Standpipe Systems for Fire Protection

Kenneth E. Isman

Standpipe Systems for Fire Protection

Springer

Kenneth E. Isman
University of Maryland
College Park, MD, USA

ISBN 978-3-319-47749-7 ISBN 978-3-319-47750-3 (eBook)
DOI 10.1007/978-3-319-47750-3

Library of Congress Control Number: 2016961307

Printed on acid-free paper

This Springer imprint is published by Springer Nature
The registered company is Springer International Publishing AG
The registered company address is: Gewerbestrasse 11, 6330 Cham, Switzerland

Preface

There have been very few books written about standpipe systems in the 150-year history of water-based fire protection system design. In 1976, Dr. John Bryan published the first edition of his book, *Automatic Sprinkler and Standpipe Systems*. This book was an important step forward because it was the first book to dedicate significant space to the subject. But half of the book was dedicated to fire sprinkler systems, so there was not enough space in the book to get into some of the more complicated issues involved in standpipe system design.

Over the years, Dr. Bryan's book was updated with more information, but it still did not have the space to tackle the real complex issues. At the same time, standpipe systems were evolving and more codes and standards organizations were getting involved in writing their own requirements, which are sometimes contradictory. The time has come for a book dedicated to standpipe systems and nothing else so that all of the complicated issues can be discussed.

This book pulls together all of the requirements for standpipe systems that can be found in the International Building Code, the International Fire Code, NFPA 14, NFPA 20, and NFPA 25 and attempts to make sense of them. In the situations where the requirements from one document contradict the rules of another, the hierarchy will be presented so that the reader will be able to determine which requirement is the one that needs to be followed.

The life cycle of standpipe systems will be covered from beginning to end. The book opens with basic information on what standpipe systems are and a short history of standpipe systems. It continues with design information and special considerations for high-rise buildings and pressure control concerns. The book will wrap up with a discussion of how to conduct the acceptance tests for a new standpipe system and how to care for an existing standpipe system with periodic inspection, testing, and maintenance.

College Park, MD Kenneth E. Isman

Acknowledgements

First and foremost, I need to thank my wife, Joan, for putting up with me and this project, which took much longer than I thought it would. She never complained, even when we ended up cancelling summer plans so that I would have more time to write.

Thanks to Paul Drougas at Springer who believed in this project from the beginning. Not only did he encourage the work, but he also made great suggestions for improvement and even provided a photo for one of the chapters when he stumbled on a situation days after we had discussed including something like it in the book. This proves that he was paying attention and really cared enough to take the picture and pass it on, and for that, I am grateful.

Nicholas Zielinski is a student in Fire Protection Engineering at the University of Maryland who proofread all of the chapters of this book for me. He found my mistakes, improved the text, and made sure that the figures and examples make sense. Thanks for his dedication and help.

Finally, thanks to Marshmallow, Contee, 2K, and Nicholas who kept me company while I was working and helped me keep pushing this project forward.

Contents

About the Author

Beginning in the fall of 2014, Kenneth E. Isman became a Clinical Professor at the University of Maryland in the Fire Protection Engineering Department, where he teaches classes in fire protection systems design and life safety analysis. He holds a Bachelor of Science degree in Fire Protection Engineering and a Master of Science in Management, both from the University of Maryland. He is a Licensed Professional Engineer in the State of Connecticut and has been inducted as a Fellow of the Society of Fire Protection Engineers.

Mr. Isman has been a member of more than 15 committees of the NFPA responsible for most of the water-based fire protection system and life safety documents since 1987. He also served on several committees of the American Water Works Association, developing recommended practices for backflow connection and cross-connection control.

Prior to coming to the University of Maryland, Mr. Isman worked for 28 years in the Engineering Department of the National Fire Sprinkler Association (NFSA), where he started as the Manager of Codes and Education and was consistently promoted over the years until he reached the position of Vice President of Engineering, which he held for 8 years. While with the NFSA, he served as the Secretary of the Engineering and Standards Committee, where he represented the interests of the fire sprinkler industry to influence changes to NFPA 14 and better define the process of designing and installing standpipe systems for fire protection.

A noted author and lecturer, Mr. Isman has written a number of publications for the NFSA, NFPA, and SFPE, including:

- The textbook *Layout, Detail and Calculation of Fire Sprinkler Systems*
- *Pumps for Fire Protection Systems*
- Chapters of the NFPA *Fire Protection Handbook*
- A chapter of the *SFPE Handbook of Fire Protection Engineering*

He has served as an international speaker at more than 500 seminars and workshops on fire protection systems design.

Chapter 1
Introduction to Standpipe Systems

History of Standpipe Systems

In the early days of fire protection (prior to 1850) buildings were limited in height because there was a limit to how high people were willing to walk up stairs. Firefighters did not ask architects to design systems into buildings to minimize the amount of equipment they had to carry during a fire because they were willing to carry their hose and equipment up the same steps that people used for everyday access to their buildings.

All that changed in 1852 when Elisha Otis invented the elevator safety brake. Prior to this, there were elevators in use, but the ropes or cables that pulled them up easily frayed and would break. When this would happen, the elevator would drop uncontrollably. Without a safety brake, people questioned the safety of elevators and generally didn't use them. Otis' safety brake would automatically deploy prongs on the side of the car that would pin the elevator car to racks on the side if the hoist rope or cable snapped. This convinced people that elevators were safe to ride and ushered in a new era of architectural design where the sky was literally the limit.

With taller buildings being constructed in the second half of the nineteenth century, firefighters started to think about ways to help them fight fire in these taller structures. Fire sprinkler systems would be invented in the latter half of the nineteenth century, but the cost of installing these systems kept them from being installed in most buildings. Instead, a piping network designed to carry water to hose connections, called standpipes, became very popular. These systems cost less than fire sprinkler systems and were marketed for both the general public to use early in a fire and for firefighters to use once they arrived on the scene.

Unfortunately, there were no national standards for the design, installation and maintenance of these early standpipe systems. Some cities developed their own requirements and some insurance companies had suggestions regarding design, installation and maintenance, but they were each different and sometimes conflicting, which caused confusion among design engineers and building owners.

K.E. Isman, *Standpipe Systems for Fire Protection*,
DOI 10.1007/978-3-319-47750-3_1

Standpipe systems, when they were installed, were not reliable enough to be consistently used.

In 1896, the National Fire Protection Association (NFPA) was formed in order to develop national standards and codes on a number of subjects. The first two documents that the NFPA produced were a standard on fire sprinkler systems and a code for electrical systems. The NFPA then began to tackle other subjects, but it would be some time before they would turn their attention to standpipe systems.

In 1903, the City of Chicago adopted their own regulations regarding the design and installation of standpipe systems. The regulations themselves were fairly basic, but one item in the report was worthy of note. A man by the name of W.C. Robinson was chair of the Advisory Board of Experts who consulted with the commission appointed by the Mayor of Chicago to write the regulations (Lockett 1903). Robinson would be an important figure in the development of standpipe system requirements in the future.

In 1911, the NFPA appointed a Committee on Standpipe and Hose Systems to study the problem and develop a standard on the subject, W.C. Robinson, from Chicago, was appointed as the committee chair given his recent experience in guiding the city of Chicago to some requirements in this field. At the 1912 annual meeting of the NFPA in Chicago, the committee reported on what they found. In part, Robinson's report reads:

> *Unfortunately, the standpipe and hose systems heretofore installed have usually only been intended for one class of service, and are frequently poorly suited for that. They have been employed for many years, and various requirements relative to their installation and use have been adopted by municipal and insurance authorities throughout the country, but far too little attention has apparently been given to their design, arrangement and equipment, and it is anticipated to their proper maintenance.*
>
> *In fact, standpipe and hose equipments have apparently fallen into such disrepute among officials and property owners in some districts that but little attention is paid to them, and, if used at all, they are only installed to fulfill the requirements of some loosely worded ordinance, or as a means of obtaining concessions in insurance rates. Standpipes are often located at remote points in the building, rendered useless by obstructions, provided with almost worthless equipment, and frequently not the slightest attempt is made to instruct those who may be called upon to use the equipment at time of fire, or to maintain the system in serviceable condition. The reason for this condition of affairs is probably another instance of our American characteristic of carelessness. Under such conditions, is it any wonder that fire departments in some of our larger cities either disregard the standpipe systems or make a practice of disconnecting the attached hose and using their own hose and equipment.*
>
> *The standpipe and hose system has been tried and found wanting by occupants of buildings so often that the general public has also lost confidence in this means of safeguarding property against loss by fire. It is particularly fitting, therefore, that this Association take up this very important subject and not only present to the public a standard so well considered that it will meet all requirements, but so prove the efficiency of this form of fire equipment that it will have a far-reaching influence, and serve as a valuable and reliable means of reducing our greatly disproportionate loss of life and property by fire (Robinson 1912)*

Robinson's report goes on to provide the committee's thoughts on the design and installation of standpipe systems and to solicit comments and questions from the membership. The report was accepted by the committee as one of progress, meaning the committee still had work to do. The committee continued working and finalized its report in 1915 with the first edition of a standard that would go on to become NFPA 14, Standard for the Installation of Standpipe and Hose Systems (NFPA 2016).

What Are Standpipes?

A traditional definition for a standpipe system, being used for fire protection purposes, would be a combination of piping, valves, and hose connections within a building that people could use for fighting fires that occur within the building manually. In other words, the standpipe system provides a mechanism for water to be distributed throughout the building with certain access points where the water can be taken from the system and discharged on a fire. In this sense, the standpipe system can be thought of as an extension of the fire hydrant system from a community into a building.

The term "standpipe" comes from the name that has been given to large vertical pipes within a building. The name comes from the idea that a large vertical pipe is thought of as "standing" up while a large horizontal pipe is thought of as "laying" down. Since the early standpipe systems were run vertically up in tall buildings, they got the name "standpipes" as a shortened version of the concept that these were pipes that were standing up. As standpipe systems evolved to include horizontal pipes in buildings as well, the name "standpipe" has remained to describe any pipe carrying water to hose connections.

Frequently, plumbers will also refer to large vertical pipes carrying water up through a building as "standpipes". Their history of the term is similar to the one for fire protection, but there are no hose connections on the plumber's standpipe. This book will ignore the plumber's standpipe systems and will only refer to those standpipes used for fire protection purposes with hose connections.

Standpipe systems may or may not have hose physically attached to the system. There have been continuing debates in the fire protection community regarding whether standpipe systems should have hose or not. On one side of the debate is the group of fire protection professionals that wants to empower people to fight fires while they are still small, before the arrival of the fire department. This group of individuals wants to provide standpipe systems with hose so that any member of the general public, whether they have been trained in how to use a hose to fight a fire or not, can grab a hose if they see a fire and begin to fight the fire. They point out that this has the potential to minimize fire damage and provide life safety by dealing with fires while they are still small and easy to control.

On the other side of the debate regarding hose are the fire protection professionals concerned with the idea of untrained individuals fighting a fire. They point out that if people find hose in a building where there is a fire, there may be a tendency to delay notification of the fire department while trying to fight that fire. They also point out that the general public will not have the protective equipment necessary to prevent burns or smoke inhalation while fighting a fire, so there is a greater chance of personal injury occurring during a fire if hose is provided. These professionals prefer to see standpipe systems reserved for fire department use without hose for the general public.

Over the more than 100 year history of standpipe systems being used in buildings, the pendulum has swung to different ends of this spectrum. During the 1950s and 1960s in the United States, there was a tendency for fire protection professionals to be on the side of wanting hose within buildings. So, the codes and standards of that time required standpipe systems to have hose attached to the system and that hose still exists in many of the buildings constructed in that time period. More recently, the pendulum has swung in the other direction. The majority of fire protection professionals now believe that fire sprinkler systems and firefighters exist to deal with the fire and the general public should just evacuate a building during a fire and call the fire department. The general public should not worry about trying to find hose to fight the fire themselves.

The Occupational Safety and Health Administration (OSHA), a branch of the United States Federal Government, has also weighed in on this subject. They have taken the position that if the hose is provided on a standpipe system, and if it is part of someone's job responsibilities to use that hose, then those people need to be trained in how to use the hose properly and they need to be provided with proper firefighting clothing and protective gear in order to use the hose safely. For example, if hose is provided on a standpipe system in a school and part of the janitor's job responsibilities include using the hose to fight a fire, then the janitor needs to be trained like a firefighter and provided with appropriate firefighting clothing and equipment. In light of this position, many companies have either removed the hose from their standpipe systems or taken firefighting responsibilities out of all employee job descriptions.

While the standpipe system is specifically for the distribution of water throughout a building in the event of fire, under normal conditions, there may or may not actually be water in the pipes. Some standpipe systems have water all of the time. Other standpipe systems are empty of water until some action causes the pipes to be filled. A third type of standpipe system is empty of water until the fire department arrives and starts pushing water into the system. There is more discussion on this topic in the next chapter of this text regarding types of standpipe systems.

What Makes a Standpipe System?

As stated earlier in this chapter, a standpipe system consists of pipe, valves and hose connections. There may or may not be a water supply connected to the piping. At one end of the system, at a location accessible by the fire department, will be a connection where the fire department will be able to put water into the system. In this book, this input point will be called a Fire Department Connection (FDC).

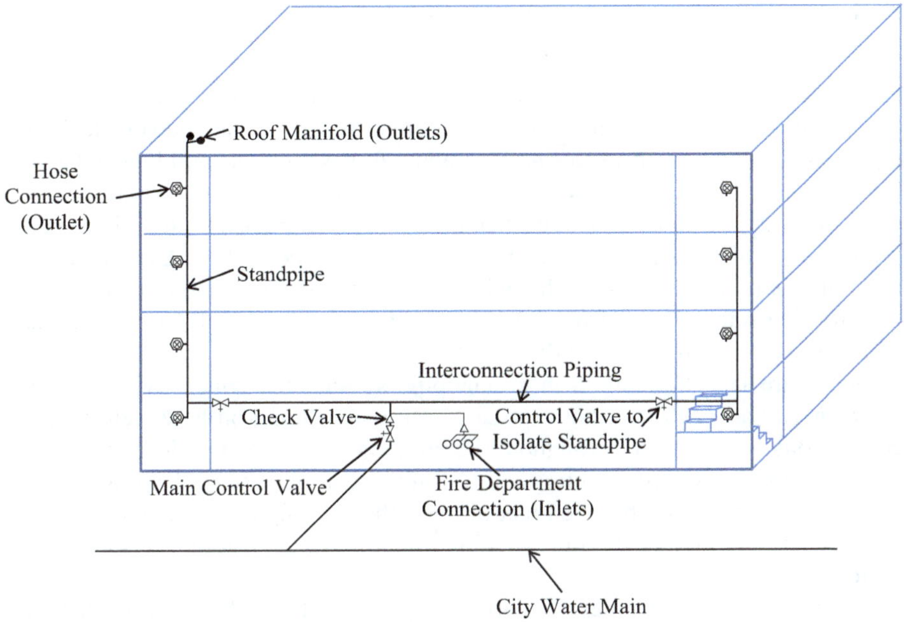

Fig. 1.1 Standpipe system overview

At other locations throughout a building will be points where water can be taken out of the system. At these locations, if there is already fire hose there, the location will be called a "hose station". If there is no hose permanently maintained at the location, then the point where water can be taken from the system will be called a "hose connection." Typically, firefighters bring their own hose when responding to a fire, so there is no need to provide hose if the only purpose for the standpipe system is to save firefighters from expending severe effort.

The piping between the input points (water supplies and fire department connections) and the output points (hose stations and hose connections) forms the basis of the standpipe system as shown in Fig. 1.1. Note that Fig. 1.1 is intended to be schematic in nature and is not intended to show every piece of equipment that may (or may not) be included in a standpipe system.

Standpipe systems typically have control valves in locations so that an individual standpipe can be isolated for maintenance without taking the whole system out of service. Other valves (check valves or backflow preventers) are added so that the water flows only in one direction from the water supply or inlet towards the output points. Depending on the water pressure in the system pressure reducing valves may also be included. There is more information on the requirements for specific types of pipe and valves in Chap. 5 of this text.

How Do Firefighters Use Standpipes?

Firefighters use standpipe systems to help them conserve energy and time during a fire. Rather than laying out hose all the way from the fire hydrant outside the building, the firefighters can bring a smaller amount of hose into the building and attach it to the nearest hose connection to the fire. While this might not sound like a big deal, consider how much time, effort and hose would be necessary to fight a fire on the top floor of a 10 story building without a standpipe system. Even if the fire hydrant was relatively close to the building, it would still take more than 400 ft of hose to get a single line up from the street and firefighters typically need more than one line to attack a fire in a building.

For a fire in a building with a standpipe system, firefighters typically need between 100 and 200 ft of hose to reach a fire, depending on how well the hose connections are scattered throughout the building. This is a significant savings in time and resources while fighting a fire. Rather than have several firefighters get worn out laying several hundred feet of hose, those firefighters can be performing other functions early in the fire (such as rescue) with a standpipe system in the building.

A firefighter is most vulnerable during the period of time when they are connecting their empty hose to the hose connection. At this time, they do not have any water available for cooling the space around them or fighting the fire. For this reason, hose connections are supposed to be placed in locations that are separated in some way from the portion of the building that is on fire. The most protected location in a building (and the least likely to have a fire) is the exit stairwell. For this reason, the most common location in a building for a hose connection is within the exit stairwell. Hose connections can also be placed in other protected locations in the building. See Chap. 3 for more information on where hose connections are required.

From time to time, it might be necessary to put a hose connection in a location in the building that is nowhere near an exit stairwell. In these situations, consideration should be given to how the firefighters will be protected from the fire while they are connecting their empty hose. Use of building features with fire resistance ratings similar to those of stairwells are helpful for locating hose connections in protected spaces for firefighters. Figure 1.2 shows a hose connection in the middle of a hotel corridor with no protection for firefighters. This is not a good location for a hose connection on a standpipe system. Note that this connection is also a significant distance above the floor, which makes it difficult to use this particular hose connection.

When standpipe outlets are installed in stairwells, firefighters use the protected location in the stairwell to connect their hose, make sure they have water, go through the fire rated door at the floor where the fire is located, and then they go to fight the fire. Once the firefighters go through the door to the stairwell, the hose prevents the door from closing completely, sometimes allowing smoke and hot gasses to get into the stairwell from the fire. For this reason, firefighters will typically connect their

Fig. 1.2 Unprotected hose connection

hose to the standpipe outlets below the fire floor so that they are not subjected to the smoke and hot gasses once the first crew goes through the door. Figure 1.3 shows a single hose connected to the standpipe outlet at the hose connection a full floor level below the fire. Figure 1.4 shows a second hose put in service from the hose connection one-half floor level below the fire, which is a more likely use of standpipe systems. The firefighters using one hose can protect the firefighters using the other hose if anything goes wrong, such as one of the hoses breaking.

The standpipe systems shown in Figs. 1.3 and 1.4 have the hose connections at the intermediate landing between floors. About half of the standpipe systems that have been installed in stairwells have been installed this way. The other half have been installed with the hose outlets at the main floor landings. More information on this difference will be included in Chap. 4 of this text.

In order to prevent smoke and hot gasses from a fire from getting into the stairwell when firefighters go through the door with their hose, some buildings are constructed with a vestibule between the stairwell and the rest of the building as shown in Fig. 1.5. The standpipe outlets can be installed in the vestibule rather than the stairwell itself. This way, the stairwell is kept smoke free during firefighting operations, which helps to evacuate the portions of the building above the fire floor. Firefighters can still walk up the stairs in a protected environment, enter the vestibule and put a hose in operation to fight the fire through the hose connection in the vestibule.

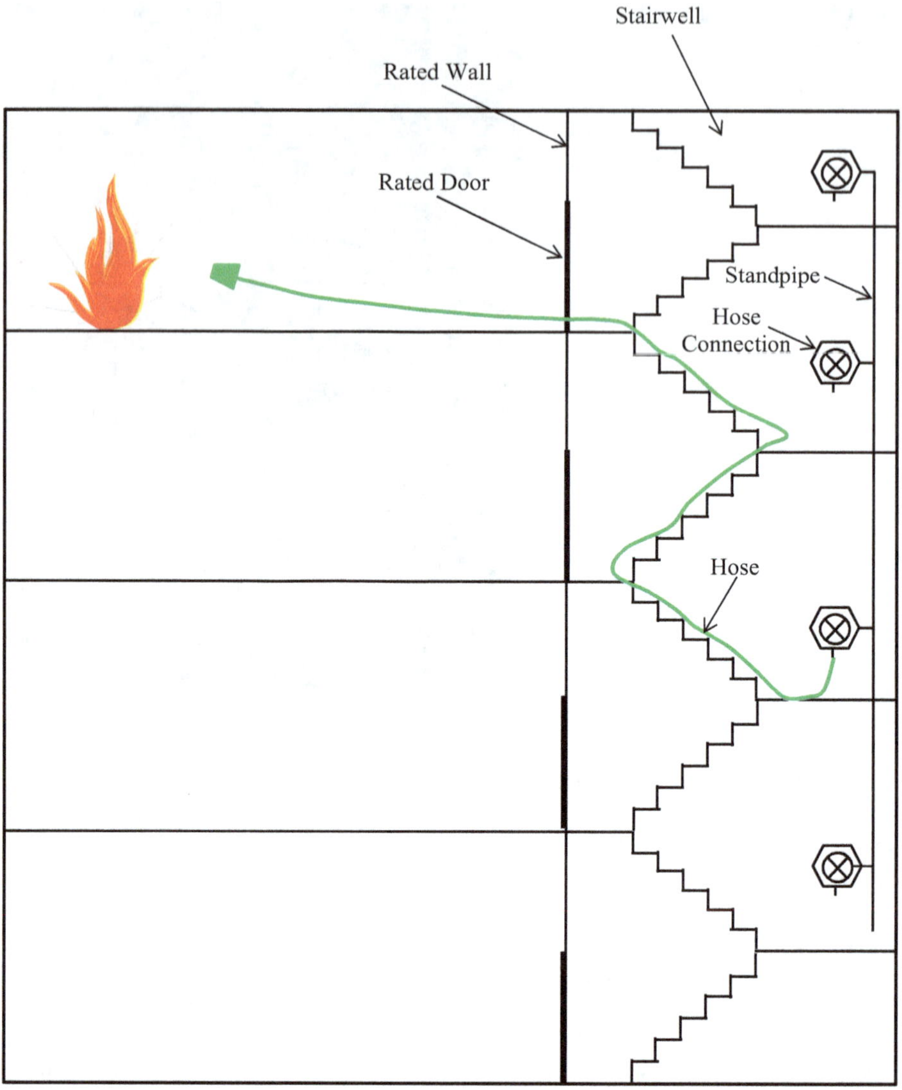

Fig. 1.3 One hose being used to fight fire

The situation shown in Fig. 1.5 still requires firefighters to climb the stairs with some hose in order to fight the fire. While this might not be too difficult for a five or six story building, it would still be a challenge in a 20 or 30 story building. On such a tall building, firefighters might take the elevators to a floor one or two below the floor where the fire is, and then go to a stairwell to get the rest of the way up to the floor where the fire is.

Typically, firefighters will bring their own hose and nozzles to connect to the standpipe system and fight the fire. Even if hose is attached to the standpipe system,

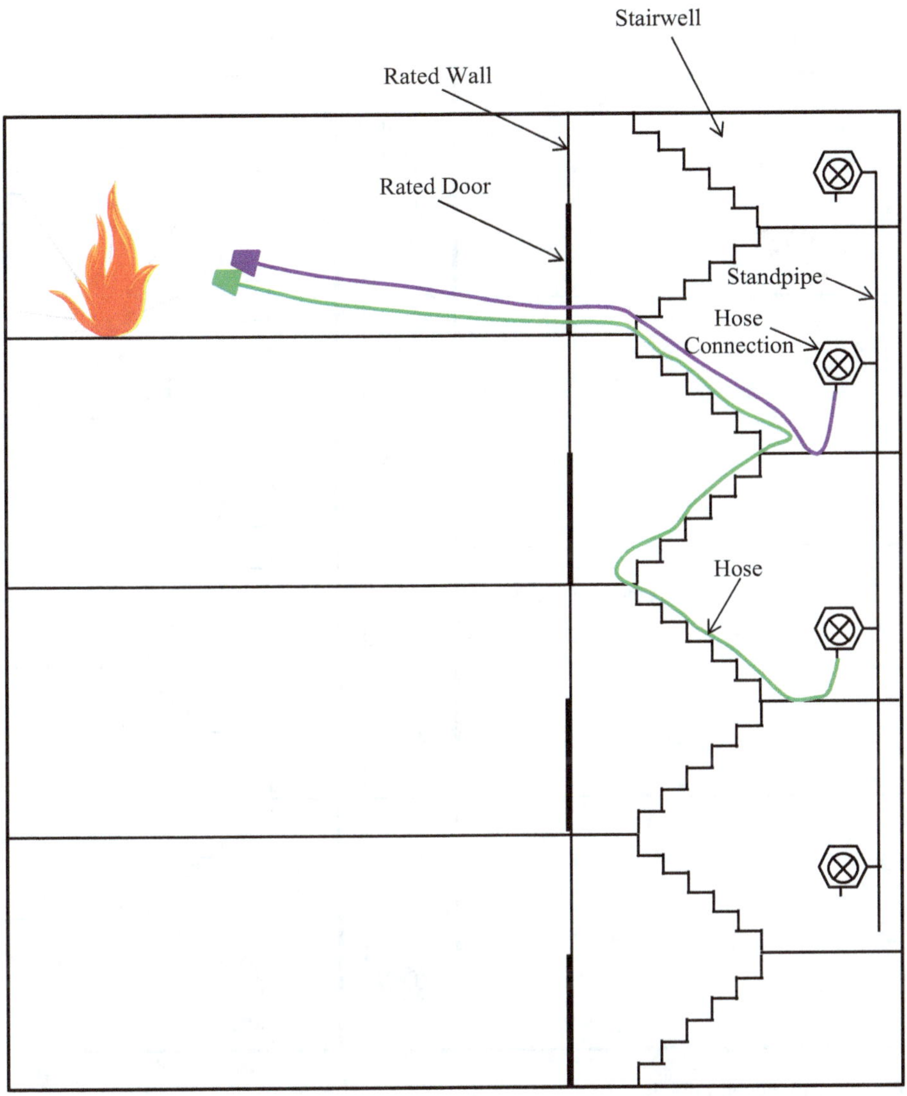

Fig. 1.4 Two hoses being used to fight fire

the firefighters will typically disconnect that equipment and connect their own. During firefighting operations, it is not worth the firefighters lives to take a chance that the hose within the building has been properly tested and maintained. The firefighters are much better off using their own equipment so that they can safely attack the fire.

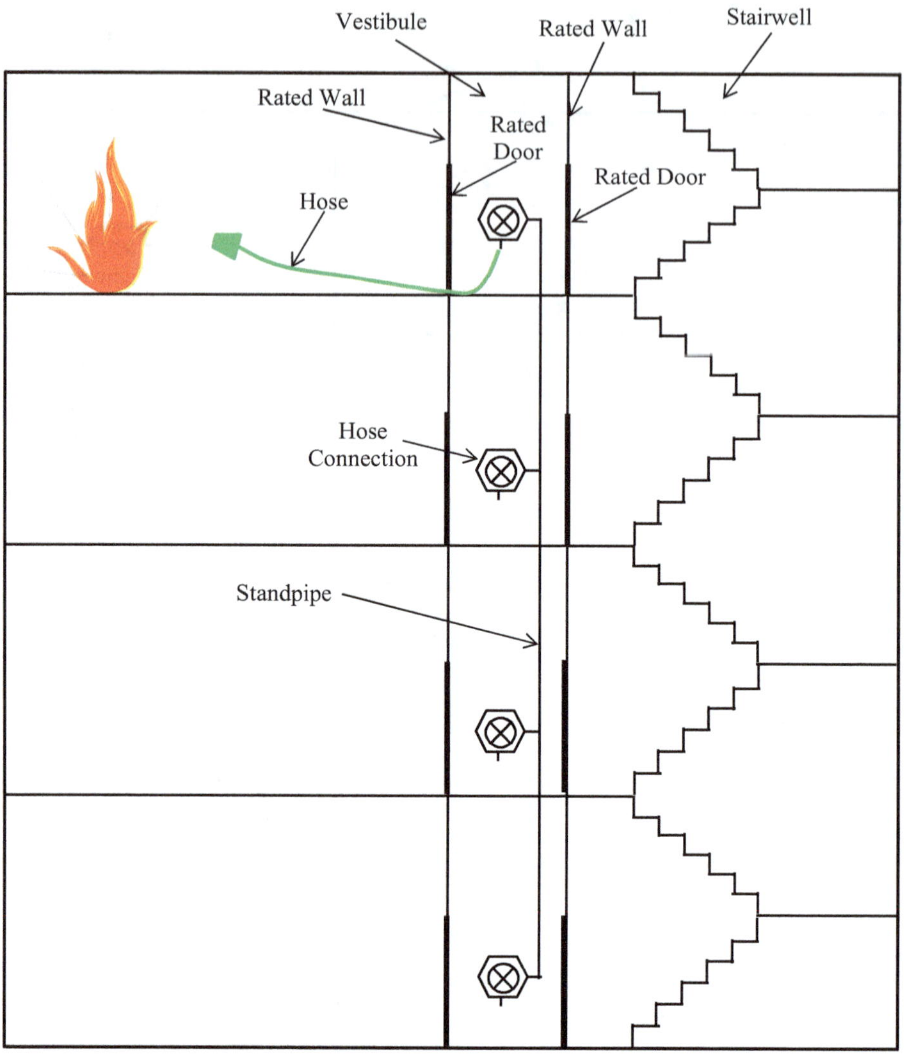

Fig. 1.5 Hose connection in vestibule

What Kinds of Buildings Need Standpipes?

In general, there are three types of buildings that need standpipe systems: those that are large, those that are tall, and those that are deep down. Large buildings are those that have a sizable area on a single floor. These buildings might only be one or two stories in height, but with a significant walking distance between the exterior door where a fire truck might pull up and a location where a fire might occur deep within the building, firefighters benefit from standpipe installation.

Tall buildings are those that have a significant vertical distance above the level where the fire truck will pull up. Firefighters benefit by having standpipe systems installed in tall buildings in order to save them from having to carry heavy hose all the way up from the ground floor where the fire truck has access near the building. Laying hundreds of feet of hose all the way up a stairwell would take considerable time during a fire and would require significant manpower. Standpipe systems save this time and manpower. Firefighters still have to carry some hose up the stairs, but only enough hose to connect to the standpipe system and go into the building to fight a fire.

Buildings that are deep down are those that are built into a hillside below the level of fire department vehicle access or those that are built underground. In these cases, standpipe systems save firefighters the effort and time of laying hose down the stairwell. While this is in the downward direction, which is slightly less difficult than laying hose up a stairwell due to gravity, there is still a significant amount of exertion in climbing down stairs, especially in full firefighter protective clothing, so standpipe systems save firefighters time and effort in these deep buildings as well.

A specific list of buildings that require standpipe system by size, height, depth and occupancy will be included in Chap. 3 of this text. This list will specifically be taken from the 2015 edition of the ICC's International Building Code and International Fire Code as well as the 2015 edition of the NFPA's Life Safety Code (NFPA 101) and Building Construction and Safety Code (NFPA 5000). While other codes or other editions of these codes might be adopted and enforced in your jurisdiction, this list is fairly representative of the types of buildings that require or could benefit from standpipes.

Installation Rules

Codes like the International Building Code (IBC) or the Life Safety Code provide the user with a list of buildings that need certain types of fire protection. For example, the IBC has a list of buildings that require standpipe systems. But the requirements for designing, installing and maintaining that standpipe system are generally not found in the Code. Instead, Codes tend to reference secondary documents, called Standards, in order to handle the design, installation and maintenance rules.

Most Codes in North America, South America, and the Middle East reference NFPA 14—Standard for the Installation of Standpipe and Hose Systems for the design and installation requirements. These same Codes tend to reference NFPA 25—Standard for Inspection, Testing and Maintenance of Water-Based Fire Protection Systems for the rules regarding maintenance of the standpipe system.

Standards are written in language that is legally enforceable and are just as much a part of the law as the Code that references them. However, since the Standard is a secondary document and its force and effect come from an initial document (the Code), the initial document (sometimes called the "Primary Legal Document") takes precedent over the Standard if there are any conflicts. Since the Code and the

Standard are both written by different organizations, there are sometimes disagreements as to how a certain situation should be handled. In these cases, the organization writing the Standard puts their requirements into their Standard, while the organization writing the Code puts their requirements into their Code. In jurisdictions enforcing both documents, the Code requirements prevail from a legal point of view. Many times, the more stringent of the two requirements is used, thereby meeting both the Code and the Standard, but in a few cases, neither of the requirements is more stringent, and the designer will need to use the rules from the Code unless a legal process is followed to allow for a different option.

In the specific case of the IBC and NFPA 14, there are a number of conflicts. The people responsible for writing the IBC either did not agree with the NFPA 14 committee, or thought that there was a chance that the NFPA 14 committee would change its mind about some aspect of standpipe system design or installation, so some of the rules of NFPA 14 have been copied into the IBC over the years. However, as one document changed, the other did not, and so some separation of the rules has occurred over the years. When it comes down to a legal position in a jurisdiction enforcing both the IBC and NFPA 14, any situation where the IBC takes a position on a subject overrides any mention of that subject in NFPA 14.

The specific requirements of NFPA 14 will be covered in the rest of the chapters of this text. In the situation where there are variations in a Code, those variations will be discussed within the chapter of this text dealing with the subject matter. For example, there are slight variations between the requirements for the location of hose connections between NFPA 14 and the IBC, those variations will be discussed in Chap. 4 of this text.

Standby Hose Stations Connected to Sprinkler Systems

This introductory chapter has spent a great deal of time talking about what constitutes a standpipe system, but it is just as important to spend some time talking about fire protection systems that are NOT standpipe systems. NFPA 13 allows small (1½ in.) and large (2½ in.) hose connections to be connected to fire sprinkler systems. Such connections are considered a part of the fire sprinkler system, not a standpipe system. These hose connections are frequently called "standby hose" connections.

Standby hose connections are intentionally different from standpipe systems. The purpose of a standby hose connection is to provide the user with some type of fire protection without providing all of the full requirements of a standpipe system. Connecting the standby hose to the fire sprinkler system allows a small amount of water to be available for firefighting (typically 50–250 gpm) without the full water supply of a standpipe system.

There are two occupancies where standby hose connections are sometimes found: storage warehouses and assembly occupancies back stage. In storage warehouses, it used to be common to find standby hose stations connected to the sprinkler system. NFPA 13 (and its predecessor documents NFPA 231 and NFPA 231C) required standby hose systems combined with the sprinkler system for many years. The original thought was that these systems could help extinguish fires in their early stages (usually prior to sprinkler activation). They were justified in storage warehouses because the ceiling is very high and the fire takes some time to grow large enough to set off sprinklers, which gives the warehouse employees time to try and extinguish the fire.

Many fire protection professionals were concerned with the thought of untrained employees in a warehouse trying to use the hose to fight a fire without any protective clothing or equipment. In addition, the Occupational Safety and Health Administration (OSHA) started to express concerns that if the hose was placed in the warehouse for employees to use, the employees needed to be provided with protective clothing, equipment and training. In the 2010 edition of NFPA 13, the requirement was changed to only have the hose stations installed when they were specifically required by the Authority Having Jurisdiction, which eliminated many standby hose systems.

The second type of standby hose systems that are connected to the fire sprinkler system are found in the back stage areas of theaters. Historically, a number of very bad fires have happened in theater occupancies. The stage itself is combustible construction and typically finished with combustible liquids. The scenery in the back stage area along with the storage of costumes and materials used for building sets are all significant fire challenges. The way that some scenery is hung from the ceiling and rigged with ropes to drop to the stage at the right time, it becomes an obstruction to the ceiling sprinklers. For all of these reasons, many fire protection professionals want to see hose connections back stage. Rather than trying to demand a full standpipe system, in a fully sprinklered building, the standby hose is seen as a supplement to the sprinkler system. Interestingly, these standby hose systems have not come under scrutiny from OSHA the way that the hose stations in storage warehouses did years ago.

This text will concentrate on full standpipe systems and will not contain design or calculation issues for standby hose systems. Other texts that focus on fire sprinkler systems can be consulted for more information on standby hose systems.

Temporary Standpipe Systems During Construction and Demolition

During the construction of a new building, the materials being used for construction, piled on the construction site, constitute a significant fire hazard all their own. Fire ignition sources are also more prevalent during building construction activities. In addition, access and egress are more difficult while a building is under construction (all staircases may not be constructed yet and elevators may not be in service). For all of these reasons, firefighters are concerned about how devastating a fire can be in a building under construction.

Fig. 1.6 Fire department connection for standpipe system in building under construction. Note: Photo taken by Paul Drougas, used with permission

If a building is going to need a standpipe system, then that system needs to be functional during the time when construction workers are on the site. The standpipe system needs to keep pace with the building construction. As the building goes up, the standpipe system has to be available at the same floors.

Access to construction sites is often difficult for fire trucks during an emergency. Construction fences designed to keep people out also make it difficult for firefighters to get in. Parking lots and driveways are not always finished when a fire occurs, so the firefighters sometimes need to leave the fire trucks in the street during a fire at a construction site. If this is the case, the standpipe system needs to have a fire department connection at the street so that the fire department can pump into the system using the truck that cannot get any closer to the building. See Fig. 1.6 for an example of a temporary fire department connection run to the street for a building under construction. After the driveway and parking lots are complete and the fire department vehicles have access to the permanent fire department connection, the temporary portion going out to the street can be removed.

If a building is being demolished, the same conditions need to apply to the standpipe system as a building under construction. The standpipe system needs to stay in service for the floors of the building that still have work going on. If access to the street is blocked, then the fire department connection needs to be temporarily extended out to the street where fire trucks can get close to the building.

More information on the requirements for standpipe systems for buildings under construction and demolition is included in Chap. 3 of this text for each of the specific building codes that have requirements in this area. Even if a building code does not have requirements in this area, it is a good idea to keep a standpipe system operational during construction or demolition.

A dramatic example of how important standpipes are during construction and demolition occurred in New York City in 2007. The Deutsche Bank building, which was heavily damaged during the attacks of September 11, 2001 on the World Trade Center, was in the process of being demolished. The building had been vacant for six years and the standpipe system was out of service. On August 18, 2007, a fire broke out on the 17th floor. This fire would eventually become a seven-alarm fire with hundreds of firefighters working to extinguish the blaze. One of the reasons the fire got this bad was that there was no standpipe to assist in getting water to the upper floors of the building. It took an hour for the fire department to establish a water supply up the outside of the building. Firefighters connected hose to the water supply and started to fight the fire, but by this time, they ran out of air. Two of these firefighters died of smoke inhalation shortly after. Before the fire was over, 115 other firefighters were injured (NIOSH 2010).

Clearly, if the standpipe system had been in service, the firefighters would not have run out of air during the beginning of their firefighting attack and would not have perished. With a reasonable water supply, the fire would have been controllable earlier in the event and would not have allowed so many firefighters to be injured.

References

Lockett, S.H. (1903). "Fire Protection in Cities", *The Spectator, An American Weekly Review of Insurance*, April 16, 1903, pp 20–23.

NFPA (2016). "Origin and Development of NFPA 14", *Standard for the Installation of Standpipe and Hose Systems*, 2016 Edition, National Fire Protection Association, Quincy, MA, 2016.

NIOSH (2010). "Two Career Fire Fighters Die Following a Seven-Alarm Fire in a High-Rise Building Undergoing Simultaneous Deconstruction and Asbestos Abatement—New York", *Death in the Line of Duty...A summary of a NIOSH fire fighter fatality investigation*, Report Number F2007-37, Washington, DC, August 5, 2010.

Robinson, W.C. (1912). Report of the Committee on Standpipe and Hose Systems at the 16th Annual Meeting of the National Fire Protection Association, Chicago, IL, May 1912.

Chapter 2
Types and Classes of Standpipe Systems

Standpipe systems are divided into two categories: types and classes. The different types describe the water supply for the system and the fluid in the piping (air or water) up to the hose connections when the system is in the ready condition (operational in case a fire occurs). The classification describes the size of the hose connection, which relates to the flow that can be expected from the connection as well as the person that is expected to be able to handle a hose attached to the connection.

Types of Standpipe Systems

Standpipes are divided into five different types considering the water supply connected to the system and whether or not the pipes are filled with water or air up to the hose connection. The five types of standpipes are as follows:

Automatic Wet Standpipe Systems

An automatic wet standpipe system is one where the water supply is permanently connected to the standpipe system and the water fills the system piping completely from the water supply all the way to the hose connections. During a fire, firefighters can connect their hose to the hose connection and open the hose connection getting water immediately at the proper flow and pressure without having to pump water into the system.

Automatic wet standpipe systems are analogous to wet pipe sprinkler systems. Both types of systems are filled with water at all times. Both types of systems also work automatically without the fire department having to bring water to the scene or having to provide any boost in pressure.

K.E. Isman, *Standpipe Systems for Fire Protection*,
DOI 10.1007/978-3-319-47750-3_2

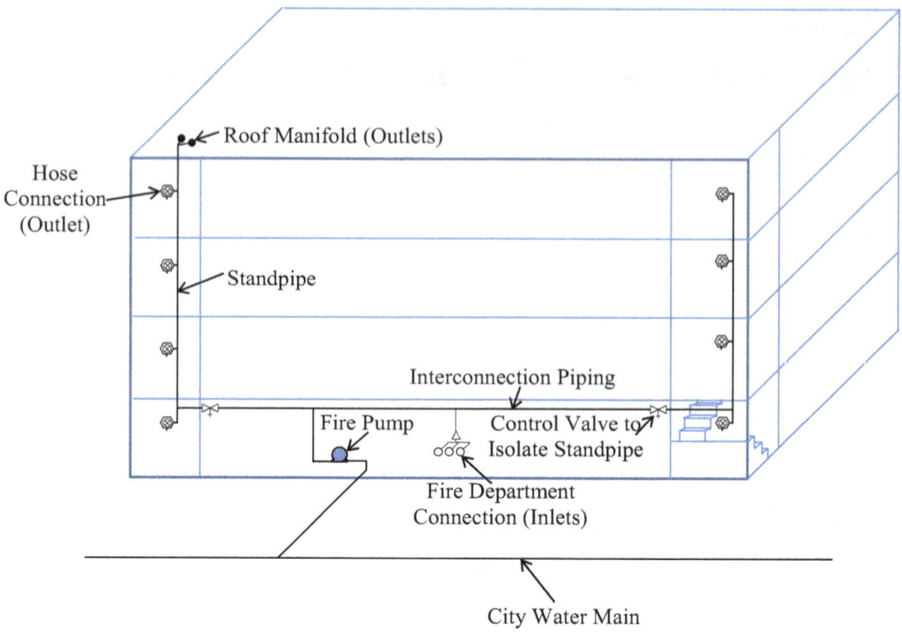

Fig. 2.1 Automatic wet standpipe system

In order for an automatic wet system to work at the proper flow and pressure for the firefighters, a pump will most likely be needed. Standpipe systems require significant pressure at the most remote portion of the system (typically 100 psi). Factoring in the friction loss between the water supply and the most remote outlet and the elevation pressure loss, the water supply pressure usually is not strong enough to meet the pressure demand. A fire pump will help boost the pressure from the water supply so that the system demand can be met without needing help from the fire department. Figure 2.1 shows the basic arrangement of an automatic-wet standpipe system with a pump. Figure 2.1 is intended to be schematic in nature and for simplicity purposes does not show all of the necessary components of a fire protection system, such as the enclosure for the pump room.

Since water fills the system piping all the way from the water supply to the hose connections, automatic-wet systems can only be installed in buildings that are heated to at least 40 °F in the locations where the pipe has been run. As Fig. 2.1 shows, it is common for standpipe systems to have outlets on the roof. In areas subject to freezing, the roof outlet will be connected to a valve at the roofline that isolates the piping above the roof and keeps the water in the heated space below the roof. This allows the system to be filled with water even though this one portion of the system is in an area subject to freezing.

Automatic wet standpipe systems are typical in high-rise buildings where the fire department wants the ability to start flowing water in buildings without having to rely on fire trucks pumping water into the system. Most codes require automatic-wet systems for the standpipe in a high-rise building.

Automatic wet standpipe systems are frequently found in buildings that also have a fire sprinkler system. Since both systems are fire protection systems that use water throughout the building, it is common to use one piping system to supply both the standpipe system and the sprinkler system. Typically, the standpipe system carries the water up through the building in the stairwells and the sprinkler system on each floor is connected to the standpipe system in the stairwell.

Automatic Dry Standpipe Systems

An automatic dry standpipe system is one where the water supply is permanently connected to the standpipe system and the water only fills the system piping from the water supply to a dry-pipe valve. Air (under pressure) fills the piping valve from the dry-pipe valve to the hose connections. During a fire, firefighters can connect their hose to the hose connection and open the nozzle on the hose. Air discharges immediately from the hose, which causes the air pressure in the piping to drop, tripping the dry-pipe valve. Water then goes from the water supply into the system piping, discharging from the hose at the proper pressure and flow for fighting the fire without the firefighters needing to pump water into the system.

The key to the operation of an automatic dry standpipe system is the operation of the dry-pipe valve. This valve acts as the point of separation between the water (which is under pressure) and the air (which is also under pressure). The surface of the dry-pipe valve in contact with the air is greater than the surface area in contact with the water. This size differential is important to the operation of the dry-pipe valve. It allows a lower air pressure to hold back a higher water pressure. Just how much air pressure can hold back the water depends on the difference between the sizes of the portion of the valve in contact with the water and the air.

Since the dry-pipe valve is manufactured with different surface areas in contact with the air and water, the valve is called a "differential dry-pipe valve". The numerical value of the differential provides the user with information of how big the difference is between surface areas. For example, a dry-pipe valve with a 5:1 (read as "five to one") differential has five times the surface area of the valve in contact with the air than it has in contact with the water. This allows 20 psi of air pressure to help back 100 psi of water pressure. Valve differentials are typically 5:1 or 6:1 although there are some dry-pipe valves with differentials as high as 17:1.

Automatic dry standpipe systems are analogous to dry pipe sprinkler systems. Both types of systems have dry-pipe valves and are filled with water from the water supply to the dry-pipe valve. Both types of systems also work automatically without the fire department having to bring water to the scene or having to provide any boost in pressure. When the system is opened (either from a sprinkler opening or a hose connection being opened) air discharges at first, causing the dry-pipe valve to trip and allowing the piping to fill with water all the way to the most remote hose connection (or sprinkler).

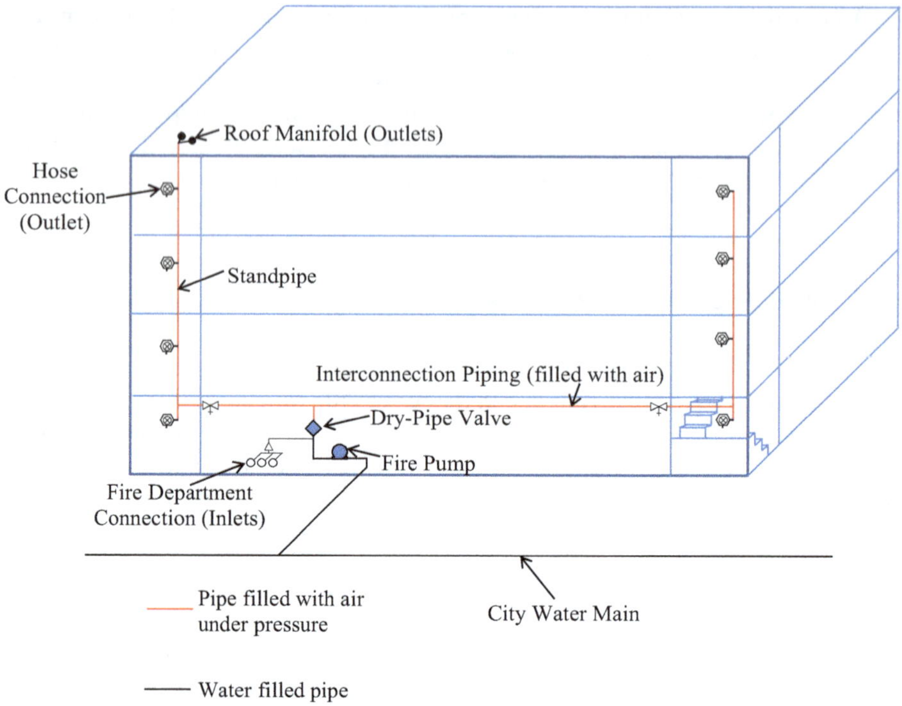

Fig. 2.2 Automatic dry standpipe system

In order for an automatic dry standpipe system to work at the proper flow and pressure for the firefighters, a pump will most likely be needed. Standpipe systems require significant pressure at the most remote portion of the system (typically 100 psi). Factoring in the friction loss between the water supply and the most remote outlet and the elevation pressure loss, the water supply pressure usually is not strong enough to meet the pressure demand. A fire pump will help boost the pressure from the water supply so that the system demand can be met without needing help from the fire department. Figure 2.2 shows the basic arrangement of an automatic-dry standpipe system with a pump. The red line represents the pipe that is filled with air under pressure during the time that the system is in service and is ready to be used. The black line represents the pipe that is continuously filled with water under pressure. Figure 2.2 is intended to be schematic in nature and for simplicity purposes does not show all of the necessary components of a fire protection system, such as the enclosure for the pump room.

Since water does not fill the system piping all the way to the hose connections, automatic dry systems can be installed in buildings that are not heated at all. The only portion of the piping that needs to be run in an area heated to at least 40 °F is the piping from the water supply to the dry-pipe valve.

Automatic dry standpipe systems are typical in high-rise open parking garages in areas of the world subject to freezing. The open parking garages make it impossible to install wet pipe systems and the fact that they are high-rise makes it a requirement in most codes to have an automatic system.

Automatic dry standpipe systems need to be inspected on a regular basis to make sure that the air pressure is holding in that portion of the system. If the air is leaking out of the piping faster than it can be replaced by the air compressor, the air pressure will drop and trip the dry-pipe valve. If the valve trips when there is no fire, the piping fills with water without people knowing it. During cold weather, the water might freeze and do damage to the piping.

A low pressure alarm can be installed on the system to provide a signal at a constantly attended location if the pressure drops to a preset value. Even with a low pressure alarm, the building owner should inspect the pressure gages and pay attention to the compressor to know whether it is cycling frequently or not. More information on inspection and maintenance of automatic dry standpipe systems is included in Chap. 14 of this text.

Semiautomatic Dry Standpipe Systems

A semiautomatic dry standpipe system is one where the water supply is permanently connected to the standpipe system and the water only fills the system piping from the water supply to a deluge valve (sometimes also called a preaction valve). Air (typically only under a small amount of pressure) fills the piping from the deluge valve to the hose connections. During a fire, firefighters can connect their hose to the hose connection. However, when the firefighters open the nozzle on the hose, the deluge valve will not trip. The deluge valve is not held closed by the air pressure. Instead, the deluge valve is connected to a separate alarm panel. Alongside the standpipe system is a separate system of initiating devices (like fire alarm pull stations) and wires also connected to the alarm panel. In order for the water to fill the system the firefighters have to push the button or pull the pull station. This sends a signal to the alarm panel, which trips the deluge valve.

Once the deluge valve is tripped, the water flows from the system automatically. The fire department does not need to augment the water supply with a pump or bring any water to the fire. The water will arrive at the hose connection at the flow and pressure necessary to fight the fire. There will be a short delay as the water flows from the deluge valve to the open hose, but as long as the nozzle is fully opened, this should be a very short period of time (less than a minute).

Semiautomatic dry standpipe systems are analogous to preaction sprinkler systems. Both types of systems have a deluge valve connected to an alarm panel. Both types of systems require a separate action in order to start the flow of water. Both types of systems also work automatically without the fire department having to bring water to the scene or having to provide any boost in pressure.

In order for a semiautomatic dry standpipe system to work at the proper flow and pressure for the firefighters, a pump will most likely be needed. Standpipe systems require significant pressure at the most remote portion of the system (typically

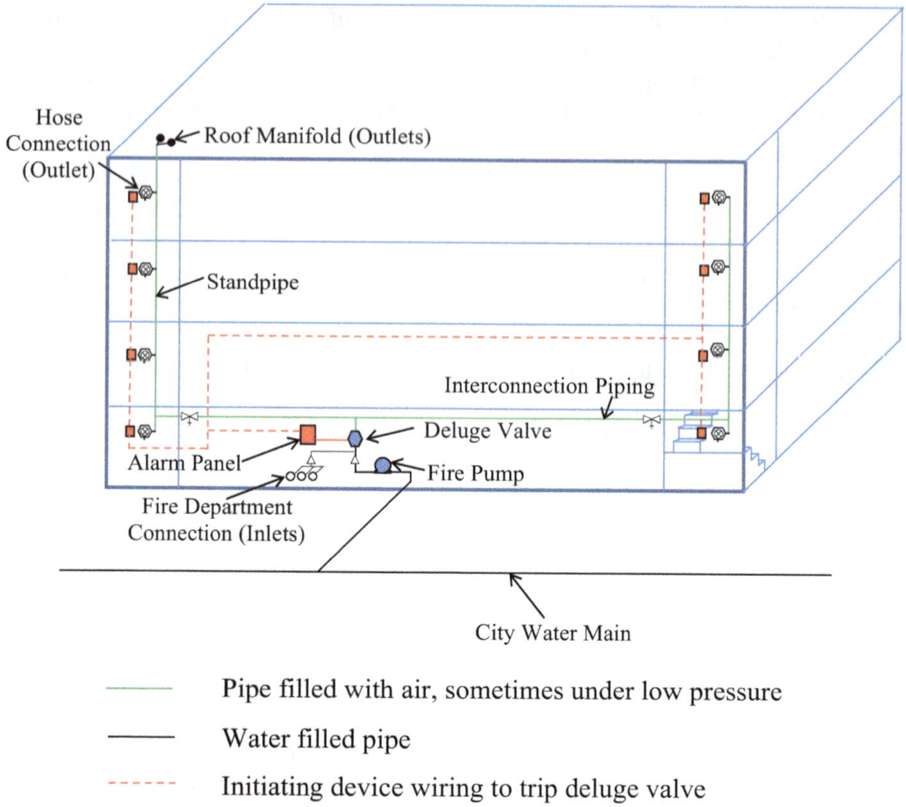

Fig. 2.3 Semiautomatic dry standpipe system

100 psi). Factoring in the friction loss between the water supply and the most remote outlet and the elevation pressure loss, the water supply pressure usually is not strong enough to meet the pressure demand. A fire pump will help boost the pressure from the water supply so that the system demand can be met without needing help from the fire department. Figure 2.3 shows the basic arrangement of a semiautomatic dry standpipe system with a pump. The green lines represent the piping that is filled with air during the time that the system is in service and is ready to be used. The black lines are the water filled portions of the system. The dashed red lines represent the wiring that connects the buttons or pull stations used to initiate the flow of water to the hose connections. Figure 2.3 is intended to be schematic in nature and for simplicity purposes does not show all of the necessary components of a fire protection system, such as the enclosure for the pump room.

Since water does not fill the system piping all the way to the hose connections, semiautomatic dry systems can be installed in buildings that are not heated at all. The only portion of the piping that needs to be run in an area heated to at least 40 °F is the piping from the water supply to the deluge valve.

Semiautomatic dry standpipe systems can be installed in any building where the owner is concerned about water damage. A drop in air pressure will not cause the system to trip. It takes the deliberate action of pushing the button or pulling the pull station to get water into the piping.

Some semiautomatic standpipe systems contain a small amount of air under pressure in order to monitor the condition of the piping. If a problem has occurred and the air is leaking faster than the compressor can keep up, the air pressure will drop in the system. The pressure can be monitored to sound an alarm if it gets down to a preset pressure. The monitoring of the air pressure in semiautomatic systems is not required, but is a feature that some building owners prefer.

Manual Wet Standpipe Systems

A manual wet standpipe system is one where the water supply is permanently connected to the standpipe system and the water fills the system piping from the water supply all the way to the hose connections. The difference between the automatic wet and the manual wet system is that the water supply is insufficient to meet the demand of the standpipe system. The fire department will have to come with a pumper and water supply to provide the flow and pressure that will be needed at the hose connections to fight the fire. The fire department will pump into the fire department connection in order to pressurize the system sufficiently to get the flows needed.

Manual wet standpipe systems are not really analogous to any type of sprinkler system. All sprinkler systems have their own automatic water supply, but a manual wet standpipe system will not have an automatic water supply. There will be a small water supply to fill the piping, but it will not provide sufficient flow or pressure to fight a fire. Manual wet standpipe systems don't have fire pumps. The pumps are not necessary because they are not intended to provide the flow and pressure on their own. Figure 2.4 shows the basic arrangement of a manual wet standpipe system. The black lines all represent water filled lines under pressure. Figure 2.4 is intended to be schematic in nature and for simplicity purposes does not show all of the necessary components of a standpipe.

Since water does fill the system piping all the way to the hose connections, manual wet systems need to be installed in buildings that are heated to at least 40 °F in all areas where there is system piping. Manual wet standpipe systems are frequently installed in buildings that also have a fire sprinkler system. The water supply is sized to automatically provide water to the sprinkler system, and is also connected to the standpipe piping. In order to save money, the standpipe piping serves the sprinkler system as shown in Fig. 2.5.

There are a number of advantages to a manual wet standpipe system. The biggest advantage is the cost. Leaving out a fire pump and a large water supply can easily save the building owner $50,000. Since the standpipe system is typically only used by the fire department and they bring water and pumps with them when they come to a fire, the pump and water supply don't need to be installed in the building.

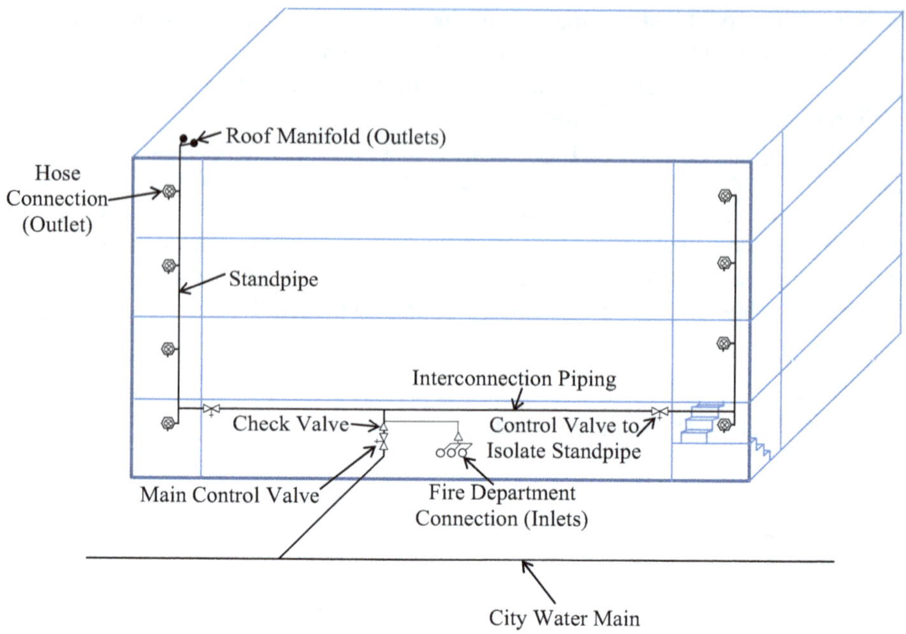

Fig. 2.4 Manual wet standpipe system

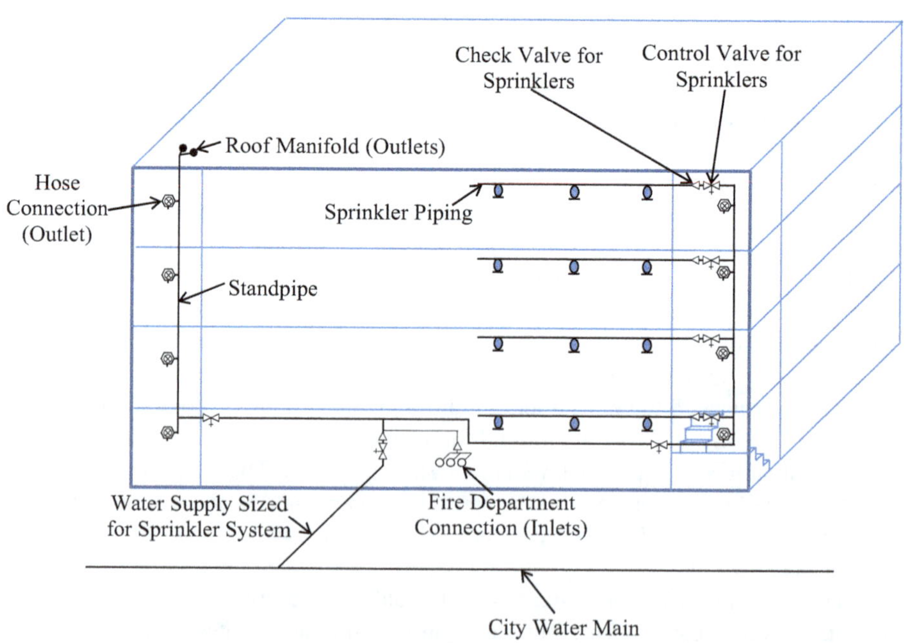

Fig. 2.5 Manual wet standpipe system combined with sprinkler system

The second advantage to the manual wet standpipe system is that the small amount of water that is in the piping helps to monitor the integrity of the piping. If there is a break in the pipe or even a small leak, the discharge of water from the system would help identify the break or leak long before a fire, helping the building owner repair the system and keep it operational in case of fire.

Firefighters are sometimes skeptical of manual wet standpipe systems because they share a similar name with manual dry standpipe systems (which are discussed next in this Chapter). While manual dry standpipe systems are a legitimate concern due to the problems with identifying broken piping, manual wet systems do not have that concern. Manual wet systems are extremely reliable.

Most codes allow manual wet standpipe systems in any building that is not a high-rise. The general feeling is that in high-rise buildings, the fire department needs the assistance of an automatic water supply to save time and effort. But for shorter buildings, the fire department does not need to spend as much effort getting up into the building to fight the fire, so they can spend a little more effort in establishing a water supply, making a manual wet system acceptable.

Manual Dry Standpipe Systems

A manual dry standpipe system is one where there is no water supply. The system is literally a empty piece of pipe. The fire department will have to come with a pumper and water supply to provide the flow and pressure that will be needed at the hose connections to fight the fire. The fire department will pump into the fire department connection in order to get any water into the system.

Manual dry standpipe systems are not really analogous to any type of sprinkler system. All sprinkler systems have their own automatic water supply. But a manual dry standpipe system will not have any automatic water supply. Figure 2.6 shows the basic arrangement of a manual dry standpipe system. The green lines show pipe that has no water, but also done not have air under pressure. The air in the pipe is just at atmospheric pressure. Figure 2.6 is intended to be schematic in nature and for simplicity purposes does not show all of the necessary components of a standpipe system.

Since there is no water in the system piping, manual dry systems are allowed to be installed in buildings regardless of the heating situation, but this is the only advantage of the manual dry system. Unfortunately, manual dry systems have a serious disadvantage. Without water or air in the piping under pressure, there is no way to monitor the integrity of the piping. At the same time, water sitting in the piping after it has been used promotes rusting of the piping. These two conditions combine to cause significantly less reliability in manual dry standpipe systems.

Every firefighter that has been around for a while has the same story to tell about using a manual dry standpipe system. The story starts with the arrival of a building on fire. The fire department starts pumping into the fire department connection on the manual dry system, but the firefighters at the hose connection at the floor where

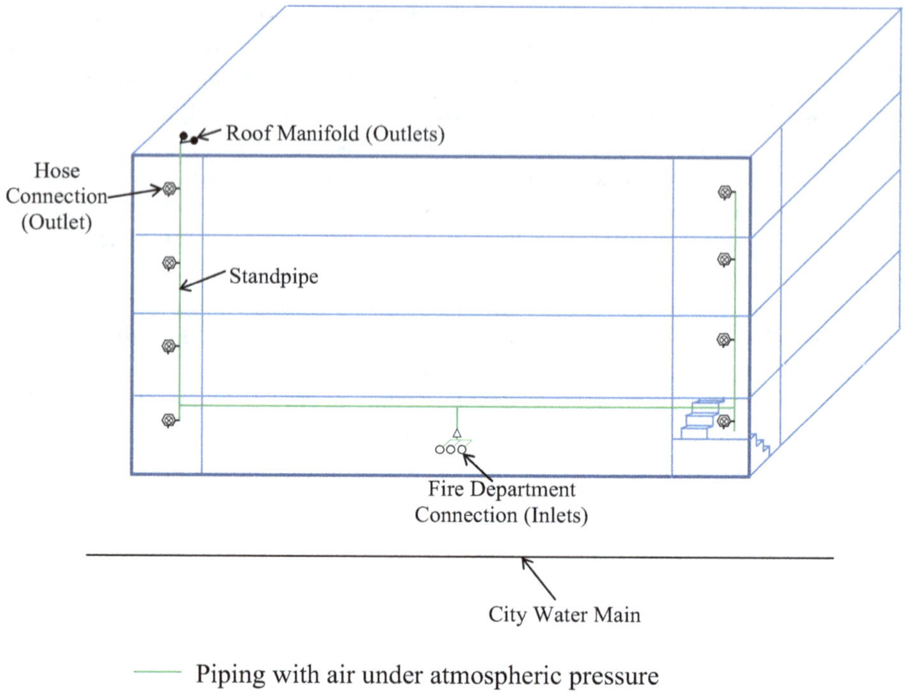

Fig. 2.6 Manual dry standpipe system

the fire is never get water. A frantic search through the building identifies a location low down in the system where the piping has rusted away and broken when water under pressure was introduced into the system. The story usually ends with the basement of the building filling with water and the fire department having to pump the water out of the basement in addition to fighting the fire.

This common story has firefighters concerned about the use of manual dry standpipe systems. These firefighters have legitimately identified the least reliable of the five different standpipe systems. As such, manual dry standpipe systems are not very common. Technically, they are permitted to be installed in non-high-rise buildings that are not heated, but there are so few of these types of building around that they are not common in new construction. There are still some manual dry standpipe systems in older systems. These systems need to be pressure tested on a regular basis to make sure that the piping is still holding together. More information on testing manual dry standpipe systems is included in Chap. 14 of this text.

Classification of Standpipe Systems

Standpipe systems are divided into three different classes of systems. The classes are completely independent of the different types discussed above. Each different class of system has different size hose connections and different expectations of flow and pressure for the water that will be necessary at the hose connection. When the different classes and types are all taken into consideration, there are theoretically 15 different combinations of type and class. For example, there could be a Class III automatic dry standpipe system or a Class I manual wet standpipe system. In reality, combinations such as Class II manual dry standpipe systems are not allowed since a manual system relies on the arrival of the fire department, but the Class II concept (as discussed below) is for use before the fire department arrives.

Class I Standpipe Systems

A Class I system is designed only for the use of trained firefighters. Large hose connections (2½ in.) are required so that the firefighters can get large volumes of water out of the connection to fight the fire. The reason that these systems are reserved for firefighters is that with the large hose connection and the significant water pressure available, the reaction forces as the water discharges from the nozzle at the end of the hose take serious effort to control. The concern is that an untrained member of the general public would not be able to hang onto the hose at the water flow and pressure that would typically discharge from the nozzle.

Class I systems never have hose permanently connected to the 2½ in. connection. It is expected that the fire department will bring the hose with them when they come to the fire. Figure 2.7 shows a common Class I standpipe system with a 2½ in. outlet in a stairwell.

Fig. 2.7 Class I standpipe with 2½ in. hose outlet

Class II Standpipe Systems

Class II standpipe systems have 1½ in. hose connections with hose permanently installed on a rack or hose reel. Class II standpipe systems were originally designed for untrained members of the general public to use, which is why the hose was already available. The hose connection is intentionally smaller than the hose connection on a Class I system so that the water flow from the system would be reduced. The assumption was that the untrained person should be able to control the nozzle with the reduced flow.

Class II systems have come under scrutiny because the hose is rarely maintained properly and there is concern about untrained people being sent to put out a fire without protective clothing. The hose could be considered an “attractive nuisance” during a fire situation and if someone gets hurt fighting the fire with the hose, the building owner could end up being held liable for providing the hose without training or protective clothing. For these reasons, Class II standpipe systems are getting more rare. In the United States, the only types of buildings that have Class II systems now are those where the building owner provides training and protective clothing for the employees in the building or where there is some expectation that trained people will be coming and using the hose. See Fig. 2.8 for an example of a Class II standpipe system.

Fig. 2.8 Class II standpipe system

Class III Standpipe Systems

For standpipe systems, one plus two really does equal three. A Class III standpipe system is supposed to be one that meets both of the sets of rules for Class I and Class III at the same time. In reality, this is getting harder to do as the rules for Class I and Class II continue to get farther apart. This subject will be discussed in more detail in Chaps. 4–6 of this text.

The original Class III systems had a 2½ in. hose connections with a reducer threaded onto a 1½ in. hose connection attached to a rack or reel of hose as shown in Fig. 2.9. If a member of the public wanted to fight the fire, they could grab the hose, pull it off of the rack or reel and take it to go fight the fire. If the fire department arrived on the scene, they could spin the reducer off of the 2½ in. connection, push the hose out of the way, and connect their own larger hose to get more water at higher pressure to fight the fire.

Another variation of the Class III standpipe system is to provide two separate connections on the same riser as shown in Fig. 2.10. One of the connections is 2½ in. in size for the firefighters to use and the other is 1½ in. in size attached to hose for building occupants. This arrangement costs more than the Class III arrangement shown in Fig. 2.9 and there is no significant advantage, so it is difficult to understand why someone would design such a system, yet they are seen around the United States, so someone must have thought it was a good idea.

In more recent times, with the hose being removed from most standpipe systems, the Class III systems remain with just a 2½ in. hose connection reduced down to a 1½ in. connection as shown in Fig. 2.11. Firefighters have their choice of using the smaller connection or the larger one, but since there is no hose permanently connected to the system, members of the general public can no longer use these systems to fight fires.

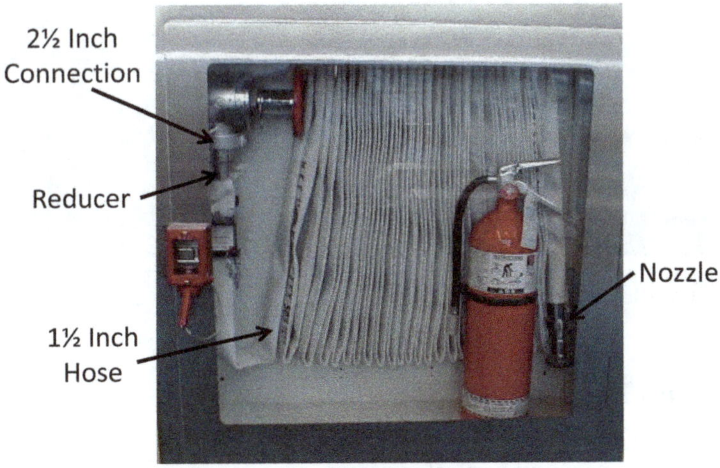

Fig. 2.9 Class III standpipe with hose

Fig. 2.10 Class III standpipe with separate connections

Fig. 2.11 Class III standpipe without hose

Vertical and Horizontal Standpipes

Most standpipe systems are vertical in nature, carrying water up through a building. The important pipes in these systems are vertical, in exit stairwells and other locations as discussed in Chap. 4 of this text. Some of the pipes might be horizontal in direction, but these pipes just join the standpipes together. Because these standpipes are so common, they are just called "standpipe systems" as opposed to vertical standpipes.

There are also standpipe systems that are primarily horizontal in nature. These tend to be in one or two story buildings that are very large in size. NFPA 14 defines a horizontal standpipe as one where the piping delivers the water supply for two or more hose connections on a single level. The installation requirements for horizontal standpipes are the same as other standpipe systems, but the calculation requirements are a bit different. The specific requirements for horizontal standpipes will be explored in Chap. 12 of this text.

Chapter 3
Buildings Required to Have Standpipes

It would be nice to keep this text as broad as possible without referencing any specific country or organization's code or standard on this subject, but then it would be impossible to discuss this subject in any more detail. Not only will the subject of which buildings need standpipe systems need to be covered, but specific rules regarding the design and installation of standpipe systems will need to be quoted in order to inform the reader of the growth, development and history behind these rules. This text will refer to the NFPA Codes and Standards, International Codes Council (ICC) Codes.

With the NFPA codes and standards being heavily adopted and enforced throughout North America, Central America, South America, and the Middle East, and with the ICC Codes being used throughout North America, referencing these codes and standards should give this text fairly wide appeal. References to the legacy codes will help to provide some perspective on how some of the rules that we currently use got to be what they are.

Unless otherwise discussed in the document, references to the ICC's International Building Code and International Fire Code will be to the 2015 edition. References to the NFPA Life Safety Code (NFPA 101) and the Building Construction and Safety Code (NFPA 5000) will be to the 2015 edition. References to the Standard for the Installation of Standpipe and Hose Systems (NFPA 14) will be to the 2016 Edition.

IBC/IFC Requirements

The ICC's International Building Code (IBC) and International Fire Code (IFC) address standpipe systems in section 905. The requirements are the same regardless of whether you look in the IBC or IFC. The requirements are debated by the Fire Safety Code Development Committee with the material that makes it through the code revision process placed in both documents using the same section number.

K.E. Isman, *Standpipe Systems for Fire Protection*,
DOI 10.1007/978-3-319-47750-3_3

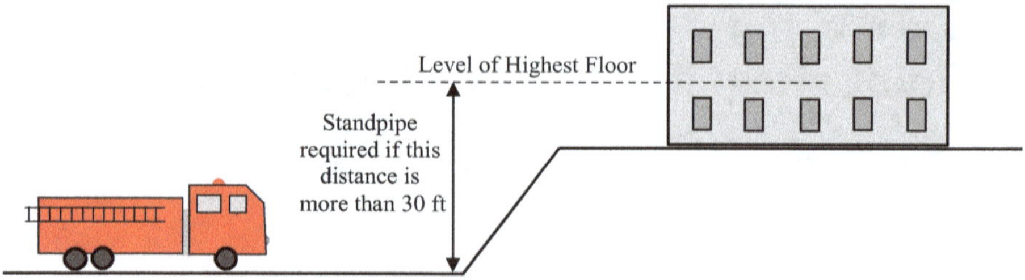

Fig. 3.1 Standpipe requirements by building height

That way, the code requirements are identical regardless of whether the building official or the fire official is the enforcement officer. The requirements for standpipe systems in the IBC and IFC are as follows:

Standpipes Required Based on Building Height

The IBC and IFC require that a standpipe system be installed in every building, regardless of the type of occupancy, where the floor of the highest story is more than 30 ft above the lowest level of fire department vehicle access as shown in Fig. 3.1. Note that the number of stories of the building is irrelevant to the discussion. Also note that the height of the roof of the building is irrelevant. The only important variable is the vertical distance from the level where the fire trucks pull up to the level of the highest floor in the building.

When trying to determine the lowest level of fire department vehicle access, the user of the IBC or IFC is not required to consider recessed loading docks for a small number of vehicles (four or fewer). With small loading docks, the fire department is not likely to use the loading dock with their trucks or to enter the building from the loading dock area. However, with larger loading docks, which you are required to consider for fire department vehicle access, the whole side of the building adjacent to the road might be taken up with the loading dock and the fire department might have to enter the building from this side. Since the code specifically says that the loading docks for four or fewer trucks can be ignored, it implies that the loading docks for five or more trucks need to be considered. Even if these larger docks are not on the street side of the building, they will be available for trucks to pull up and firefighters might need to access the building from these docks.

The user is also not required to take into consideration the topography around a building where the fire department vehicles will not be able to actually drive. Such areas of lawn, woods, or bodies of water around a building make it impractical or impossible to drive a fire truck around the building. Since the firefighters can't get to the building from these locations, they can be ignored when determining the vertical distance for compliance with the code.

Standpipe systems are also valuable for firefighters when the building is lower than the location where fire trucks can pull up, such as buildings constructed into a

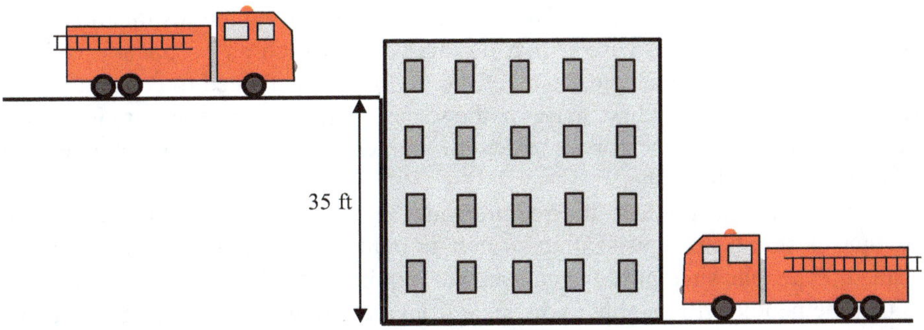

Fig. 3.2 Standpipe system required due to level of lowest floor below fire department vehicle access

hill and buildings with deep basements. The IBC and IFC also require standpipe systems in every building, regardless of the type of occupancy, where the floor of the lowest story is more than 30 ft above the highest level of fire department vehicle access. Figure 3.2 shows a building constructed into a hill that requires a standpipe system even though every floor is within 30 ft of one of the locations where fire trucks can pull up.

The base requirement in the IBC and IFC is for the standpipe system to be a Class III system, which needs to be one with an automatic water supply (automatic wet, automatic dry or semiautomatic dry), at least for the Class II portion. However, there are a number of exceptions. Class I standpipe systems are permitted in the following circumstances:

1. Buildings that are fully sprinklered in accordance with NFPA 13 or NFPA 13R.
2. Open parking garages where the top floor is within 150 ft of the lowest level of fire department vehicle access. In this case, the Class I system is permitted to be a manual system.
3. In open parking garages, manual dry standpipes are allowed if the garage is subject to freezing. The outlets for this Class I system need to be closer together per the distance requirements for Class II systems. The distance requirements between outlets will be discussed in more detail in Chap. 4 of this text.
4. In basements equipped with automatic sprinkler systems.

Standpipes for Assembly Occupancies

The IBC and IFC require standpipe systems in all assembly occupancies that have an occupant load greater than 1000 people regardless of the vertical distance from where the firefighters arrive and the highest floor level in the building. It is quite possible that a single story building as small as 7000–15,000 ft^2 could hit this threshold depending on how tightly packed the architect or engineer believes the occupancy will be. While a building between 7000 and 15,000 ft^2 is generally not considered overly large, especially when it is a single story, the justification for a standpipe system in this case is that during a fire in such an occupancy with an occupant load of 1000 people, many of the firefighters will need to concentrate on rescue efforts when they first arrive on the scene, so the standpipe system is helpful in saving effort and time for those few firefighters that start manual suppression when they get to the scene.

There is an exception to the requirement for a standpipe system in assembly occupancies with open air seating spaces. In outdoor assembly spaces with more than 1000 people, standpipe systems are not required at all because the assumption is that the people will be able to disburse from any fire without needing assistance from the firefighters. The assumption is also that the firefighters will be able to easily set up and fight the fire without needing standpipe systems. However, if the venue is a long distance away from water supplies or fire hydrants, a standpipe system might be helpful, even if it is not required by the Code.

The type of standpipe system required for these assembly occupancies is a Class I system. In buildings that are not high-rise, the Class I system is allowed to be automatic dry, semiautomatic dry, or manual wet. Manual dry systems are not permitted to meet the requirement for a standpipe system in assembly occupancies due to the concerns regarding the reliability of these systems discussed in Chap. 2. Automatic wet standpipe systems are required if the assembly occupancy is in a high-rise building because it will be more difficult for the firefighters to provide the pressure necessary to fight the fire on the higher levels of the building from the street level. For high-rise buildings, the ability to generate significant flow at a significant pressure at the higher floors of the building needs to be handled within the building.

A high-rise building is defined by the IBC and IFC as a building with an occupied floor more than 75 ft above the lowest level of fire department vehicle access. Note that it is not the height of the building that counts, nor is it the distance from the lowest floor to the highest floor. The important distance is the vertical measurement from the location outside where the fire department can pull up their trucks to the highest floor in the building.

Stages

In buildings that are unsprinklered or partially sprinklered, all stages that are greater than 1000 ft^2 in area are required to have a Class III standpipe system installed. The system is required to have both 1½ and 2½ in. outlets on both sides of the stage. The system in this case needs to meet all of the requirements for a standpipe system.

In buildings that are fully sprinklered, a standpipe system is still required if the stage exceeds 1000 ft^2, but the requirement for the size of the outlets changes to just 1½ in. hose connections on both sides of the stage. The designer in these cases is allowed to have a full Class II standpipe system installed, or the designer is allowed to connect the hose stations to the fire sprinkler system in the building. If the option is selected to install the hose connections to the sprinkler system, the hose connections will be considered a part of the sprinkler system and will not technically be considered a standpipe system, so none of the rules of NFPA 14 will apply.

Regardless of whether the building is sprinklered or not, hose is required to be connected to the 1½ in. outlet. The hose needs to be long enough to get water from one of the connections to every part of the stage area. The hose is required to be in a rack or cabinet. A nozzle is required at the end of the hose.

The 1000 ft^2 threshold for the standpipe or hose connection requirement is somewhat arbitrary. There has been a significant history of fires in theaters and the backstage workshops, storage areas and on the stage itself. Set construction includes extremely flammable materials and once these sets are finished, portions get hung from the ceiling to be dropped in at specific times during the play. These hanging portions of the set can be obstructions to ceiling sprinklers, making it more difficult to control a fire on the stage.

During the debate regarding the requirement for standpipes and hose connections for stages, the point was made that there are some small stages that typically don't use much in the way of scenery. These stages tend not to have much in the way of storage backstage or flammable materials. Rather than require standpipes for all stages, a threshold of 1000 ft^2 was established. While it was somewhat arbitrary, it was the size at which the members of the committee responsible were comfortable and it has remained the threshold for more than a generation with no real concern from stage owners.

Shopping Malls

The IBC and IFC require standpipe systems in all shopping malls. The malls that meet the height requirements previously discussed will be required to have full Class I or Class III systems. For those malls that have less height, Class I systems are required. Rather than a full standpipe system, the Class I outlets are permitted to be connected to the sprinkler system. In these cases, the water supply is not required to meet the full requirements of NFPA 14. More information on these special requirements is contained in the hydraulic calculation portion of this text in Chap. 6.

The IBC and IFC also modify the requirements for where the Class I outlets are required to be installed in these shopping malls that do not require a full standpipe system. The outlets only need to be located in five places:

1. Within the mall at the entrance to each exit passageway or corridor.
2. In the interior exit stairways that open directly on the mall. The outlets need to be installed at each floor-level landing.
3. For covered mall buildings, at exterior public entrances.
4. For open mall buildings, at public entrances along the perimeter line.
5. In any location within the mall that is more than 200 ft from one of the hose connections required above. Every location within the mall needs to be within 200 ft from a hose connection as measured along the path of travel.

Underground Buildings

All underground buildings are required to have a Class I standpipe system. The IBC and IFC allow the standpipe system to be an automatic wet or a manual wet standpipe system. There is no specific definition for what constitutes an "underground" building, but it is generally accepted that this is a building where methods of escape through windows is not possible due to the ground covering the windows. Even though windows are not used as means of egress, they do serve as last-ditch methods to save people during fires. Windows also serve as mechanisms for firefighters to ventilate buildings during a fire. Without these windows, fighting a fire in the building can be more difficult, so standpipe systems can help firefighters.

Since the lack of windows is what distinguishes an underground building, a legitimate question to ask would be whether a building that is above ground without windows or without windows that open might need standpipes as well. Such a discussion is extremely logical and would make a great deal of sense. Even though the IBC and IFC do not have a specific requirement for such windowless buildings or buildings with unopenable windows, it would be a good idea to include a standpipe system in such buildings.

Heliports

In buildings designed for helicopters to land on the roof, the standpipe system is required to be extended to the roof. The IBC and IFC permit the standpipe system to be a Class I or Class III system. The outlets on the standpipe system need to be within 150 ft of every portion of the heliport portion of the roof.

The way that this requirement is written, for the standpipe system to be extended to the roof, assumes that the building will have a standpipe system in the first place. Most buildings with heliports on the roof will be tall enough to require a standpipe system anyway. But if a heliport is designed for a short building that is not otherwise required to have a standpipe system, one will have to be installed due to the heliport.

The need for a standpipe system on the heliport is fairly obvious. Helicopters can break down and crash during landing or taking off. In addition, helicopters can be challenging to land in difficult weather conditions and heliports are generally installed on buildings like hospitals where they are needed in emergency conditions, making their use during bad weather something that occurs often enough that there needs to be a plan in place for how to deal with a crash if it does occur.

The requirement for there to be enough outlets so that every portion of the heliport is within 150 ft of an outlet generally comes from the idea that the firefighters will connect 100 ft of hose to the outlet. The firefighters will stretch that hose as far as they can, and the water will flow in a stream from the nozzle. At the flow and pressure typical to a Class I or Class III system, the water should go at least 50 ft from the nozzle. This is referred to in the industry as, "100 ft of hose and 50 ft of 'throw'" to justify the distance from any point in the building to a hose connection.

Marinas

NFPA 303 contains the requirements for which marinas require a standpipe system. The 2016 edition of NFPA 303 requires Class I standpipe systems for all piers and marina buildings in two situations:

1. Where the distance that firefighters need to lay the hose from the apparatus (at the closest place they can pull up) to any portion of the structure is more than 150 ft.
2. Any building for the rack storage of boats.

Once it has been determined that the marina needs a standpipe system, the outlets need to be located within 150 ft of every part of the marina. The distance of 150 ft was determined using the same 100 ft of hose and 50 ft of throw discussed under heliports. In addition, flags or other obvious means of identifying the outlet locations are required. The flags or identifying locators need to be visible from all parts of the pier.

In general, the standpipe systems for piers and marinas follow the same installation rules as other standpipe systems, but NFPA 303 has some special allowances. First, hose, hose racks, and hose reels are not required. It is assumed that firefighters will bring whatever hose they need and that nobody will need to try and fight any fire before firefighters arrive.

Second, the supply piping needs to be sized for a minimum flow rate of 300 gpm. This varies from NFPA 14 by 50 gpm since that standard requires a minimum flow of 250 gpm for pipes that only serve a single hose connection. What is not clear is whether the intent of the 300 gpm rule in NFPA 303 is meant to nullify the requirement in NFPA 14 to size the pipe for flows of 500 gpm for pipes that serve two hose connections or 750 gpm for pipes that serve three or more hose connections. When performing hydraulic calculations for standpipe systems on piers or marinas, it is probably best to calculate the system for 300 gpm if it serves a single outlet, 500 gpm if it serves two outlets and 750 gpm if it serves three or more outlets.

NFPA 303 specifically allows manual dry standpipe systems. It is important for NFPA 303 to come right out and say this since many firefighters have very bad experiences with manual dry standpipe systems as described in Chap. 2 of this text.

NFPA 303 also allows the authority having jurisdiction to permit unlisted flexible connections on floating piers. These flexible connections are important in allowing the standpipe system to move differentially as the floating pier moves up and down with the tide. Note that the designer is not permitted to use the flexible connections in all cases. Instead, the designer is required to get the permission of the authority having jurisdiction prior to deciding to use the flexible connections.

Buildings Under Construction or Demolition

Section 3311.1 of the IBC requires a portion of the standpipe system to be in service during the construction of a building if the building is one of those that requires a standpipe system once it is built. During construction, only one of the standpipes needs to be in service. That standpipe needs to be in service by the time that the construction has exceeded 40 ft in height above the lowest level of fire department vehicle access. The hose connections (outlets) need to be near the usable stairways at accessible locations. As construction continues upward, the standpipes need to be extended with no more than one secured deck or floor above the highest operational standpipe outlet.

In the case of a building being demolished, the situation is the same in reverse. If the building has a standpipe system before demolition starts, then the standpipe system needs to remain in service as the building is being taken down. Only one floor is permitted above the highest operational outlet on the standpipe system.

The standpipes being used construction or demolition can be permanent or temporary. They are allowed to exist without a water supply as long as the fire department can pump into the standpipe at the level of fire department vehicle access and get water out on the floors of the building.

NFPA 101 Requirements for Standpipe Systems

The Life Safety Code (NFPA 101) is focused on minimum building requirements for maintaining life safety during fires and other similar events. There aren't many requirements for standpipe systems in the Life Safety Code because standpipe systems are seen as tools for firefighters to use when they arrive on the scene of a fire, which is typically 15 or 20 min after ignition. By this time, most fire deaths will have occurred, so standpipes are not a primary tool used by the Life Safety Code.

But there are a few circumstances where the Life Safety Code does require standpipe systems. Like the IBC and IFC, they tend to be for tall buildings or large buildings where firefighters need help advancing hose deep into a building. The Life Safety Code is somewhat unusual as codes go because it applies to both new and existing buildings. In the requirements listed below, a distinction will be made between the requirements for new buildings and the requirements for existing buildings. Standpipe systems are required in the following types of buildings:

Regular Stages: A regular stage is one where the vertical distance from the lowest level of the stage floor to the highest level of the roof or floor above is 50 ft or less. All new and existing regular stages over 1000 ft^2 or more in area are required to have 1½ in. outlets on both sides of the stage. The hose connections can be extended from the sprinkler system or can be full Class II or Class III standpipe systems at the discretion of the designer (sections 12.4.6.12 and 13.4.6.12).

Legitimate Stages: A legitimate stage is one where the vertical distance from the lowest level of the stage floor to the highest level of the roof or floor above is more than 50 ft. All new legitimate stages are required to have 1½ in. outlets on both sides of the stage. All existing legitimate stages that are more than 1000 ft^2 in area are required to have 1½ in. outlets on both sides of the stage. The hose connections can be extended from the sprinkler system or can be full Class II or Class III standpipe systems at the discretion of the designer (sections 12.4.6.12 and 13.4.6.12).

Air Traffic Control Towers: A Class I standpipe system is required in all new air traffic control towers where the floor of the cab is more than 30 ft above the lowest level of fire department vehicle access. The system is allowed to be a manual system as long as this is permitted by the authority having jurisdiction (section 11.3.4.5.3).

Detention and Correctional Occupancies: A Class I standpipe system is required for all new and existing detention and correctional occupancies (such as prisons and jails) that are three or more stories in height. This requirement is based on the assumption that the facility is fully sprinklered (as is required by the Life Safety Code for these occupancies). However, many operators of jails and prisons do not like fire sprinklers, so they apply for code variances to get out of putting sprinklers in their facilities. In recognition of this, the Life Safety Code specifically requires that a Class III standpipe system be installed if the facility is not fully sprinklered (sections 22.3.5.5 and 23.3.5.5).

High-Rise Assembly Occupancies: A Class I standpipe system is required for all new high-rise assembly buildings and all new assembly occupancies in the high-rise portion of another occupancy (section 12.4.4).

High-Rise Educational Occupancies: A Class I standpipe system is required for all new high-rise educational buildings (section 14.4.2).

High-Rise Day Care Occupancies: A Class I standpipe system is required for all new and existing buildings that have day care occupancies in the high-rise portion of the building (section 16.4.2 and 17.4.2).

High-Rise Health Care Occupancies: A Class I standpipe system is required in all new high-rise health care occupancies (section 18.4.2).

High-Rise Ambulatory Health Care: A Class I standpipe system is required in all new high-rise ambulatory health care occupancies (section 20.4.2).

High-Rise Hotel and Dormitories: A Class I standpipe system is required in all new high-rise hotel and dormitory occupancies (section 28.4.1.1).

High-Rise Apartments: A Class I standpipe system is required in all new high-rise apartments (section 30.4.1.1).

High-Rise Board and Care Occupancies: A Class I standpipe system is required in all new high-rise residential board and care facilities. If the roof is sloped more than 3 in 12, the standpipes are not required to be extended to the roof (sections 32.3.3.9.1, 32.3.3.9.2, and 32.3.4.1).

High-Rise Mercantile Occupancies: A Class I standpipe system is required in all new high-rise mercantile occupancies (section 36.4.2).

High-Rise Business Occupancies: A Class I standpipe system is required in all new high-rise business occupancies (section 38.4.2).

High-Rise Industrial Occupancies: A Class I standpipe system is required in all new high-rise industrial occupancies (section 40.4.2.1).

High-Rise Storage Occupancies: A Class I standpipe system is required in all new high-rise storage occupancies (section 42.4.2).

Building Rehabilitation: Rehabilitation of a building is defined as repair, renovation, modification, reconstruction, change of use, change of occupancy classification, or an addition to the building. The Life Safety Code requires that any building with a standpipe system has to keep the standpipe system functional up to the highest point of any rehabilitation work for all of the time that the rehabilitation work is going on. In fully sprinklered buildings, the standpipe is not required to have a fire pump during rehabilitation work as long as the system can deliver 250 gpm at 65 psi to the top floor. If the building is not fully sprinklered, the standpipe is not required to have a fire pump during rehabilitation work as long as the system can deliver 500 gpm at 65 psi to the top floor (sections 43.6.4.4 and 43.6.4.6).

NFPA 5000 Requirements for Standpipe Systems

The NFPA Building Construction and Safety Code (NFPA 5000) is a traditional building code. It only applies to new buildings. Although it has not been adopted by many jurisdictions within the United States, it has seen significant use in the Middle East and Central and South America. The requirements for standpipe systems in NFPA 5000 are very similar to those for new occupancies in the Life Safety Code. Standpipes are required by NFPA 5000 in the following situations:

Regular Stages: A regular stage is one where the vertical distance from the lowest level of the stage floor to the highest level of the roof or floor above is 50 ft or less. All regular stages over 1000 ft^2 or more in area are required to have 1½ in. outlets on both sides of the stage. The hose connections can be extended from the sprinkler system or can be full Class II or Class III standpipe systems at the discretion of the designer (section 16.4.6.10).

Legitimate Stages: A legitimate stage is one where the vertical distance from the lowest level of the stage floor to the highest level of the roof or floor above is more than 50 ft. All legitimate stages are required to have 1½ in. outlets on both sides of the stage. The hose connections can be extended from the sprinkler system or can be full Class II or Class III standpipe systems at the discretion of the designer (section 16.4.6.10).

Air Traffic Control Towers: A Class I standpipe system is required in all new air traffic control towers where the floor of the cab is more than 30 ft above the lowest level of fire department vehicle access. The system is allowed to be a manual system as long as this is permitted by the authority having jurisdiction (section 31.6.13.5.3).

Assembly Occupancies: Standpipe systems are required in the following assembly occupancies:

- Class I standpipe systems are required in all buildings four or more stories in height (section 16.3.5.2.1).
- Class I standpipe systems are also required in all buildings where at least one occupiable level is more than 30 ft above the level of fire department access (sections 16.3.5.2.2).
- Class I standpipe systems are required in all buildings having four or more basement levels (section 16.3.5.2.1).
- Class I standpipe systems are also required in all buildings where at least one occupiable level is more than 30 ft below the level of fire department access (section 16.3.5.2.2).
- Class I standpipe systems are required in all buildings not protected throughout by a complete, electrically supervised sprinkler system in where there is an occupiable area more than 150 ft from the closest point of fire department entry into the building (section 16.3.5.2.3).

Educational Occupancies: Standpipe systems are required in the following educational occupancies:

- Class I standpipe systems are required in all buildings four or more stories in height (section 17.3.5.6.1).
- Class I standpipe systems are required in all buildings having four or more basement levels (section 17.3.5.6.1).
- Class I standpipe systems are required in all buildings not protected throughout by a complete, electrically supervised sprinkler system in where there is an occupiable area more than 150 ft from the closest point of fire department entry into the building (section 17.3.5.6.2).

Day Care Occupancies: Standpipe systems are required in the following day care occupancies:

- Class I standpipe systems are required in all buildings four or more stories in height (section 18.3.5.6.1).
- Class I standpipe systems are required in all buildings having four or more basement levels (section 18.3.5.6.1).
- Class I standpipe systems are required in all buildings not protected throughout by a complete, electrically supervised sprinkler system in where there is an occupiable area more than 150 ft from the closest point of fire department entry into the building (section 18.3.5.6.2).

Health Care Occupancies: Standpipe systems are required in the following health care occupancies:

- Class I standpipe systems are also required in all buildings where at least one occupiable level is more than 30 ft above the level of fire department access (sections 19.3.5.8).
- Class I standpipe systems are also required in all buildings where at least one occupiable level is more than 30 ft below the level of fire department access (section 19.3.5.8).
- Class I standpipe systems are required in all buildings not protected throughout by a complete, electrically supervised sprinkler system in where there is an occupiable area more than 200 ft from the closest point of fire department entry into the building (section 19.3.5.8).

Ambulatory Care Facilities: Class I standpipe systems are required in the following ambulatory care facilities (section 20.3.5.4):

- All buildings four or more stories in height.
- All buildings more than 50 ft above grade plane with intermediate stories or balconies.
- All buildings with more than one story below grade plane.
- All buildings more than 20 ft (6100 mm) below grade plane.

Detention and Correctional Occupancies: Standpipe systems are required in the following detention and correctional occupancies:

- Class I standpipe systems are required in all buildings four or more stories in height (section 21.3.5.5.1).
- Class I standpipe systems are also required in all buildings where at least one occupiable level is more than 30 ft above the level of fire department access (sections 21.3.5.5.1).
- Class I standpipe systems are required in all buildings having four or more basement levels (section 21.3.5.5.1).
- Class I standpipe systems are also required in all buildings where at least one occupiable level is more than 30 ft below the level of fire department access (section 21.3.5.5.1).
- Class I standpipe systems are required in all buildings where there is an occupiable area more than 200 ft from the closest point of fire department entry into the building (section 21.3.5.5.1).
- Class III standpipe systems are required to be provided in all nonsprinklered buildings three stories or more in height.

Hotels and Dormitories: Class I standpipe systems are required in the following hotels and dormitories. If the roof slopes more than 3 in 12, the standpipes are not required to be extended to the roof (section 24.3.5.10):

- All buildings four or more stories in height.
- All buildings more than 50 ft above grade plane with intermediate stories or balconies.

- All buildings with more than one story below grade plane.
- All buildings more than 20 ft (6100 mm) below grade plane.

Apartments: Class I standpipe systems are required in the following apartment buildings. If the roof slopes more than 3 in 12, the standpipes are not required to be extended to the roof (section 25.3.5.11):

- All buildings four or more stories in height.
- All buildings more than 50 ft above grade plane with intermediate stories or balconies.
- All buildings with more than one story below grade plane.
- All buildings more than 20 ft (6100 mm) below grade plane.

Residential Board and Care Facilities: Class I standpipe systems are required in the following residential board and care facilities. If the roof slopes more than 3 in 12, the standpipes are not required to be extended to the roof (section 26.3.3.9):

- All buildings four or more stories in height.
- All buildings more than 50 ft above grade plane with intermediate stories or balconies.
- All buildings with more than one story below grade plane.
- All buildings more than 20 ft (6100 mm) below grade plane.

Mercantile Occupancies: Class I standpipe systems are required in the following mercantile occupancies (section 27.3.5.4):

- All buildings four or more stories in height.
- All buildings more than 50 ft above grade plane with intermediate stories or balconies.
- All buildings with more than one story below grade plane.
- All buildings more than 20 ft (6100 mm) below grade plane.
- For mall buildings with no standpipe system in accordance with one of the above situations, 2½ in. hose connections are required and are permitted to be supplied by the sprinkler system in the mall. The system is required to provide at least 250 gpm to the most hydraulically remote single outlet. The outlets are required within the mall at the entrance to each exit passage or corridor, at each floor level landing within enclosed stairways opening directly onto the mall, and at exterior public entrances to the mall.

Business Occupancies: Class I standpipe systems are required in the following business occupancies (section 28.3.5.2):

- All buildings four or more stories in height.
- All buildings more than 50 ft above grade plane with intermediate stories or balconies.
- All buildings with more than one story below grade plane.
- All buildings more than 20 ft (6100 mm) below grade plane.

Industrial Occupancies: Class I standpipe systems are required in the following industrial occupancies (section 29.3.5.2):

- All buildings four or more stories in height.
- All buildings more than 50 ft above grade plane with intermediate stories or balconies.
- All buildings with more than one story below grade plane.
- All buildings more than 20 ft (6100 mm) below grade plane.

Storage Occupancies: Class I standpipe systems are required in the following storage warehouses (section 30.3.5.4):

- All buildings four or more stories in height.
- All buildings more than 50 ft above grade plane with intermediate stories or balconies.
- All buildings with more than one story below grade plane.
- All buildings more than 20 ft (6100 mm) below grade plane.
- Where the storage occupancy is an open parking structure, the system is allowed to be a manual dry system.

Building Rehabilitation: Rehabilitation of a building is defined as repair, renovation, modification, reconstruction, change of use, change of occupancy classification, or an addition to the building. The Life Safety Code requires that any building with a standpipe system has to keep the standpipe system functional up to the highest point of any rehabilitation work for all of the time that the rehabilitation work is going on. In fully sprinklered buildings, the standpipe is not required to have a fire pump during rehabilitation work as long as the system can deliver 250 gpm at 65 psi to the top floor. If the building is not fully sprinklered, the standpipe is not required to have a fire pump during rehabilitation work as long as the system can deliver 500 gpm at 65 psi to the top floor (section 15.6.2.5.5).

Chapter 4
Installation Rules for Hose Connections

Hose connections are the outlets of the standpipe system. If the hose connection already has hose attached, it is called a hose station. Hose stations will be addressed in the Class II discussion later in this Chapter. For hose connections that don't normally have hose, a protective cap is threaded over the couplings. During a fire, the firefighters unthread the protective cap and thread their hose onto the coupling. Once the hose is attached, the firefighters can open the valve, make sure they have water, and go fight the fire.

The proper location for hose connections is important because the firefighters use the hose connection as a starting point. They make their plans on how much hose they will carry in their packs and how they will attack the fire based on the location of the hose connections. Therefore, the key to standpipe system design is knowing where to put the hose connections.

The contents of this chapter will mostly come from the 2016 edition of NFPA 14—Standard for the Installation of Standpipe and Hose Systems. NFPA 14 is referenced by most building codes and fire codes in the United States as well as the Life Safety Code, which is used internationally and has been adopted as a minimum requirement by a number of U.S. Federal Agencies including the Centers for Medicare and Medicaid Services (CMS) for health care occupancies. The IFC and IBC do make a few modifications to the requirements of NFPA 14, which will be discussed in this chapter. Previous editions of NFPA 14 had similar requirements. Some of the more notable rules in previous editions that were different than the 2016 edition will be discussed in this chapter in the context of explaining how the rules have evolved.

K.E. Isman, *Standpipe Systems for Fire Protection*,
DOI 10.1007/978-3-319-47750-3_4

Location of Class I Hose Connections

The hose connections on Class I systems have 2½ in. threads so that firefighters can attach large hose and attack the fire with large quantities of water. These hose connections are only for firefighter use, so they will not have hose under normal conditions. Firefighters will bring the hose to the fire when it occurs.

For more than 100 years, there has been a standardized hose thread so that manufacturers of equipment can easily supply couplings for hose connections and so that fire departments can respond to fires outside their immediate jurisdiction and provide assistance. Unfortunately, fire departments have not all agreed to use the standard thread. In 2016, there are more than 100 fire departments in the United States that are using a unique thread that is different from the standard thread and different from all other fire departments. This means that the designer of a standpipe system needs to make sure that the couplings on the hose connection have a compatible thread with the local fire department.

There are five different locations within a building where hose connections will be required. The general concept when placing hose connections is to remember that the firefighters need to stay protected until they have attached their hose to the hose connection. Once they have the hose filled with water, they can protect themselves from the fire and it is safe to go fight the fire. Therefore, the hose connection needs to be in a location where the firefighter is protected. If the designer is ever confused about where the hose connections need to go, remembering this simple concept of keeping the firefighters protected until they have water in the hose should help in the decision making.

For each of the locations where Class I hose outlets are installed, they need to be at the proper distance from the floor. They should not be too close to the floor because it is difficult for firefighters to bend over and connect the hose. They should also not be too high above the floor so that firefighters do not have to reach too high over their heads or stand on ladders just to connect their hose. A good comfortable distance above the floor would be waist level for the average firefighter. NFPA 14 calls for the distance from the floor to the middle of the valve to be at least 3 ft and no more than 5 ft (section 7.3.1.1).

Outlets in Exit Stairwells

The most common place to find Class I hose connections is in exit stairwells. NFPA 14 requires the hose connections to be installed in every exit stairwell, regardless of how closely spaced the exit stairwells might be. Even though exit stairwells might be spaced closely together enough that firefighters could reach every part of the building with their hose from only a single stairwell, the hose connections need to be placed in all of the exit stairwells. The reason for this is so that firefighters do not have to guess at which stairwell has the hose connections. During a fire, the firefighters need to know that they can go into any stairwell and find hose connections that they can use to fight the fire.

There are two types of stairs that can be designed by an architect. The first is where the stairwell goes straight up from one floor to another with no intermediate landings or changes of direction. The second is a stairwell where the stairs change direction and at least one (sometimes more than one) intermediate landing exists between floor levels of the building. The decision will be made by the architect based on how much room there is in the building for the stairwell and where the users of the stairs need to come out at each floor level.

Regardless of which stair design has been decided on by the architect, the standpipe hose connections need to be placed at one of the landings. The hose connections should not be placed adjacent to the stairs themselves because firefighters need a stable platform to connect their hose and start their work. It would be too difficult for them to wrestle a stubborn cap off of a hose connection and open a tight valve when their feet are at different levels on different steps. So, hose connections have always been placed at a landing.

For stairs where there is no intermediate landing, the standpipe hose connections will be placed at the main floor landings. There really isn't any other option. The hose connection needs to be placed so that it is out of the way of egress in the stairwell, but so that the firefighters can still connect their hose and turn the valve without hitting the wall with the hose or their hands. Figure 4.1 shows a plan view of a portion of a stairwell with the egress path highlighted and the standpipe connection well out of the way of egress, but still accessible to firefighters.

Note that in Fig. 4.1, the hose connection comes out of the standpipe parallel to the wall, but still far enough away from the wall so that the firefighter can get a hand

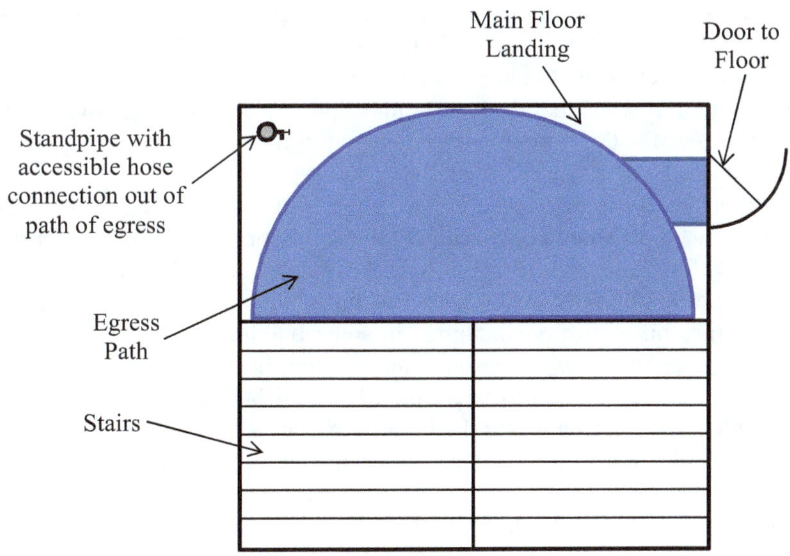

Fig. 4.1 Hose connection is accessible and clear of egress path

Fig. 4.2 Hose connection at main floor landing

around the hand wheel and turn it. Turning the wheel is how the firefighter opens the connection and gets water in the hose. The designer and installer need to make sure that there is sufficient clearance between the wall and the hand wheel for the firefighter, with a glove on, to get their hand around the wheel. Recent editions of NFPA 14 require a minimum clearance of 3 in. around the hand wheel so that the firefighter can access the wheel (see section 4.7.5). Figure 4.2 shows a standpipe at a main floor landing, accessible to firefighters and out of the path of egress. This hand wheel is about as close as should be installed and still provide access to the firefighters. Also note in Fig. 4.2 that the hose connection is at a slight angle to the vertical, making it easier for the firefighters to connection their hose and stretch it up the stairs or out the door without kinking the hose.

For stairs that have intermediate landings there is a continual debate as to whether the hose connections should be placed at the main floor landing or the intermediate floor landing. There are pros and cons to each location. In the 23 years from 1990 to 2013, the rules in NFPA 14 changed several times on this issue, so it is common to find standpipes on main floor landings in some buildings and intermediate floor landings in other buildings in the same jurisdiction. In the 2016 edition, NFPA 14 requires the hose connection to be at the main floor landing (section 7.3.2.1), but then provides an exception where the fire department is allowed to require the hose connections at the intermediate floor landing (if there is more than one intermediate floor landing between stories of the building, the hose connection is required at the highest intermediate floor landing) in section 7.3.2.1.1.

If a fire department is going to require hose connections at the intermediate floor landings, they need to do so through an official process where the intent to make the rule is publicized and the public has a right to provide input. After that, if the fire department still wants to pass the requirement, they should do it as an official part of the fire code and make the change available to the installing contractors that are doing business in their community. The rise of the Internet has helped to provide fire departments with a fair method of posing such local requirements in a location where everyone has access. If these local requirements get posted in a prominent location on a fire department's website, it helps contractors that do not frequently do work in a specific jurisdiction get their design and installation correct in the rare occasions that they do.

The advantage of putting hose connections on the main floor landings start with the fact that the main landings are generally larger than the intermediate landings and allow the firefighters more room to prepare to fight the fire. Another advantage of the main floor landing is that the riser is in a good position to also serve the fire sprinkler system. See Fig. 4.3, which shows a fire sprinkler system connected to the same standpipe riser that also serves the Class I hose connection.

Having the hose connection at the main floor landing can also save hose for firefighters if the standard operating procedures for the fire department call for putting two hose lines in operation as shown in Fig. 4.4 with the hose connected at the floor of the fire (green hose) and at the floor below (purple hose). This saves firefighters from having to run extra hose up some of the stairs, however, some fire departments do not like to connect hose right at the floor of the fire because smoke getting into the stairwell when firefighters open the door might affect a firefighter standing by

Fig. 4.3 Standpipe riser serving Class I hose connection and sprinkler system

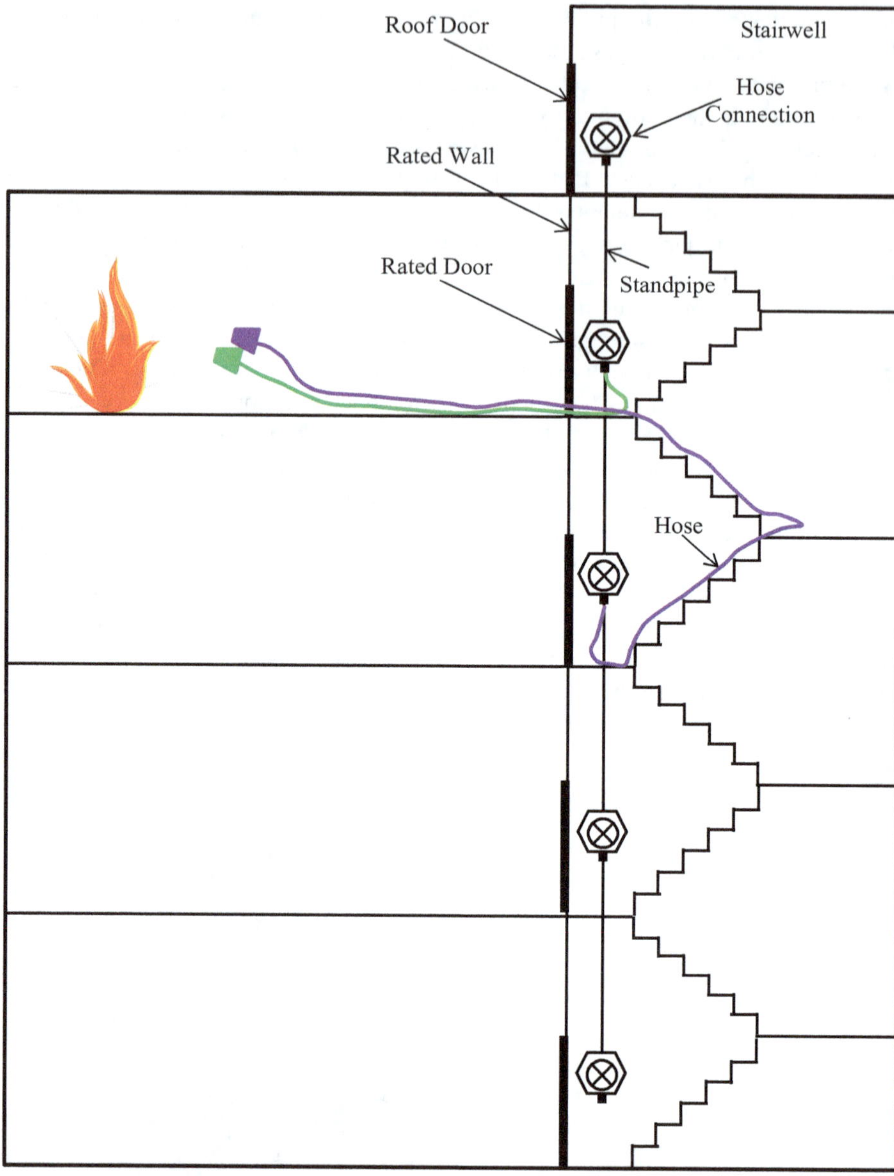

Fig. 4.4 Standpipe with hose connections at main floor landings and hose in service from the fire floor and the floor below

the hose connection. In these cases, the fire department's standard operating procedure might call for using the two connections below the fire floor, which then takes an additional length of hose as shown in Fig. 4.5.

Firefighters that are opposed to putting hose connections at main floor landings generally raise three issues in the discussion. First, they point to the difficulty of

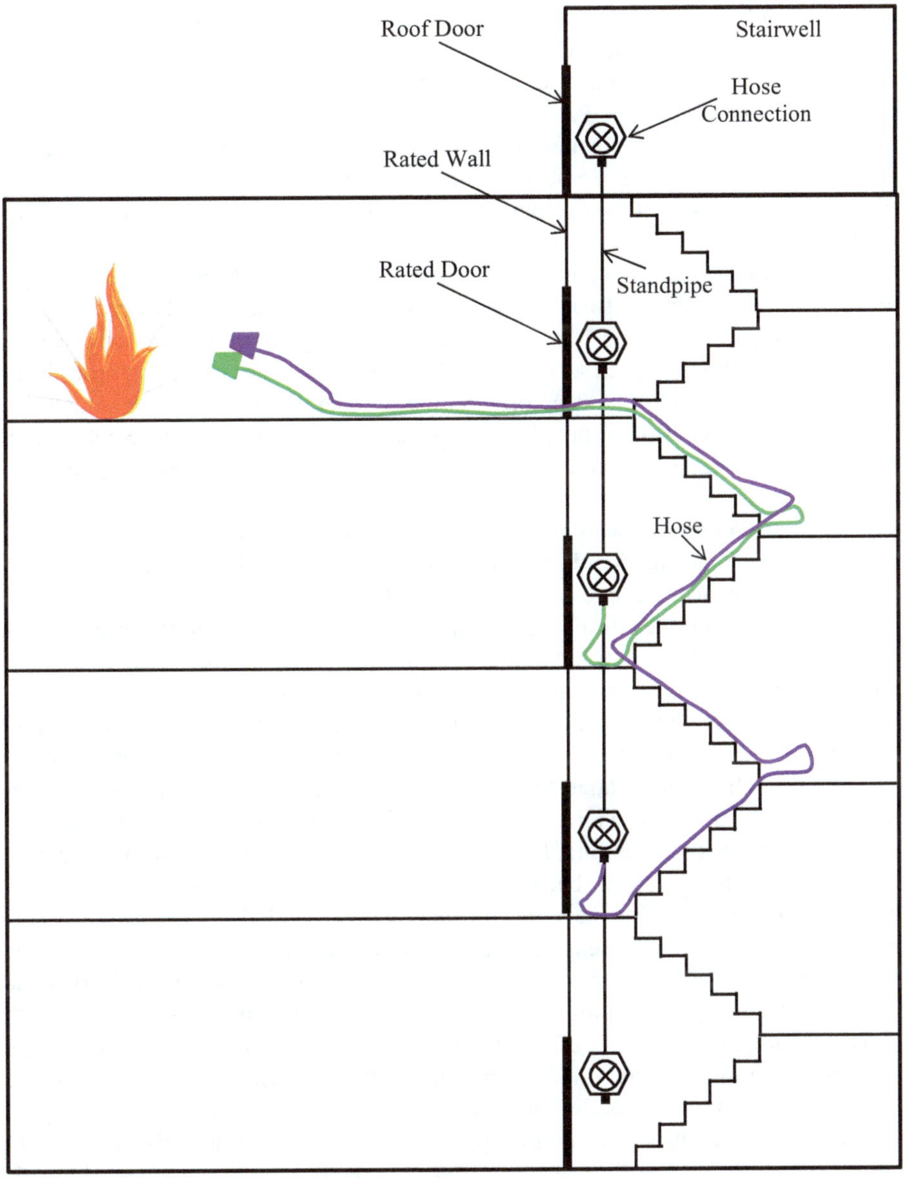

Fig. 4.5 Standpipe with all hose connected below fire floor

connecting 100 ft or more of hose at the main landing as shown in Fig. 4.4 and then trying to go right through the adjacent door. Large hose, when filled with water tends to be very stiff and it is difficult to get the hose to bend from the connection, which is typically close to the door, and go right through the door.

The second argument that they raise when discussing this issue is that since the Americans with Disabilities Act (ADA) was passed in 1990, a place of refuge for people who can't use the stairs (such as people in wheelchairs) needs to be provided on each floor of a building. Typically, this place of refuge is the main floor landing in a stairwell. People who cannot use the stairs go to this place of refuge and await firefighter rescue. If there are people waiting on the main floor landing for rescue, they will be in the way if the firefighters also need to put hose in service on this floor landing.

The argument that people using the stairwell as a place of refuge seems to be a weak argument. If one or two people are in such a position (the ADA requires the place of refuge to be sized for two people), the firefighters could easily evacuate them before putting hoses in service. Fire department standard operating procedures call for evacuation of civilians before firefighting operations commence anyway, so it would not matter if these people were at the same landing as a hose connection because the firefighters would get them out of the way before connecting hose to fight the fire anyway.

The third argument that is used against the hose connection being at the main floor landing is that the position of the hose at the intermediate landing saves the firefighters from using some hose. This is only true if the operations of the fire department call for the use of all hose connections to be below the fire floor as shown in Fig. 4.5. If the fire fighters use the hose connections as shown in Fig. 4.4, having the hose connections on the main landing saves hose.

If the hose connections are placed at the intermediate landings, as shown in Fig. 4.6, then it does allow the firefighters to stretch out the hose, go up the stairs and straight through the door rather than having to make a sharp turn immediately after the hose connection. This is the most logical argument for having the hose connection at the intermediate landing. However, this makes the standpipe riser difficult to use for the sprinkler system because the connection for the sprinkler system will need to be higher than fitters can reach from the floor and ladders (the height that is necessary to use to reach the point that will be needed for the connection) can't be used safely in stairs. In order to connect sprinkler systems to standpipe risers at the intermediate level, scaffolding typically needs to be built in the stairwell, which adds considerably to the cost of construction. Figures 4.7 and 4.8 show a standpipe system and sprinkler system in the same stairwell using separate risers because the standpipe system has the outlets at the intermediate landings and it was too expensive to use the standpipe riser for the sprinkler system, so a completely independent sprinkler system riser was added with connections at the main landing. This ultimately cost the building owner more than if the standpipe outlets were at the main floor landing and the standpipe riser was combined with the sprinkler riser.

Although the requirements have gone back and forth over the years, the current preference of NFPA 14 is to put the standpipe outlets at the main floor level. Fire departments are allowed to pass rules that require the outlets at the intermediate landings, so before a designer finalizes the standpipe system design, they should check the requirements of the local fire department. Regardless of whether the hose connections are at the main floor landing or the intermediate landing, there will be hose connection required for Class I systems in the exit stairwells.

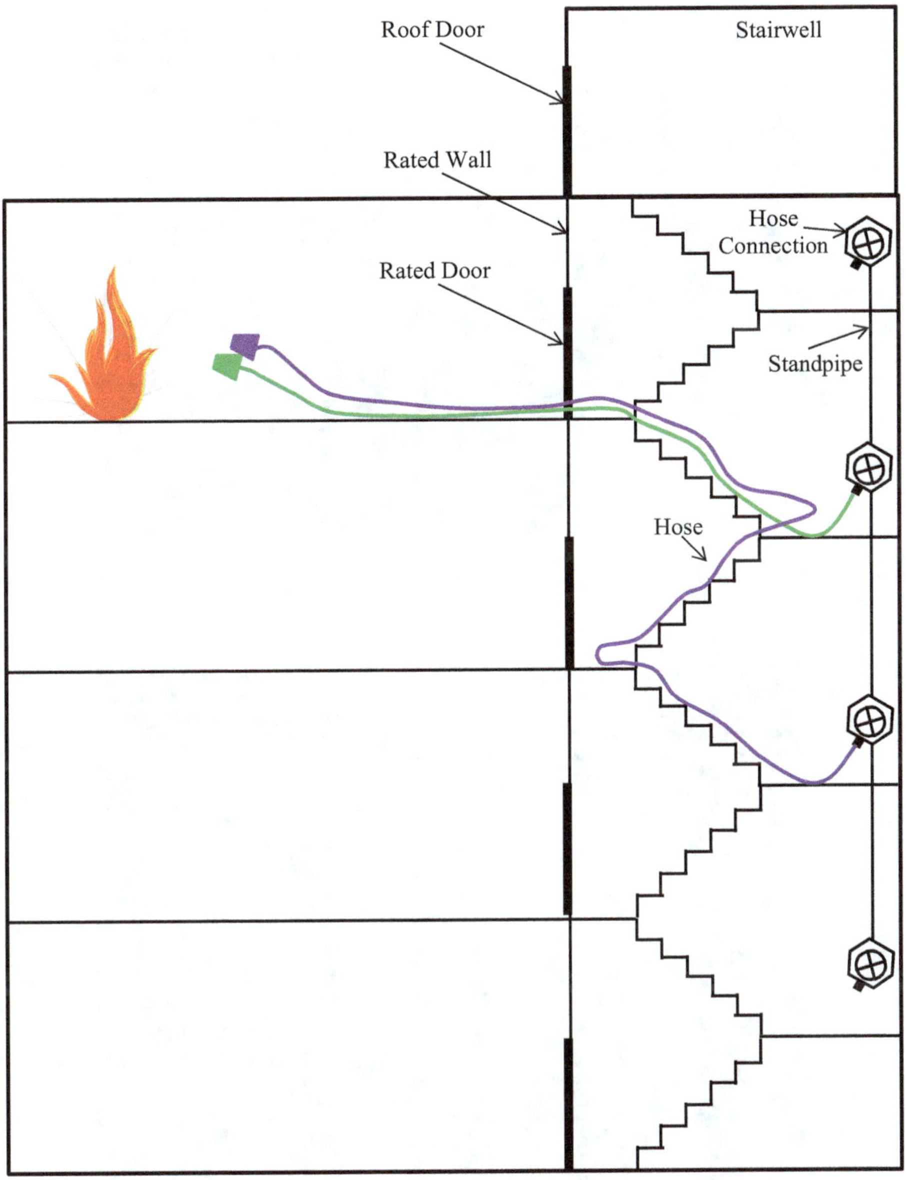

Fig. 4.6 Hose connections at intermediate landings

The International Building Code has come to the opposite conclusion of NFPA 14 regarding the landing at which the Class I outlet needs to be installed. Section 905.4 requires the Class I outlets to be at the intermediate landings unless otherwise approved by the authority having jurisdiction (AHJ). Legally, the IBC rules take precedence over the NFPA 14 rules. Since both documents specifically allow the

Fig. 4.7 Standpipe riser with hose connections at intermediate landings and no sprinkler connection

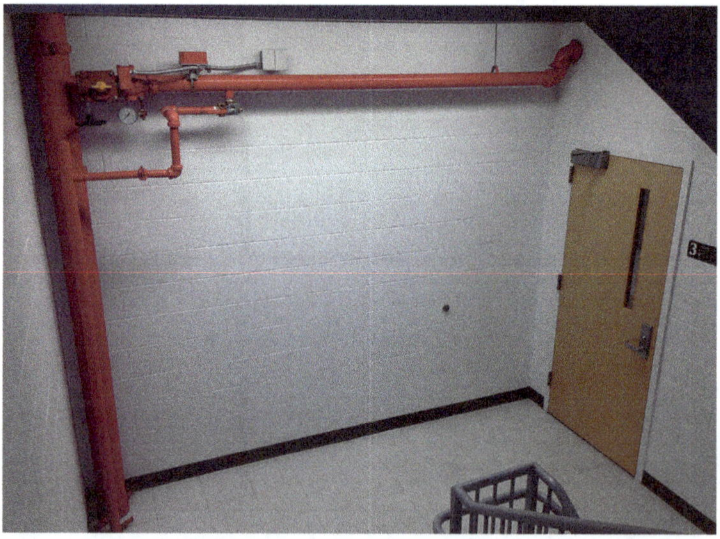

Fig. 4.8 Separate riser for sprinkler system in same stairwell as standpipe system shown in Fig. 4.7

AHJ to make the final decision, you should always check with the AHJ before finalizing any decision on the location of your Class I outlets.

Many building codes and the Life Safety Code allow convenience stairs between two floors of a building. Such stairs are there for the building owner (or tenant) to save time in walking around the building, but are not a part of the egress system for the building. Such stairs are typically not enclosed and are not expected to be used during a fire. These convenience stairs are not required to have standpipe systems or hose connections because they are not protected locations and would not afford firefighters any safety in connecting their hose and preparing to fight a fire.

Outlets at Horizontal Exits

When a person sees the term "Horizontal Exit", they might think that the simple definition of the term is any exit that exists in the horizontal direction, but this would be incorrect. The concept of a horizontal exit is much more challenging. The Life Safety Code defines a Horizontal Exit (in section 3.3.83.1) as:

> A way of passage from one building to an area of refuge in another building on approximately the same level, or a way of passage through or around a fire barrier to an area of refuge on approximately the same level in the same building that affords safety from fire and smoke originating from the area of incidence and areas communicating therewith.

The idea is that the horizontal exit produces the same level of safety as the exit stairwell. Rather than passing from the portion of the building where there is a fire to an exit stairwell, the person evacuating the building passes through a fire barrier of the same fire rating as the stairwell into an adjacent building or into an area in the same building that is separated from the fire. Once they are in this relatively safer area, they still need to eventually get to an exit.

The concept of a horizontal exit was developed to save the building owner from having to put a tremendous number of stairwells in a building. Consider the floor plan of a multiple story building shown in Fig. 4.9. Without horizontal exits, the building would require six exit stairs to handle the occupant load (based on use of the occupancy and capacity of the exit stairs) of the upper floors. The black circles in each stairwell represent the standpipe location with Class I outlets. But with the use of a horizontal exit, the owner only has to build four stairwells, as shown in Fig. 4.10. In this case, the horizontal exit saves the building owner from the cost of constructing two stairwells and the floor space where these stairwells would have gone is now usable (or rentable) space, helping to make the building owner money.

If a fire occurs in the north side of the building shown in Fig. 4.10, people can travel through the horizontal exit and be safe on the south side. They can eventually go down stairwells 3 or 4 to the ground level and exit the building. If the fire is on the south side of the building, people can go through the horizontal exit to the north side of the building and be safe, eventually going down stairwells 1 or 2 to the ground level. The wall containing the horizontal exit needs to have the same fire resistance rating as the stairwells in the building. For a building with four or more levels being served by the stairwell, this would typically be a 2-h fire resistance rating. The exit doors in a 2-h horizontal exit wall would typically be rated for 1.5 h (see Table 8.3.4.2 of NFPA 101).

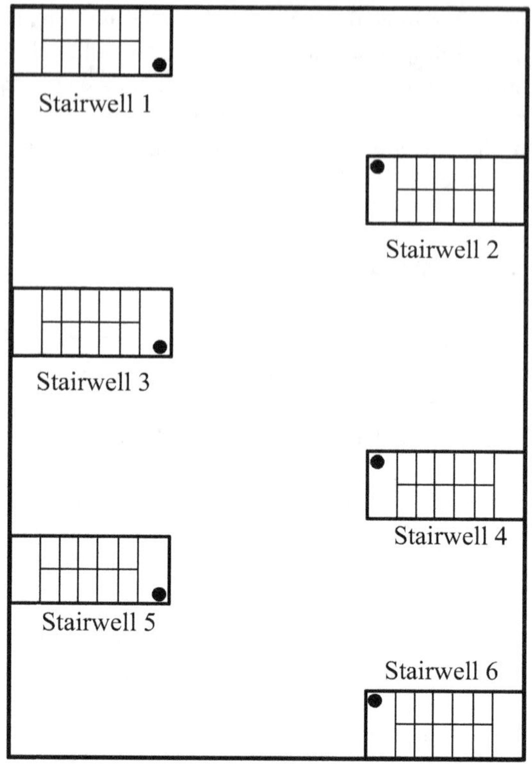

Fig. 4.9 Floor plan of multi-story building without horizontal exits

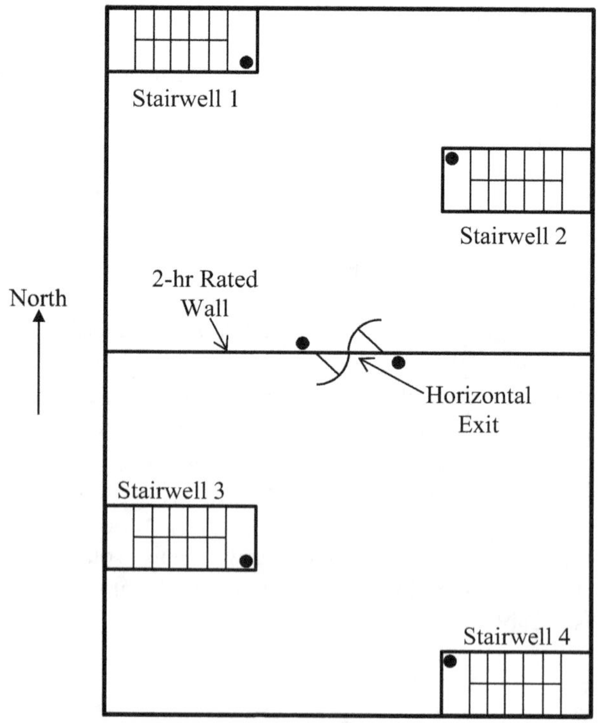

Fig. 4.10 Floor plan of multi-story building with a horizontal exit

NFPA 14 requires Class I outlets adjacent to each side of a horizontal exit. When the standard talks about "each side", it means on both sides of the fire rated wall. This is consistent with the philosophy that the firefighters stay in a protected space until they have water in their hoses to go fight a fire. If a fire were to happen in the north side of the building shown in Fig. 4.10, the firefighters would go up stairwell 3 or stairwell 4 (on the protected side of the building from the fire) and connect a hose to the Class I outlet on the south side of the horizontal exit (represented by the black circle adjacent to the horizontal exit). Once the firefighters had water, they could go through the door to fight the fire. Likewise, if the fire was on the south side of the building, the firefighters could go up stairwell 1 or stairwell 2 and stay on the protected side of the horizontal exit while the connected hose on the north side. Once they had water, they could go through the door of the horizontal exit and fight the fire on the south side of the building.

One important characteristic of a horizontal exit is a door that swings in each direction. You never know which side of the building is going to have a fire. The horizontal exit needs to provide protection for people in both directions. Since doors are required to swing in the direction of exit travel, and since you don't know which way the doors are going to be needed, the typical design for a horizontal exit is with a door that will swing in both directions, sometimes called a "double swinging door". Figure 4.11 shows a horizontal exit from one building to another. Note the double swinging doors and the sign announcing that the user is going from one building with one address to another building with a different address (7840). The Class I hose connection for the building in which the photographer is standing is on the left, just prior to the door. The Class I hose connection for the Building 7840 side is on the right side of the corridor, just past the door.

Fig. 4.11 Horizontal exit and Class I standpipe hose connection

On the architectural plans for a building, the architect rarely puts the words "horizontal exit" and rarely puts a big arrow pointing to the horizontal exit. So, how is a designer of standpipe systems supposed to know that a horizontal exit is present? It certainly helps to understand occupant loads and egress capacity so that you can have a sense for the number and size of exits from different occupancies. But even without a close familiarity with egress rules, it can be pretty easy to spot a horizontal exit. Any time there is a wall with a significant fire resistance rating going across a building and double swinging doors in that wall, the odds are that you are looking at a horizontal exit. Anytime you are in doubt, it never hurts to pick up a phone and call the architect (or the person that designed the egress system for the building) and ask whether they have designed a horizontal exit or not.

Earlier in this chapter, the statement was made that NFPA 14 requires Class I hose connections on "each side" of a horizontal exit and the concept of "each side" was defined as having an outlet for the standpipe system on both sides of the building separated by the horizontal exit. There are some plan review authorities that want to interpret the words "both sides" as meaning to the left and right of the doors, both on one side of the wall. This would be an incorrect interpretation of the rules of NFPA 14. Consider the building shown in Fig. 4.10. If the hose connections were shown on the left and right side of the doors, both on the south side of the building, and the fire was also on the south side of the building, how would the firefighters have a safe place to connect their hose before getting water to fight the fire? They would have to go into the south side of the building to connect their hose, which would expose them to the smoke and heat from the fire before they had water to fight the fire. The language in NFPA 14 is to put the outlets on both sides of the horizontal exit, not both sides of the doors. The horizontal exit is formed by the wall separating the building, so one outlets need to be on each side of the wall.

In the 2013 edition of NFPA 14, a new section was added to allow one of the outlets at a horizontal exit to be eliminated if the hose connection from a stairwell on the opposite side of the horizontal exit could be used to reach all of the portions of the building. The new language was inserted in section 7.3.2.2.1, which reads as follows:

> Where all floor areas are reachable from an exit stairway hose connection on the same side of a horizontal exit within the distances required by 7.3.2.2.1.1 or 7.3.2.2.1.2 as applicable, the hose connection on the other side of the horizontal exit shall be permitted to be omitted.

Sections 7.3.2.2.1.1 and 7.3.2.2.1.2 go on to say that the maximum travel distance (also the maximum amount of hose that would be needed to reach all portions of the building on the side of the horizontal exit in question) would be 200 ft in a fully sprinklered building or 130 ft in an unsprinklered or partially sprinklered building. Even though NFPA 14 allows the hose to be 200 ft long in a fully sprinklered building to use this concept to eliminate a hose outlet at the horizontal exit, the IBC does not. The IBC only allows the outlet at the horizontal exit to be eliminated if the area can be covered by 100 ft of hose and 30 ft of throw from the nozzle coming from an outlet in one of the exit stairwells.

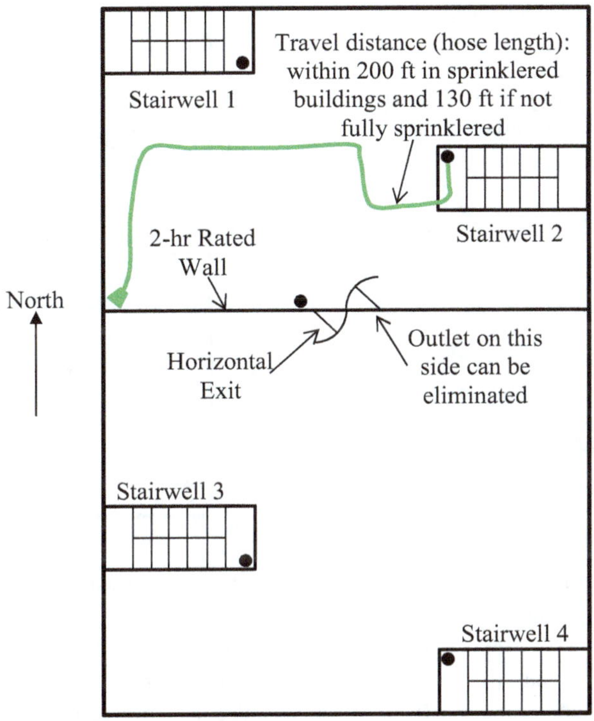

Fig. 4.12 Elimination of a hose connection at a horizontal exit

While the language in NFPA 14 is somewhat complex, it can be simplified by using a drawing. Consider Fig. 4.12, which shows a building with one of the exit stairwells close to the horizontal exit on the north side. If the travel distance from the hose connection in this exit to all portions of the building that would have been reached from the outlet on the south side of the horizontal exit (following the path of travel along corridors and through rooms) is within 200 ft in sprinklered buildings and within 130 ft in unsprinklered or partially sprinklered buildings, then the hose connection on the south side can be eliminated because the hose connection in the stairwell serves the same function.

Outlets in Exit Passageways

An exit passageway is a specific component of a means of egress. The term is not officially defined in the Life Safety Code, but it is defined in section 3.3.5.1 of NFPA 14 as, “Hallways, corridors, passages, or tunnels used as exit components and separated from other parts of the building in accordance with NFPA 101.” This means that every corridor and every passageway is NOT an exit passageway. Only those corridors that have a specific separation from the rest of the building that are

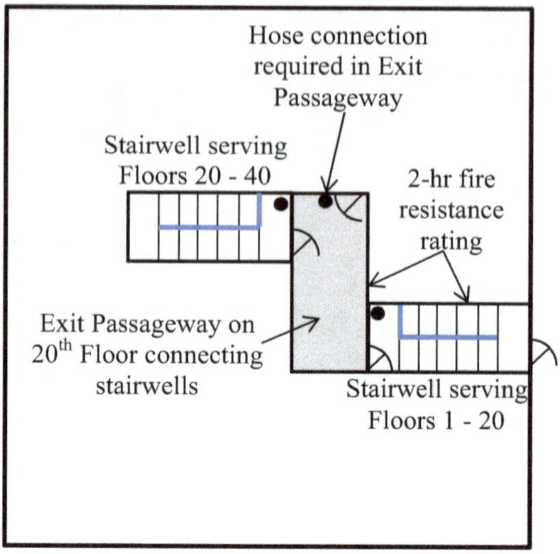

Fig. 4.13 Plan view example of exit passageway joining exit stairs in different locations in a building

intended to be a specific part of a means of egress are considered "exit passageways."

Section 7.2.6 of the Life Safety Code (NFPA 101) lays out the requirements for a separation for an exit passageway. In short, the separation needs to match that of the exit stairs. If the stairs serve four or more stories, the exit passageway needs to be separated from the rest of the building by construction with a 2-h fire resistance rating. If the exit stair serves three or fewer stories, the separation is permitted to have a 1-h rating. For situations where the fire resistance rating of the wall needs to be 2-h, the doors all need to be at least 1.5-h rated.

There are two conditions where exit passageways are used as components in a means of egress. The first is to provide continuity to exit stairwells. For example, in a tall building, it might not be possible to line up the exit stairwells from the top to the bottom of the building. In such situations, exit passageways can be used to connect the discharge from one stairwell with the entrance to another as shown in Fig. 4.13, which is an example of a 40 story building with one exit stair serving floors 20 through 40 and a different stair serving floors 1 through 20 and an exit passageway on the 20th floor connecting the discharge from one stair to the other. The exit passageway is the shaded portion of the floor area.

Another issue regarding continuity of an exit stairwell is the situation where the exit stairwell is not near on the exterior wall of the building, so the people coming down the exit stairs don't get to the outside of the building. In such conditions, an exit passageway can connect the bottom of the exit with a discharge to the outside as shown in Fig. 4.14. The exit passageway is shown as the shaded portion of Fig. 4.14.

The second condition where exit passageways are used as a component of a means of egress is to help with the problem where a building is so large that there are portions of the building that exceed the maximum travel distance. With each

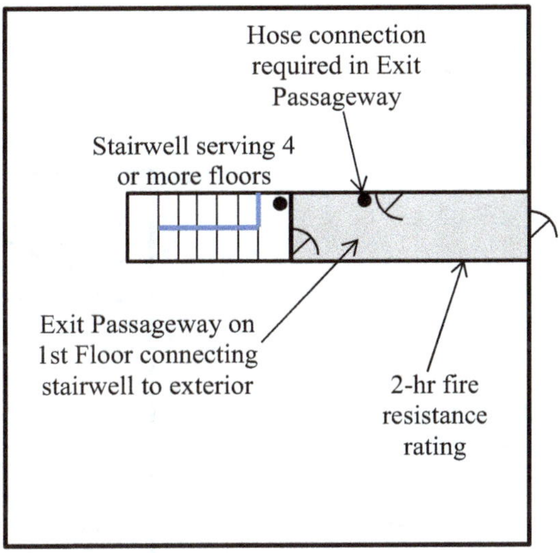

Fig. 4.14 Plan view of first floor of building with exit stairs with exit passageway to the exterior

occupancy, building codes, fire codes and the Life Safety Code require that all portions of the building be within a certain distance of an exit. An exit passageway can count as an exit in these circumstances as long as it eventually leads to an exit.

Figure 4.15 shows an example where the distance from the remote portion of a room to the actual exit stairs exceeds the maximum allowable travel distance. Stairwell 2, at the bottom of the figure can handle the rooms at that end of the building with the red line showing the maximum allowable travel distance. For the upper portion of the building, the blue line shows the maximum allowable travel distance for the occupancy. The blue line comes up short of stairwell 1, so an exit passageway (the shaded portion of the corridor) was constructed to make the situation code compliant. The exit passageway is like a protected tunnel into the building. The exit passageway is permitted to have openings into other building areas, but they need to have appropriately rated doors, which are 1.5-h rated doors for a 2-h rated wall.

NFPA 14 requires Class I standpipe hose connections at the entrance to the building. In Fig. 4.13, there is a door from the 20th floor into the exit passageway, which then leads to the stairwell down to the first floor. A hose connection is required at this door within the exit passageway. During a fire on the 20th floor, firefighters can attach hose to the connection, make sure they have water while they are in the protected passageway, then go through the door and fight the fire.

In Fig. 4.14, the exit passageway goes from the bottom of the exit stair to the outside of the building. There is also an entrance into the exit passageway from another portion of the first floor. A hose connection needs to be installed at the door from the rest of the first floor into the exit passageway as shown in the figure. In case of a fire on the first floor, the firefighters can come into the exit passageway from the outside, attach the hose to the connection and then go fight the fire when they have water.

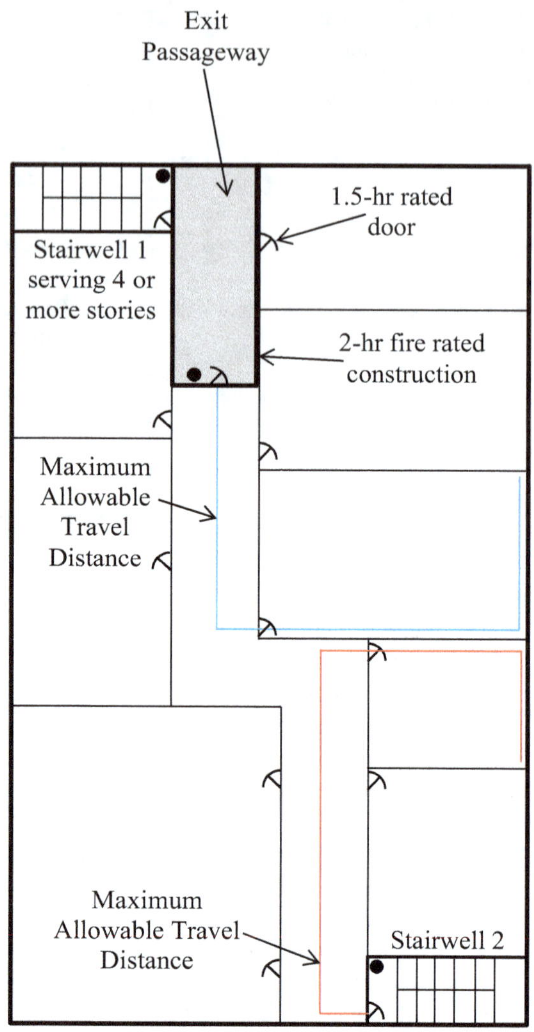

Fig. 4.15 Floor plan of building where exit passageway helps comply with travel distance requirements

One of the interesting discussions regarding the situation in Fig. 4.14 is that there isn't always a door from the first floor into the exit passageway. Sometimes, the passageway goes directly from the bottom of the stairs directly to the outside. In these conditions, NFPA 14 does not have an exception, and a hose connection is still required in the exit passageway. In these conditions, the best place to put the hose connection would be at the door to the exit stairwell in the passageway.

In Fig. 4.15, the exit passageway extends from stairwell 1 into the building. NFPA 14 requires a hose connection at the door inside the passageway as shown in the figure. In case of fire, the firefighters can come up stairwell 1 and into the exit passageway. They are still protected in the passageway as they connect their hose and make sure they have water. Then they can leave the exit passageway and go fight the fire.

Although NFPA 14 does not allow outlets in exit passageways to be eliminated if the area covered by the outlet can be covered from other outlets in exit stairwells, but the IBC does. If the area can be covered by 100 ft of hose and 30 ft of water discharging from the nozzle from a hose connection in an exit stairwell (regardless of whether the building is sprinklered or not), the outlet in the exit passageway can be eliminated according to section 905.4(3) of the IBC.

Outlets in Covered Mall Buildings

The rules for Class I hose connections in covered mall buildings are very similar to those for other occupancies with a few additional locations where hose connections need to be installed. Covered mall buildings are typically very large and represent situations where firefighters might have to lay a large amount of hose to fight a fire deep in the mall, even if the mall is only one or two stories tall.

The first additional location that covered mall buildings need hose connections is at public entrances to the mall from the exterior (not at public entrances from anchor stores). In these cases, the hose connection is required at the interior of the mall adjacent to the entrance door (see section 7.3.2.6 of NFPA 14 for the requirement). See Fig. 4.16 for an example of hose connection placement. The reason that firefighters need help in these circumstances is that fire hydrants tend to be a long distance from mall entrances. While these hose connections are not in protected locations, the height of a covered mall is usually tall enough to collect the smoke and heat from a fire and not let it spill down on the firefighters. The assumption is that the covered mall will also be fully sprinklered, limiting the size of the fire that would likely occur. The black circles in Fig. 4.16 show where the Class I hose connections are required.

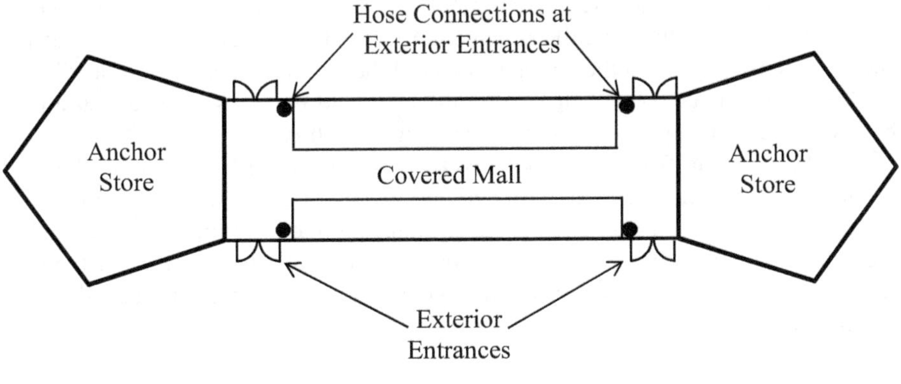

Fig. 4.16 Hose connections at exterior mall entrances

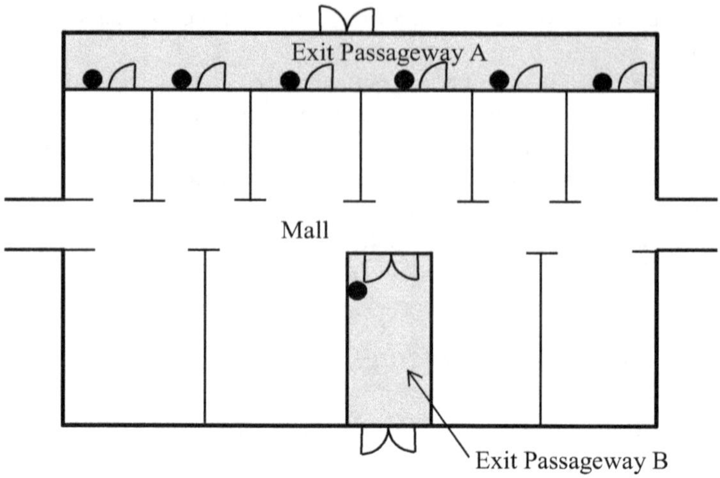

Fig. 4.17 Hose connections at exit passageways in a portion of a covered mall

Another location where Class I hose connections are required for covered mall buildings is at the entrance to each exit passageway (see section 7.3.2.6 of NFPA 14). While this may sound the same as the discussion for exit passageways in other buildings, the committee responsible for NFPA 14 thought that it was important enough to repeat it in the standard since exit passageways are so common in covered malls. Most of the time, the covered mall buildings need to use the exit passageways to overcome the maximum travel distance rules. See Fig. 4.17 for an example of two different types of exit passageways in a portion of a covered mall building.

Figure 4.17 shows two different types of exit passageways. The first is labeled Exit Passageway A and helps extend the travel distance out the back of the stores in the mall. Since each of these doors enter from the exit passageway into individual stores, a Class I hose connection is required at each of the entry points. The thick line around the passageway represents a 2-h fire resistance rated wall. If a fire occurs in one of the stores, the firefighters can enter the exit passageway from the outside and work their way down the passageway until they get to the entry door leading to the fire. They can connect their hose in the protected passageway, make sure they have water, and then go through the door to fight the fire.

Exit Passageway B in Fig. 4.17 helps the travel distance for people in the mall portion of the building. The thick line around the exit passageway represents a 2-h fire rated wall. NFPA 14 requires the hose connection at the entrance to the mall in the exit passageway. If a fire occurs in the mall, the firefighters can enter the passageway, connect their hose, make sure they have water and go through the door and fight the fire.

Outlets on Roofs

There are three reasons why standpipe hose connections are valuable on the roof for firefighters. First, there is the obvious concern that a fire on the roof might need to be extinguished. Second, there is the strategic ability to fight a fire on an adjacent building from the roof. Firefighters can direct water down directly at a fire next door, or they can spray water between the buildings in an effort to absorb the radiant heat from the fire next door to prevent the fire from jumping to the next building. Finally, the roof also makes a good platform for flow testing the standpipe system. Roof drains can be sized to handle the flow or the discharge can be directed over the side of the roof. Note that in order to use roof drains, they need to be properly designed to handle the flow and pressure from the test.

Most drain designers do not consider pressure when designing a drainage system since the water starts out at atmospheric pressure, but in a very tall building, the drain pipes will back up since the water generally does not flow out as fast as it flows in (standpipe tests can be run at flows up to 1250 gpm). The rising column of water develops an elevation pressure due to gravity of 0.433 psi per ft of elevation. In a building that is 100 ft high, that would only be 43 psi, but in a building 500 ft high, the elevation pressure would be 216 psi, which would potentially break pipe and fittings rated at 175 psi, so make sure that the drainage system can handle the flow and the pressure before counting on the drainage system to handle the flow. Some fire protection contractors wait until a rainy day is in the forecast to flow some (or all) of the discharge from the test over the side of a building. If this is going to be your plan, make sure that the space below the building is blocked off and that the water has somewhere safe to go. A flow of 1250 gpm can be significant off the side of a building.

The basic philosophy of NFPA 14 is that firefighters can get water to the roof in two ways. One is to place outlets directly on the roof. The other is to provide outlets in stairwells that have access to the roof. Placing the outlets in stairwells is less costly because the stairwells tend to be heated and tend to keep the standpipe equipment protected from the weather, which saves on maintenance costs.

Section 7.3.2.9 of NFPA 14 requires that at least one Class I hose connection be placed on the roof if the roof is not accessible from stairwells. The requirement for only one outlet is relatively new. Older editions of NFPA 14 required at least two outlets to facilitate testing. The two outlets were frequently extended to the roof from the most remote standpipe riser because this is where the flow needs to be created during the test of the standpipe system. These two outlets were typically called a "roof manifold" and consisted of two outlets side by side at a slight angle to allow for hoses to be tightened by the firefighters hands and tools without interference from the adjacent hose on the connection. See Fig. 4.18 for an example of a roof manifold.

There were several editions of NFPA 14 that required two outlets on the most remote standpipe to facilitate testing. These two outlets were required to be at the most remote location, which would typically be on the roof or in the stairwell

Fig. 4.18 Roof manifold

Fig. 4.19 Two outlets at a remote location to facilitate testing

adjacent to the roof as shown in Fig. 4.19. The purpose of the two outlets is to make sure that 500 gpm can be provided for the test. Typically each 2.5 in. outlet can allow about 250 gpm to discharge. Rather than trying to force 500 gpm from a single outlet, two outlets are provided and 250 gpm can be flowed from each. This practice is no longer required by NFPA 14, but the reader is likely to still see them on existing systems. It is expected that in newer systems, a single outlet at the most remote location is all that is needed because a hose can easily be stretched from the second most remote location, up the stairs, and to a place where the water can be discharged and 250 gpm can still be flowed from each of the two outlets using a little hose.

Even older editions of NFPA 14 required all standpipe risers to be extended to the roof. This would give firefighters the ability to fight fires on the roof or to extend hose to the edge of the building to fight fires in adjacent buildings without having to worry about bringing lots of hose to the roof. Over time, the requirement has been cut back due to maintenance concerns for roof outlets in freezing weather and the fact that few roof outlets are being used these days. With the rise of fire sprinkler systems in almost every building of considerable size, fewer fires are getting out of control to the point where they need to fight the fires from adjacent roofs and so few fires occur on roof structures these days, it is difficult to justify the expense of extending so many standpipes to the roof.

The previous discussion on only providing a single outlet only applies to a roof with no occupancy or use. A roof with a swimming pool, social gathering area or even just a garden is a space that is being used and needs more outlets. See the discussion later in this chapter regarding travel distance to every location on the roof from hose connections on the roof or hose connections in stairwells accessible to the roof to determine how many outlets will be needed.

In situations where roof outlets are installed and the climate is subject to freezing, care needs to be taken to install the equipment in such a way that it water will not freeze in the piping and do damage. There are a number of options for such systems. If dry systems (dry automatic, semiautomatic dry, or manual dry) systems are allowed, the whole system could be installed as a dry system to avoid the problem of the roof outlet. This would be unusual since the maintenance problems associated with dry systems would not warrant the installation of a dry system just for the roof outlets. A second solution would be the installation of a dry-pipe valve in the riser to the roof outlet(s). This too would be unusual since the maintenance for a dry-pipe valve would also be more than what is warranted for roof outlet(s).

The most common solution to prevent roof outlets from freezing is to use a valve in a heated space below the roof to control the flow of water to the hose connection above the roof as shown in Fig. 4.20. The valve will need an extended stem that comes up above the roof so that water can be made available to the outlet directly adjacent to the roof hose connection. After the roof connection, a drain will be needed in order to clear the piping of water after the roof outlet is used. The drain should be checked just prior to freezing weather to make sure that water has not been left in the piping after a test.

If the stairs permit access to the roof, specific outlets are not required on the roof (see section 7.3.2.9.2). Firefighters that want to stretch hose lines to the roof, in order to fight a fire on the roof, to fight a fire in an adjacent building, or to test the system can just stretch the hose from the stairwell onto the roof. In this case, the hose connection needs to be at the highest main floor landing, or the highest intermediate landing if the outlets are being installed at intermediate landings. A question is frequently asked regarding how "accessible" the stairways need to be to the roof. If the stairway has steps that go all the way up to the roof and a full size door that opens directly onto the roof, as shown in Fig. 4.6, there is no question that the roof is accessible from the stairwell. Less clear is a situation where the stair terminates at a ladder that goes through a roof hatch. Some fire protection experts feel that the

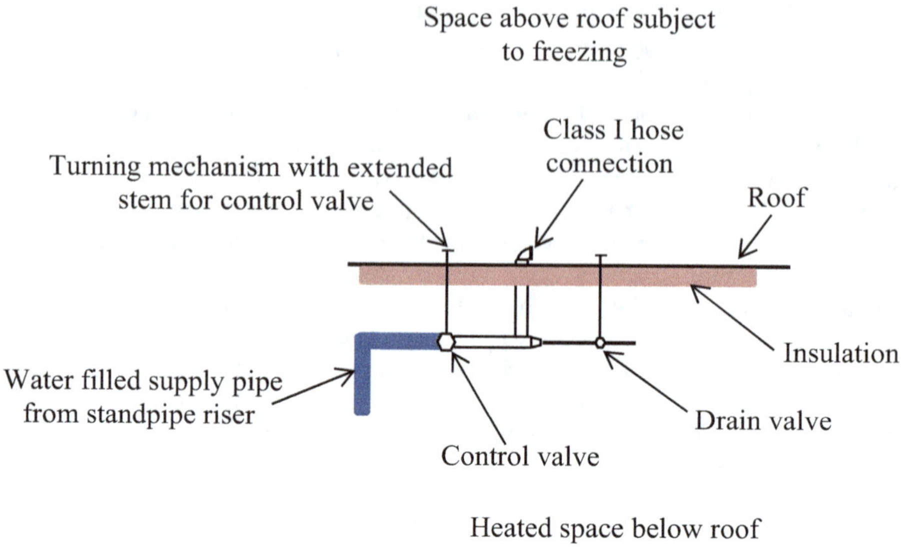

Fig. 4.20 Roof outlet protected from freezing

ladder and roof hatch constitute sufficient "accessibility" to warrant omission of the hose connections on the roof. Others feel that the ladders are too difficult for firefighters in full gear (with boots) to climb and that the access hatches are too small for a firefighter to get through with protective gear including an air pack. NFPA 14 does not specifically address the issue, so be sure to contact the local plan review officials if you have a ladder with an access hatch and want to omit roof hose connections.

While the three reasons mentioned above are compelling for putting outlets on the roof, they are only valuable if the firefighters can actually stand on the roof. Some buildings have a roof that is sloped so significantly that the firefighters cannot stand on the roof safely. In such a situation, roof outlets are not advisable. NFPA 14 used to be silent on this issue, which is why many of the building codes, fire codes and the Life Safety Code specifically talk about the issue in several of the occupancies discussed in Chap. 3 of this text and exempt buildings with a significant roof slope from the requirement of a roof outlet. The current editions of the IBC and NFPA 14 permit the roof outlets to be omitted when the roof slope exceeds 4 in 12 (a vertical rise of 4 units for every horizontal run of 12 units). See section 7.3.2.9.1 of NFPA 14 and section 905.4(5) in the IBC.

The rules discussed in this section apply to roofs without heliports. If the building in question is being designed with a heliport on the roof, standpipe outlets will be needed. See the discussion in Chap. 3 of this text regarding heliports for the requirements regarding how many hose connections will be needed for the roof.

Section 905.3.8 of the IBC requires outlets on the roof of any building with a rooftop garden or landscaped roof. The IBC is not clear whether just a single outlet would meet this requirement or whether each of the standpipes needs to be extended to the roof. The way that the section is written, using the phrase, "shall have the standpipe system extended to the roof level" implies that the whole system needs to be extended to the roof and not just a single outlet.

Outlets Based on Travel Distance

Once all of the Class I outlets have been placed in the building exit stairwells, horizontal exits, exit passageways, and mall exterior entrances, the designer then needs to make sure that every point in the building is within a certain distance from a hose connection. This is the maximum amount of hose that the firefighters will need to bring into a building to fight a fire in that building. The maximum allowable distance from an outlet depends on the type of standpipe system in the building and whether the building is sprinklered or not.

In a sprinklered building, every point in the building needs to be within 200 ft of a hose connection. In the application of this rule, NFPA 14 allows buildings that have been protected in accordance with either NFPA 13 or NFPA 13R to meet the definition of a fully sprinklered building. The original rule for 200 ft was adopted from the building codes and the Life Safety Code, which allows a travel distance of 200 ft from any point in a sprinklered building to the exit door at the entrance to an exit stairwell for the following occupancies:

- Educational
- Day Care
- Health Care
- Ambulatory Health Care
- Detention and Correctional
- Enclosed Parking Structures

For these occupancies, the intent of the standpipe committee was that the designer of the standpipe system should be able to just use the exit stairwells and not worry about putting outlets in hallways or other unprotected spaces. After all, the stairways are the most protected space in the building and the safest place for firefighters to connect to the standpipe system. For many years, the term "travel distance" was used in NFPA 14 to describe this requirement. Section 7.6 of the Life Safety Code tells the user that "travel distance" is measured from the most remote portion of the building to the centerline of the exit door. The problem is that the standpipe outlets are not at the centerline of the exit door. The outlets are in the stairwell at the main floor landing or at the intermediate floor landing. This adds between 5 and 10 ft to the distance from the most remote portion of the building to the standpipe outlet.

For many years, this was not a problem because the distance was measured from the exit to the most remote portion of the building, not from the hose connection. But in the 2013 edition of NFPA 14, the language changed slightly. Section 7.3.2.2 was rewritten to say that the measurement needed to be made from the hose connection to the most remote portion of the building. In the 2016 edition, this concept was moved to section 7.3.2.11.1 and cleaned up so that there is no mistaking the committee's intention. The most recent position of the committee is that the maximum allowable distance is required to be measured from the most remote portion of the building to the standpipe outlet in the stairwell.

In a fully sprinklered building, where the architect has taken full advantage of the 200 ft travel distance permitted by the Life Safety Code with the distance measured from the most remote portion of the building to the center of the door in the exit stairwell, the standpipe outlet in the stairwell will be insufficient for meeting the intent of NFPA 14 (2013 or 2016 edition). The distance from the most remote portion of the building to the standpipe outlet will be 205 ft or 210 ft, which will mean that another standpipe outlet will have to be located somewhere in the building outside of a protected exit stairwell. The designer should try and find a protected place that is accessible by the firefighters to put this additional outlet.

In buildings where the Life Safety Code, or the building code, allows a travel distance greater than 200 ft (such as Business Occupancies where the maximum allowable travel distance is 300 ft in a sprinklered building) it is common to need additional standpipe outlets beyond those located in the exit stairs. Such locations are not as protected as the exit stairs. Care should be used in finding locations that the firefighters can use. Figure 4.21 shows a Class I hose connection in the middle of a hallway that is not protected, but was installed in a sprinklered building because the travel distance from the outlets in the stairwells to the most remote portion of the building exceeded 200 ft.

In unsprinklered buildings, Class I outlets need to be spaced closer together. NFPA 14 requires that every portion of the building needs to be within 150 ft of a Class I outlet (see section 7.3.2.10). If the building is a parking garage with a manual

Fig. 4.21 Class I outlet in hallway because outlets in stairs were too far from remote points in building

Table 4.1 Travel distance to Class I hose connections

Type of building	Type of sprinkler system	Type of standpipe system	Maximum allowable distance from any point in the building to the Class I hose connection (ft)
Anything except a parking garage	NFPA 13	Automatic wet, automatic dry, semiautomatic dry, manual wet, or manual dry	200
	NFPA 13R	Automatic wet, automatic dry, semiautomatic dry, manual wet, or manual dry	200
	Partially sprinklered, but not meeting NFPA 13 or NFPA 13R	Automatic wet, automatic dry, semiautomatic dry, manual wet, or manual dry	150
	Not sprinklered	Automatic wet, automatic dry, semiautomatic dry, manual wet, or manual dry	150
Parking garage	NFPA 13	Automatic wet, automatic dry, semiautomatic dry, or manual wet	200
	NFPA 13R	Automatic wet, automatic dry, semiautomatic dry, or manual wet	200
	Partially sprinklered, but not meeting NFPA 13 or NFPA 13R	Automatic wet, automatic dry, semiautomatic dry, or manual wet	150
	Not sprinklered	Automatic wet, automatic dry, semiautomatic dry, or manual wet	150
Parking garage	Sprinklered or not	Manual dry	130

dry standpipe system, the outlets need to be even closer together with a maximum of 130 ft from every point in the garage to a Class I hose connection. In most occupancies, the Life Safety Code allows the travel distance to be more than 150 ft, even when the building is unsprinklered, so the Class I outlets will typically have to be installed in unprotected corridors or other accessible locations to the firefighters in addition to the stairwells. Table 4.1 summarizes the travel distance requirements for Class I outlets in all buildings with all types of sprinkler systems and standpipe systems.

As discussed earlier in this chapter, the roof is not required to follow the travel distance requirements as long as the roof is not being used for any type of occupancy (see section 7.3.2.11.3 of NFPA 14). If the roof is being used for any type of activity, including assembly (such as cocktail receptions or social events) then the minimum travel distance rules do apply.

The travel distance for standpipe outlets is measured using the same technique as the measurement of travel distance for egress purposes. See section 7.6.1 of the Life Safety Code (NFPA 101) for more information on how to measure travel distance. Basically, you start at the most remote corner of a room. A case could be made that the Code allows you to start 1 ft away from each of the walls rather than starting exactly in the corner because section 7.6.1(2) allows you to stay 1 ft away from obstructions and the walls in the room can be considered obstructions.

You then continue along the normal walking path through the room(s) out to the corridor and towards the exit door, keeping in the center of the walking surface. If the path requires going up or down stairs, the travel distance is measured along the plane parallel to the nose of the stairs. The travel distance for standpipe purposes terminates at the hose connection rather than at the exit door. It is important to consider the need to get around furniture and other obstructions in a room, so you are not allowed to measure the diagonal distance across a room. Instead, you should consider the distances at right angles as shown in Fig. 4.21, without doubling back on the path in the room.

NFPA 101 does not require the measurement of travel distance to be made with sharp turns. Instead, an arc with a 1 ft radius is allowed when changing directions. When changing direction 90° (making a left or right turn), this arc results in a path that is 1.6 ft in length ($2\pi r/4$) rather than the 2 ft associated with going right up to the point of a turn and making the turn. While the difference is only 0.4 ft, depending on how many turns are between the remote portion of a building and the hose connection, this might make the difference between a compliant hose connection in a stairwell and having to add a hose connection in a corridor. The red line in Fig. 4.22 shows the travel distance from the point in a room 1 ft from each of the walls near a far corner, through the room, out the door (through the center of the doorway), to the center of the corridor, down the hall, through the center of the doorway to the stairs, around the landing and down the stairs to the outlet on the intermediate floor landing.

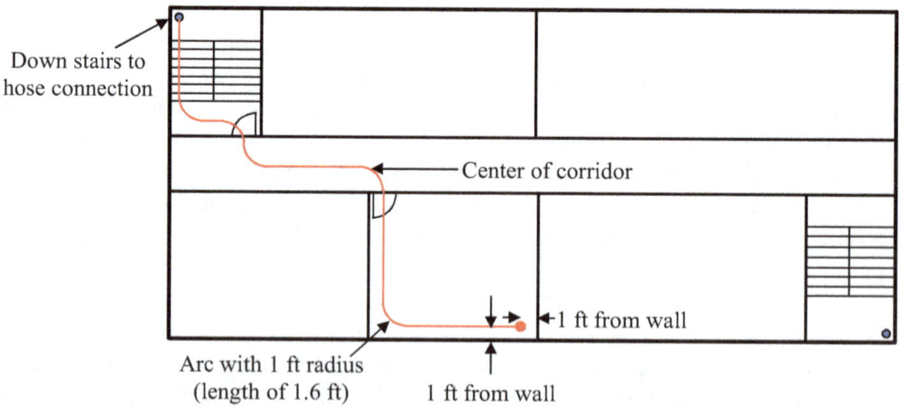

Fig. 4.22 Travel distance measurement for standpipe system

Outlets and ADA

The Americans with Disabilities Act (ADA) was a relatively simple piece of legislation that passed in 1990. The ADA did not contain any specific rules for building construction and is not written like a building code. Instead, the ADA states that buildings need to be accessible to everyone. In order to help architects and engineers design buildings, a number of ADA guidelines have been written and code groups have addressed accessibility issues by putting specific rules into building and fire codes. The advantage to having these accessibility codes and guidelines is that they help architects and engineers think about situations that might never have come to their attention. Most architects and engineers want to do what is right for the building occupants, but some of the situations just have not been something that they have thought about.

One of the issues that few architects and engineers think about is the difficulty that a blind person experiences walking in a means of egress on their own with a cane. The blind person uses the cane to feel out the pathway in front of them. A projection into a means of egress, even if it leaves an acceptable width for exit capacity, can possibly represent a hazard to a blind person walking with a cane. A Class I standpipe hose connection can be one of these hazards. The vertical pipe only sticks into the walking path (corridor or stairwell landing) a little bit, and the cane should alert the blind person to the presence of the pipe, but the hose connection sticks even further into the walking path and does not extend to the floor, so a person walking with a cane might not know that it is there until they bump into it.

In new construction, objects that are between 27 and 80 in. high are not supposed to protrude into a circulation path more than 4 in. (see section 204 of the ADA Standards). Standpipe outlets certainly have the potential to do that, so the designer needs to recognize the situation. Figure 4.23 shows both a plan view and an elevation view of the area of concern. Objects less than 27 in. from the floor are not considered a problem because the person walking with a cane should be able to detect them. Objects more than 80 in. high are generally not a concern because they are above the level of most people's heads (6 ft–8 in.). However, tall blind people might need to be aware of the limit so that they don't bump their heads. Note that the protrusion of the hose connection is measured from the standpipe, not from the wall. The standpipe will be noticed by the cane, but the hose connection is the concern. Most hose connections should be okay, but the designer of the standpipe system should be aware of the situation so that they do not create a barrier to the disabled.

Special IBC Rules

The International Building Code (IBC) has a number of special requirements that override the requirements of NFPA 14. As the primary legal document, the building code has the right to change the requirements, or add to the requirements, of its

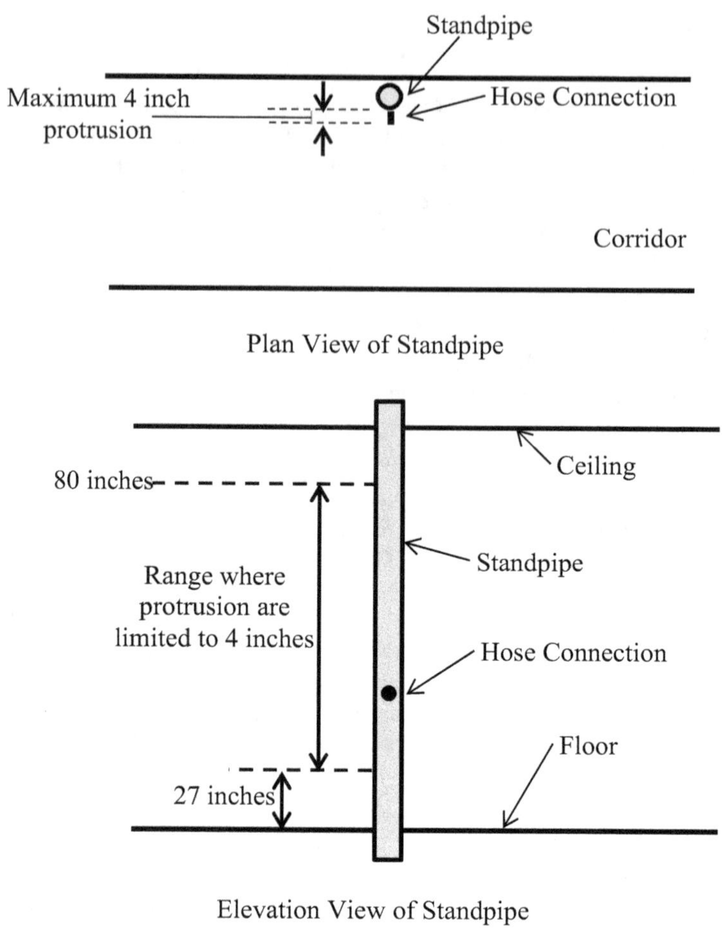

Fig. 4.23 Range of concern for protrusions due to ADA

referenced standard. The following is a list of the changes that the IBC has made to NFPA 14:

Malls: As explained in Chap. 3 of this text, section 905.3.1 of the IBC has requirements for full Class I standpipe systems to be installed in certain malls that are both covered and open. These full standpipe systems need to comply with NFPA 14. However, not all malls are required to have full standpipe systems. Section 905.3.3 requires that the malls that do not have full standpipe systems need to have modified standpipe systems. The IBC does not use the term "manual wet" standpipe system, but the description in section 905.3.3 sounds like one. These modified standpipe systems need Class I hose connections in the following five locations:

1. Within the mall at the entrance to each corridor or exit passageway.
2. Within interior exit stairways at each floor landing opening directly onto the mall.

3. For covered malls, at the exterior public entrances.
4. For open malls, at the public entrances along the perimeter line.
5. After all hose connections have been placed in accordance with the four items above, additional outlets need to be added so that all parts of the mall are within 200 ft of a hose connection.

There are special hydraulic calculation requirements for these modified standpipe systems. They only need to be sized to flow 250 gpm to the most remote standpipe outlet. In addition, the maximum pressure loss allowed (due to friction loss and elevation loss) between the most remote outlet and the fire department connection is 50 psi with the 250 gpm flowing. The goal is typically to get 100 psi at the most remote outlet, so the limit in this section of the IBC is assuming that the fire department can pump into the fire department connection at 150 psi. With the water losing 50 psi in the piping due to friction loss and elevation, the water will arrive at 100 psi at the most remote outlet. The IBC makes no specific mention of a water supply. Since it specifically contains a requirement related to the fire department connection, it makes sense that this is just intended to be a manual wet system.

Class I Hose Connections: Section 905.4 of the IBC contains a list of locations where Class I hose connections need to be installed. For the most part, this list matches the list discussed above from NFPA 14. However, there are a few notable exceptions:

1. In exit stairwells, the outlet needs to be installed at the intermediate landing unless the fire code official approves an alternate location (which would usually be the main floor landing).
2. The hose connection required at the horizontal exit can be eliminated, but only if the area that would be covered by the hose connection can be covered from an outlet in an exit stairwell with 100 ft of hose and 30 ft of water discharge from the nozzle at the end of the hose.
3. The hose connection required in the exit passageway can be eliminated, but only if the area that would be covered by the hose connection can be covered from an outlet in an exit stairwell with 100 ft of hose and 30 ft of water discharge from the nozzle at the end of the hose.

Rooftop Gardens: Section 905.3.8 specifically requires that the standpipe system be extended to the roof when there is a rooftop garden or a landscaped roof. The concern is that without regular rain or watering, such gardens could produce combustible situations that could result in a serious fire on the roof.

Location of Class II Hose Connections

As defined in Chap. 2 of this text, Class II hose systems are those that have smaller (typically 1.5 in.) connections. Under limited circumstances, the connection might be as small as 1 in. There may or may not be hose actually attached to the system. If there is hose attached, the connection is actually called a "hose station".

NFPA 14 starts out in section 5.3.2.1 requiring 1.5 in. hose for the hose stations. However, section 5.3.2.2 allows the hose to be reduced to 1 in. in size if the building can be classified as a light hazard occupancy and if the local authority having jurisdiction agrees. Using smaller hose can be advantageous because it lowers the water pressure at the nozzle (by creating friction loss in the hose), reducing the reaction forces and making it easier for an untrained (or lesser trained) individual to use the hose.

Section 4.6.2 of NFPA 14 requires the 1.5 in. hose to be lined, but allows this size hose to be either collapsible and stored on a rack or non-collapsible and stored on a reel. Where 1 in. hose is installed, it is required to be non-collapsible and stored on a continuous reel. When hose is stored on a rack, all of the hose needs to be taken off of the rack before you open the valve and fill the hose with water because the water filled hose won't fit into the rack and the rack won't support the weight of the water filled hose. When using the hose on a continuous reel, the reel has been designed to hold the weight of the water and the hose is not collapsible, so there is already room for the water filled hose. Therefore, you do not need to pull all of the hose off of the reel in order to use the hose.

Section 4.6.5 of NFPA 14 requires that a label be installed on the rack or reel that says, "FIRE HOSE FOR USE BY TRAINED PERSONNEL". This section also requires that operating instructions be included on the label. This is an interesting contradiction of ideas. One could argue that if the hose is only for trained personnel that they should not need the operating instructions. They should know how to operate the hose from their training. The label reflects the growing trend to make sure that building occupants evacuate a building if there is a fire instead of trying to fight it themselves. The requirement for the operating instructions might tempt an untrained person to try to use the standpipe rather than evacuate.

Section 4.6.2.1 of NFPA 14 requires hose stations be provided with a maximum of 100 ft of hose. It is interesting that this distance is considered a maximum, since a little more hose might be helpful in some situations, especially when you consider the spacing requirements that will be discussed next in this chapter. But the reason for the maximum of 100 ft has to do with the fact that all of the hose needs to be pulled off of the rack before the water can be turned on to fill the hose. If there were more than 100 ft of hose, it would take longer to get the hose in service. Also, there is a concern that someone might not be able to handle a longer run of hose because it takes significant physical effort to wrestle water filled hose into place.

NFPA 14 requires that Class II hose stations with 1.5 in. hose be located so that every point within the building is within 130 ft of a hose station (section 7.3.3.1). If the hose station has 1 in. hose, this distance is reduced to 120 ft. Given that the paragraph above says that the maximum amount of hose provided at the hose station is 100 ft, how is it that the standard allows locations in the building that are 120 or 130 ft away from a hose station? The answer is that NFPA 14 counts on the fact that water will travel a good distance (at least 30 ft) after it discharges from the nozzle. The phrase 100 ft of hose and 30 ft of "throw" is common in the industry.

The challenge when designing Class II systems is remembering this last 30 ft of throw. Water does not bounce around corners very well, so if the last 20 or 30 ft has to go around a corner, it might be better to move the hose station closer to this point so that water can get to a fire that might occur at this location in the building. The IBC specifically requires the user to consider this last 30 ft of nozzle discharge. Rather than require that every point in the building be within 130 ft of a hose connection, the IBC specifically says that every point in the building has to be within 30 ft of a nozzle attached to 100 ft of hose. This implies that the last 30 ft cannot be around a corner or a bend in the building's design.

Location of Class III Hose Connections

Class III systems are simply a combination of Class I and Class II. However, it is getting more difficult to design a single system of outlets that meets all of the requirements of both Class I and Class II. For example, the hose connections for a Class I system in a sprinklered building are based on a travel distance requirement of 200 ft while the hose connections for Class II are based on a travel distance of 130 ft. This means that for a Class III system, the outlets will have to be spaced on the lesser dimension (travel distance of 130 ft), which will put some outlets in unprotected positions for firefighters.

Another challenge in designing a single system that meets all of the rules of Class I and Class II at the same time is the water pressure at the outlet. Class I systems require a minimum pressure of 100 psi while Class II systems only allow a maximum of 100 psi at the outlet. It is extremely difficult to design a single system that delivers exactly 100 psi to every outlet in the system. More information on pressure control in standpipe systems will be included in Chap. 8 of this text.

Given the difficulty in designing a single system that will meet the requirements of both Class I and Class II, many designers are now complying with the requirements for Class III systems by installing two separate systems. This makes a great deal of sense. Several cycles ago, the idea of eliminating Class III from NFPA 14 was proposed to the committee. While the NFPA committee recognized the challenges associated with a single Class III system, they decided not to eliminate the classification of standpipe systems from the standard because it is often referenced by building codes. But the committee did recognize that separate hose connections might be necessary to meet the rules.

In fully sprinklered buildings (meeting the requirements of NFPA 13R or NFPA 13), the need for hose to be attached to the hose connections is diminished. The fire will be controlled by the sprinkler system and there is less of a need to fight the fire in its early stages manually. Building occupants can spend their time on evacuation and firefighter notification. In recognition of this fact, NFPA 14 allows the user to eliminate the hose from Class III systems. The Class III system then just becomes a Class I system with a 2.5 in. outlet and a 2.5–1.5 in. reducer. The firefighters can then connect any size hose that they want when they get to the scene of a fire. It seems like a copout to call a system with no hose a Class III system, but that is how NFPA 14 is written. See section 7.3.4.1 of NFPA 14 for more information on this unusual allowance.

Hose Connections on Sprinkler Systems That Are Not Standpipes

For many years (1960s, 1970s, 1980s and 1990s), NFPA 13 (and its predecessors NFPA 231 and NFPA 231C) required hose connections for first aid firefighting in most storage warehouses. In 2007, the rules changed to only require the hose connections when the authority having jurisdiction specifically mandated them. These hose connections where technically not standpipe systems. Instead, these hose connections were directly connected to the sprinkler system and were specifically covered by NFPA 13 as a part of the sprinkler system rather than be considered a separate standpipe system.

Most of these systems utilized 1.5 in. connections and hose. Section 8.17.5.1 of NFPA 13 covers the installation rules for such small hose connections that are directly connected to the sprinkler system. The connections are required to be spaced around a building so that all portions of the building can be reached within 100 ft of hose and 30 ft of throw. The connections are permitted to be made directly to wet pipe sprinkler system piping of the following sizes:

- 2.5 in. piping on a tree sprinkler system (one flow path from the water supply to the connection).
- 2 in. piping on a looped or gridded sprinkler system (more than one flow path from the water supply to the connection).
- For piping serving a single hose connection:
  - Minimum 1 in. pipe for horizontal runs up to 20 ft.
  - Minimum 1.25 in. pipe for horizontal runs up to 80 ft. Note that the whole run needs to be 1.25 in. pipe. You are not allowed to switch to 1 in. pipe for the last 20 ft.
  - Minimum 1.5 in. pipe for horizontal runs over 80 ft. Note that the whole run needs to be 1.5 in. pipe. You are not allowed to switch to smaller pipe sizes towards the end of the run.
  - Minimum of 1 in. for vertical runs.
- For piping serving multiple hose connections, a minimum of 1.5 in. pipe is required.

For these hose connections in storage warehouses, the connections need to be made in such a way that the closure of a sprinkler system valve does not take all of the fire protection out of service at the same time. Protection in storage warehouses is accomplished with fire sprinklers at the ceiling, hose connections around the warehouse, and sometimes with in-rack sprinklers if the warehouse has racks and if the in-rack sprinklers have been selected as a part of the design by the design professional. NFPA 13 provides the user that is protecting a storage warehouse with the following options for connecting the hose stations to a water supply:

1. A separate piping system for just the hose stations. This would have its own connection to the water supply with its own control valve and closing any valve on the fire sprinkler system would have no effect on whether water could get to the hose stations.

2. A connection to the sprinkler riser upstream of the control valve. This is really the same as option 1 above.
3. An adjacent sprinkler system as long as the minimum pipe sizes discussed earlier are met. This is an interesting solution that can save quite a bit of piping. Feeding the hose stations in one part of a warehouse with the ceiling sprinkler piping from another part of the warehouse makes sure that when one valve is closed, the protection is not completely eliminated from each area of the building. Closing the valves to the ceiling sprinklers in one area will take out the hose stations in another area, but in that area the ceiling sprinklers should still be active. In the area where the ceiling sprinklers were shut down, the hoses will be fed from a different sprinkler system, so they should still be active.
4. In rack storage warehouses where the decision has been made to install in-rack sprinklers, the hose stations are permitted to be fed from the ceiling sprinklers in the same area as the hose stations as long as the in-rack sprinklers have a completely different control valve. In this way, at least one of the three types of fire protection in an area will be active at all times.

Another type of occupancy where hose stations are frequently connected directly to the sprinkler systems are back stage areas of theaters. Many building and fire codes require that these areas have 1.5 in. hose stations, but the codes allow them to be directly connected to the sprinkler systems rather than have them installed as separate standpipe systems. The general rules for the installation of these hose systems is the same as discussed above for warehouses with one exception. The hose stations for non-storage situations are allowed to be directly connected to the ceiling sprinkler system in the same portion of the building as the hose stations. See section 8.17.5.1.3(6) for this allowance in NFPA 13.

Test Yourself

4.1. What is the difference between a hose station and a hose connection?

4.2. What size is the connection for a Class I outlet?

4.3. What size is the connection for a Class II outlet?

4.4. What size is the connection for a Class III outlet?

4.5. At what five locations does NFPA 14 require Class I outlets?

(a) ______________________________
(b) ______________________________
(c) ______________________________
(d) ______________________________
(e) ______________________________

4.6. What is the maximum allowable travel distance from any point in a building to a Class I outlet in the following circumstances:

(a) An unsprinklered office building with a manual wet standpipe system?

(b) A sprinklered hospital (NFPA 13) with an automatic wet standpipe system?
(c) A sprinklered parking garage (NFPA 13) with a dry-pipe sprinkler system and an automatic dry standpipe system?
(d) An unsprinklered parking garage and a manual dry standpipe system?

4.7. Given the plan view of an unsprinklered office building shown in Fig. 4.24, determine where the hose connections would be required for a Class I standpipe system. In this building, both of the stairwells provide access to the roof. The fire department has stated that they require the hose connections at intermediate landings in stairwells in accordance with NFPA 14 and the IBC. Assume that all of the doors shown are 3 ft wide and are installed 6 in. away from the wall next to where they open up. This will make the path through the center of the door 2 ft from the wall.

4.8. Given the plan view of a sprinklered office building shown in Fig. 4.25, determine where the hose stations would be required for a Class II standpipe system. Assume that all of the doors shown are 3 ft wide and are installed 6 in. away from the wall next to where they open up. This will make the path through the center of the door 2 ft from the wall.

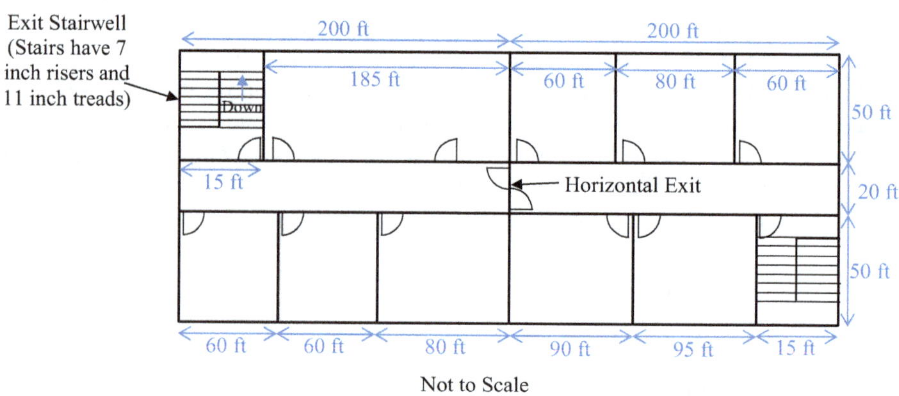

Fig. 4.24 Floor plan of office building for Class I standpipe system in question 4.7

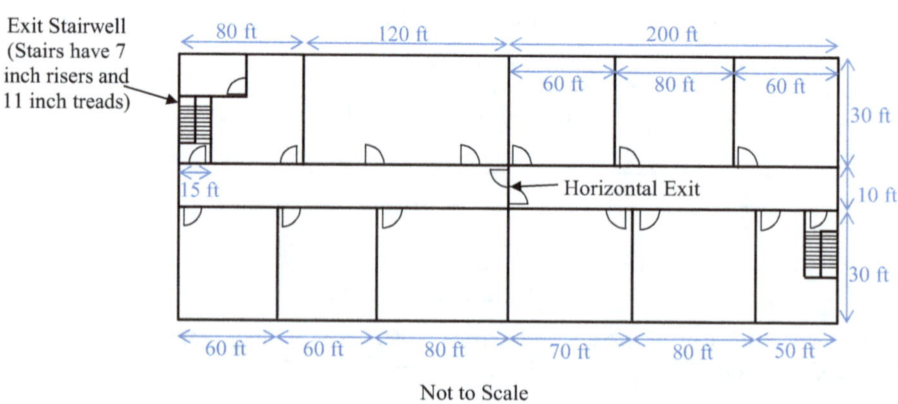

Fig. 4.25 Floor plan of office building for Class II standpipe system in question 4.8

Chapter 5
Installation Requirements for the Rest of the Standpipe System

The previous chapter covered the installation rules for the standpipe hose connections. This chapter will cover the installation requirements for the rest of the system components including pipe (underground and above ground), valves, alarms, drains, fire department connections, and gauges. In addition, this chapter will include some special requirements for assembling these parts into systems including the special requirements that apply to dry systems (automatic, semiautomatic and manual).

In addition to the rules discussed in this chapter, there are requirements in NFPA 14 for protecting the pipe from gravity, mechanical damage, fire damage, and seismic events. These requirements will be discussed in detail in Chap. 11 of this text.

Pipe and Fittings: Above Ground

There is technically a difference between the terms "pipe" and "tube" that has to do with the manufacturing process that is used to make the product, however, the NFPA codes and standards tend to use the terms interchangeably. Basically, we are talking about the hollow product that allows water to flow through the center of the product and travel from one location to another. Typically, the term "pipe" is used for steel and ductile iron products while the term "tube" is used for copper, so this text will try and stick with these breakdowns, but a mistake might creep into the language by mistake. Ultimately, it should not matter since both pipe and tube serve the same function.

NFPA 14 starts out be requiring that the above ground pipe be ductile iron, steel or copper (Table 4.2.1). But then the standard allows other options if the pipe is specifically listed for use in a standpipe system. At this time, there are no other materials specially listed for standpipe systems, but the reality is that some other type of pipe could achieve a special listing at some time in the future. It is worth noting that the term "above ground" applies to any pipe installed in a structure that is not buried in dirt. A parking garage might have levels for parking underground,

K.E. Isman, *Standpipe Systems for Fire Protection*,
DOI 10.1007/978-3-319-47750-3_5

but the piping at these levels would be considered "above ground" because it is exposed to the air and not buried in dirt.

The most common material for above ground pipe used in standpipe systems is steel because ductile iron is a brittle product and more susceptible to potential mechanical damage if it gets struck and copper is extremely expensive in the sizes required for standpipe systems. If ductile iron pipe is used in a standpipe system, it needs to be internally lined with cement mortar in order to prevent corrosion by adding a layer of separation between the iron and the water (see section 4.2.2 of NFPA 14). While the cement mortar lining helps to control the corrosion situation, it decreases the internal diameter of the pipe, which needs to be taken into account in the hydraulic calculations. See Chap. 6 of this text for more information about hydraulic calculations of standpipe systems.

Copper tube is permitted to be Type K, Type L, or Type M. The difference between these three types of copper tube is the thickness of the wall of the tube. The thicker the walls of the tube, the more pressure the tube can withstand. However, the thicker the wall, the heavier the tube, which adds to the installation cost. At the same time, the thicker tube wall is less hydraulically efficient, also adding to the cost of the system.

Steel pipe is available in a number of different wall thicknesses. The thicker the wall, the more pressure the pipe can hold and the longer it will take for corrosion to become a problem. Just like copper tube, the thicker wall steel pipe is heavier, which adds to the installation cost and is less hydraulically efficient. So, selecting a thickness of steel pipe for any specific standpipe system is a function of balancing the concerns of corrosion and water pressure against the cost of installation and the strength of the available water supply.

Steel pipe thickness is expressed in terms of a "Schedule". Standards have been developed by the industry to define a specific thickness of pipe for each wall thickness for different Schedules of pipe. Each different Schedule is given a number with the lower numbers indicating thinner wall pipe and the higher number indicating thicker wall pipe. The standard wall thickness is considered Schedule 40 pipe. Other defined Schedules that are thinner than Schedule 40 are 10 and 30. Schedule 80 pipe is thicker than Schedule 40 and is typically only used if very high pressures are expected in the system.

See Table 5.1 for specific pipe diameters and wall thicknesses for Schedule 10, 30, 40 and 80 in the sizes common to standpipe systems. The 1.5 and 2 in. sizes are typically only used in Class II systems while the larger sizes are typically used in Class I and Class III systems. The 2.5 in. size is only typically used in branch lines to a single outlet. Note that for each nominal size of pipe, the outer diameter of the pipe is the same. With the wall thickness changing, this changes the inside diameter of the pipe, increasing the efficiency of the water flow in the pipes with thinner walls. The abbreviation OD stands for outside diameter, the abbreviation ID stands for inside diameter, and the abbreviation WT stands for wall thickness. The abbreviation NTU stands for not typically used, which might mean that the pipe is not manufactured, or that it is manufactured for other industries, but has not been made available for fire protection because it is not desirable from a cost, strength or corrosion perspective. The values in Table 5.1 are in inches.

Table 5.1 Diameters and wall thickness for steel pipes

Nominal diameter (in.)	OD	Schedule 10 WT	Schedule 10 ID	Schedule 30 WT	Schedule 30 ID	Schedule 40 WT	Schedule 40 ID	Schedule 80 WT	Schedule 80 ID
1½	1.9	0.109	1.682	NTU	NTU	0.145	1.61	0.2	1.5
2	2.375	0.109	2.157	NTU	NTU	0.154	2.067	0.22	1.935
2½	2.875	0.12	2.635	NTU	NTU	0.203	2.469	0.28	2.315
4	4.5	0.12	4.26	NTU	NTU	0.237	4.026	0.34	3.82
6	6.626	0.134	6.357	NTU	NTU	0.28	6.065	0.43	5.766
8	8.625	0.188	8.249	0.277	8.071	0.322	7.981	0.5	7.625
10	10.75	0.188	10.37	0.307	10.14	0.365	10.02	0.59	9.57

Note that in 8 and 10 in. nominal sizes, both Schedule 30 and Schedule 40 pipe are manufactured, but at these sizes, Schedule 30 is used much more often than Schedule 40. The Schedule 30 pipe is much lighter than its Schedule 40 counterpart of the same nominal size and the walls are thick enough that pressure and corrosion are not a significant concern. Using Schedule 30 pipe in the 8 in. and larger sizes rather than Schedule 40 makes the system more efficient from both a cost and hydraulics perspective.

Most components of fire protection systems are required to "listed", meaning that they have been evaluated for service in the specific fire protection system (in this case a standpipe system) by a laboratory in accordance with published testing standards and that the laboratory monitors the manufacturing process to make sure that the quality of the product is as least as good as what was submitted for evaluation. But for standpipe systems, some types of steel pipe are an exception to this rule (see sections 4.1.2 and 4.2.1 of NFPA 14). The following types of steel pipe systems are not required to be listed as long as the pipe is manufactured in accordance with ASTM A53, ASTM A135 or ASTM A795:

- Steel pipe joined by welding as long as the water pressure is expected to be 300 psi or less and the pipe wall is at least as thick as required for Schedule 10.
- Roll-grooved steel pipe as long as the water pressure is expected to be 300 psi or less and the pipe wall is at least as thick as required for Schedule 10.
- Threaded steel pipe as long as the water pressure is expected to be 300 psi or less and the pipe wall is at least as thick as required for Schedule 40 pipe up to 6 in. in size and Schedule 30 pipe for sizes 8 in. and larger.
- Cut-grooved steel pipe as long as the water pressure is expected to be 300 psi or less and the pipe wall is at least as thick as required for Schedule 40 pipe up to 6 in. in size and Schedule 30 pipe for sizes 8 in. and larger.

There are four different types of joining techniques discussed above: welding, roll-groove, threaded, and cut-groove. This text will not attempt to explain these joining techniques in sufficient detail for the reader to become a standpipe system installer, but it is important for the design professional to understand some basics regarding these joining techniques.

Welding is a process where a large amount of energy is focused on a very small segment of the pipe, melting some of the material and allowing another piece to be joined to the melted piece before it cools. Then, as the material cools, the two pieces of pipe become one. While welding is a very secure process for joining pipe, the process of using such high energy has the concern of potentially starting fires in a building during installation. Therefore, most codes and standards prohibit the use of welding in the field except for emergency repairs and spot welding of small tabs for connection of some earthquake braces. When welding does need to be performed, small pieces of pipe are welded together in a shop where the hazards can be controlled and the process can be done safely. The small pieces of welded pipe are then shipped to a job site and installed to other pieces of pipe using one of the other methods discussed here.

Roll-grooving is the process of pressing an indentation into the pipe near the end of each of the two pieces that need to be joined. The pieces of pipe are then held close together and a fitting goes over the pipe, fitting into the groove and holding the two pieces of pipe together. Since the grove is pressed into the pipe, the wall thickness of the pipe does not change, but an indentation forms on the inside of the pipe as shown in Fig. 5.1. The indentation caused by the roll-groove has been the source of some concern in recent years for dry pipe systems. While some engineers like the roll-grove because it does not change the pipe wall thickness (keeping the strength of the pipe and the ability of the pipe to resist corrosion through the pipe wall), other engineers are concerned about using roll-groove pipe for dry systems because each time the water is drained from the system, water collects at the indentation and does not leave the system. Once the system is put back in service with air in the piping, the combination of water and air at the indentation can cause excess localized

Fig. 5.1 Roll-groove pipe

Fig. 5.2 Cut-groove pipe

corrosion right at this point. It is not surprising to see the phrase "no roll groove pipe" in the specifications for some dry (automatic, semiautomatic or manual) standpipe systems.

Cut-groove pipe is very similar to roll-groove except the indentation is cut into the pipe instead of pressed. The cutting process takes the pipe wall away and does not leave an indentation on the inside of the pipe as shown in Fig. 5.2. Many design professionals are beginning to choose cut-groove pipe for dry standpipe systems because it does not leave an indentation on the inside of the pipe and can be completely drained after the system trips (assuming that the piping is correctly pitched), eliminating the concern about localized corrosion near the end of the pipe where it will be joined to the next piece. Since the cut-groove takes away some of the pipe wall, there is a small concern that this eliminates some of the strength of the pipe, which is why NFPA 14 requires that unlisted cut-groove pipe have a greater minimum wall thickness than roll-groove pipe as discussed above.

Threaded pipe is a process by which the outer wall of the pipe is cut away in a twisted pattern so that the pipe can be screwed into a threaded fitting. Another piece of pipe is then threaded into the other end of the fitting, joining the pipe. Threaded pipe is the oldest form of joining piping for fire sprinkler systems. Since the process of threading the pipe removes some of the pipe wall, NFPA 14 requires that the minimum wall thickness of unlisted pipe be thicker than welded or roll-groove pipe. Threaded Schedule 40 pipe is considered the standard from an acceptable pressure and corrosion resistance point of view. Other methods of joining pipe are sometimes compared to the performance of threaded Schedule 40 when trying to estimate how long the pipe will last.

NFPA 14 does allow other types of pipe besides the three discussed above (ductile iron, steel and copper). These other types of pipe could take three potential forms: other materials, other methods of joining the three permitted materials, and traditional methods of joining the three permitted materials with thinner walls. The first of these possibilities, using other materials, potentially opens the door to pipe materials such as plastic, copper-nickel and stainless steel. Each of these materials might have advantages of corrosion resistance in certain circumstances, but in order to use them in accordance with NFPA 14, the product would have to be specifically listed for use in standpipe systems. At the publication of this book, none of these products have been listed this way. There are limited circumstances where copper-nickel or stainless steel products have been used in standpipe systems, but these have been done as equivalencies based on the performance-based aspects of these materials as compared to the steel pipe produced in accordance with standards like ASTM A53, A135 and A795.

In addition to welding, roll groove, cut groove and threading, there are other methods of joining steel pipe. These other methods would be permitted by NFPA 14 as long as the method is listed for use in standpipe systems. The NFPA Committee on Standpipe and Hose Systems does not want to limit ingenuity, but does not want to be put in a position to have to establish equivalent performance from some new joining technique with those that have been in use for decades, so they leave this evaluation to a listing laboratory, who not only investigates the performance of the product, but the manufacturing process to make sure that each product continues to be made with the quality of the original samples that were investigated by the lab.

As discussed above, it is also possible to use threaded or cut groove pipe with a thinner wall than required by NFPA 14. For example, it is possible for a manufacturer to achieve a listing for a threaded steel pipe that has a pipe wall thinner than Schedule 40. This is acceptable to use in a standpipe system as long as the listing is specifically for standpipe systems and the pipe is used and installed in accordance with all of the special requirements of the listing. This could be done by having the manufacturer control the tolerances better on the threading process, or by the manufacturer making the pipe stronger in some other way.

Copper standpipe systems are rare, but when they are installed, the tube is required to be brazed as its joining technique. While NFPA 13 allows soldering of copper tube under some low hazard situations, NFPA 14 does not have the same allowance. Standpipe systems need to be brazed, which takes a higher temperature to achieve. The thought is that brazed tubes will be less likely to fall apart from the heat of a fire than soldered tubes. Since the process of brazing the tubes takes heat, and has to be done during the installation of the system in the building, the installer needs to take care not to start a fire while installing the standpipe system.

While the types of pipe discussed above are generally rated for 300 psi, the fittings that join them together are not always allowed to be used for such a high water pressure. NFPA 14 limits the use of standard pattern fittings based on the material that the fitting is made from and the water pressure that is expected in the fitting. Table 5.2 summarizes the limitations that NFPA 14 imposes on fittings in section 4.3.3 and its subsections.

Table 5.2 Use of fittings to join steel pipe

Type of fitting	Weight pattern of the fitting	Nominal size	Water pressure allowed
Cast iron	Standard	2 in. and smaller	300 psi
	Standard	2½ in. and larger	175 psi
	Extra heavy	2½ in. and larger	Over 175 psi up to rating of fitting
Malleable iron	Standard	6 in. and smaller	300 psi
	Standard	8 in. and larger	175 psi
	Extra heavy	8 in. and larger	Over 175 psi up to rating of fitting

Interconnection of Standpipes

Where a building has two or more standpipes, all of the standpipes are required to be interconnected with piping (NFPA 14 section 7.5.1). In this way, a single water supply, or a single fire department connection can provide water to any standpipe and any outlet on a standpipe system. Where the water supply for the standpipe system is in the lower portion of the building, such as an underground water main or a fire department connection near ground level, the interconnection should be in the lower portion of the building in order to save friction loss as shown in Fig. 5.3. Note that the requirement is not to have the interconnection piping at the very bottom of the system, but just somewhere near the bottom. If the water supply is higher up in the building, like a tank on the roof of the building, then the interconnection is required by NFPA 14 to be at the top of the building (section 7.5.2) as shown in Fig. 5.4. This interconnection piping is sometimes called a “feed main”.

From time to time, it may be advantageous to interconnect the standpipes at both the top and the bottom. For example, in the situation discussed above, when NFPA 14 requires the system to be interconnected at the top, it might save pipe to interconnect the piping at the bottom so that the pipe from the fire department connection can feed each of the standpipes without having the water go all the way to the top and feed back down again. When this happens, section 7.5.3 of NFPA 14 requires the installation of check valves in the standpipes so that the standpipe system does not become a loop. The check valve arrangement is shown in Fig. 5.5. Without the check valves, the flow of water through the system would be more like a grid, which might tempt the designer to decrease the pipe size, which is exactly what the committee responsible for NFPA 14 does not want to happen. So, to prevent the piping from acting like a grid, the check valves are required.

Dry standpipes (automatic, semiautomatic or manual) are not permitted to be interconnected at more than one level (section 7.5.4). If a dry standpipe was interconnected at more than one level, the possibility exists for air to get trapped in the system and delay the delivery of water to any specific hose connection. This is similar to the prohibition of gridded dry-pipe sprinkler systems in NFPA 13.

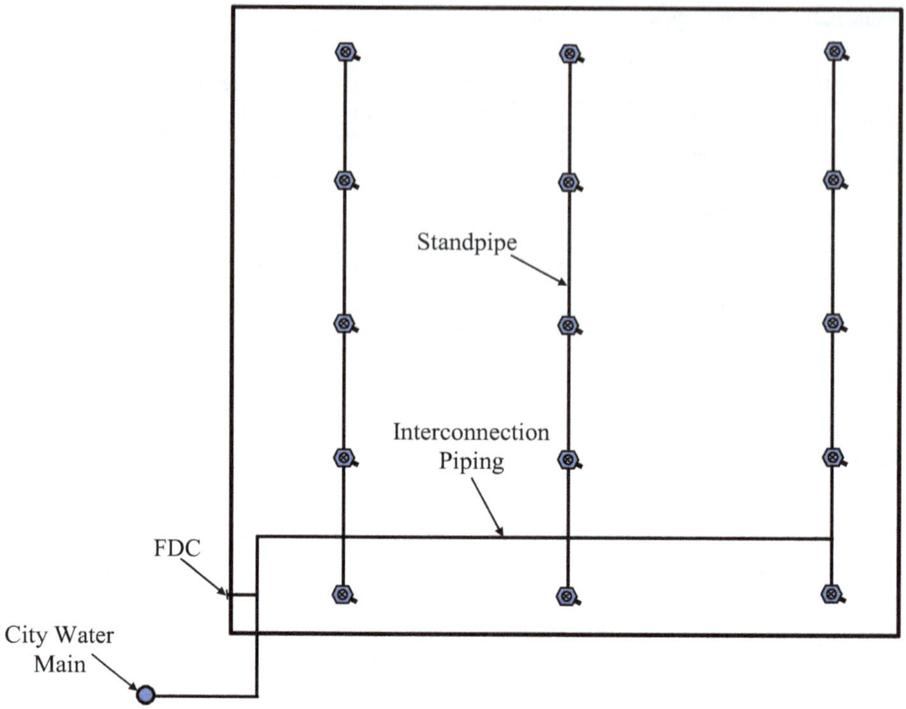

Fig. 5.3 Interconnection of standpipes near the bottom of the building

Control Valves

Where control valves are installed on standpipe systems, they need to be listed indicating valves such as outside screw and yoke (OS&Y) valves or butterfly valves. The OS&Y valve has a threaded stem that pokes out of the valve when the valve is open. If you can see the valve stem poking out of the valve, then you know that the valve is open. Figure 5.6 shows an OS&Y valve that has been taken out of a system. With the valve stem out of the valve, the gate that blocks the flow of water is stored in the gate housing, completely out of the flow of water, allowing water to flow unimpeded through the system. When it is time to shut the valve, the wheel is turned, pushing the valve stem into the valve and pushing the gate down until it blocks the flow.

Butterfly valves have a gate inside the valve that is connected to a wheel. The connection is typically in the form of a gear, so they are sometimes called butterfly gear valves. When the wheel is turned, the gear rotates the gate. When the gate is parallel to the pipe, water can flow in the pipe. When the gate is turned perpendicular to the pipe, the flow is blocked. Unfortunately, we can't see the position of the gate at any given time. So, to indicate that the valve is open, the gate is connected to

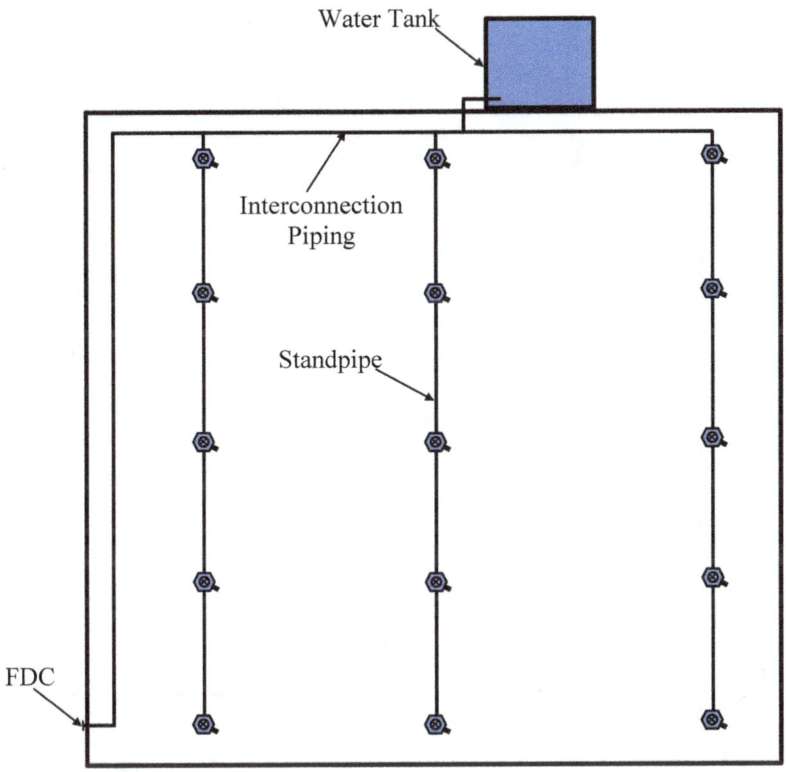

Fig. 5.4 Interconnection of standpipes at the top of a building

a paddle outside of the valve. When the paddle is parallel to the pipe (as shown in Fig. 5.7) the valve is open. When the paddle is perpendicular to the pipe, as shown in Fig. 5.8, the valve is closed and water cannot flow in the pipe. The gate on a butterfly valve can be longer than the body of the valve, so when a butterfly valve is designed or installed in a system, care needs to be taken to make sure that there is enough room for the valve to fully open and close. Do not design or install an elbow, tee, or other device close to the butterfly valve that would interfere with the ability of the valve to open or close.

One of the important features of a listed valve is that the valve must close slowly. NFPA standards require that the valve take at least 5 s to close (section 4.5.2). The purpose of the slow closing valve is to make sure that flow through a system is not abruptly stopped, especially large water flows. Standpipe systems have the potential for large flows of water. If this flow is stopped abruptly, a pressure wave can occur in the piping that bounces back and forth. This pressure wave is called "water hammer" and has the potential to do serious damage in the piping system. The best way to reduce the possibility of water hammer is to shut down water flow slowly.

Control valves are required at three different locations in standpipe systems: connections to water supplies, isolation of standpipes, and connections to sprinkler

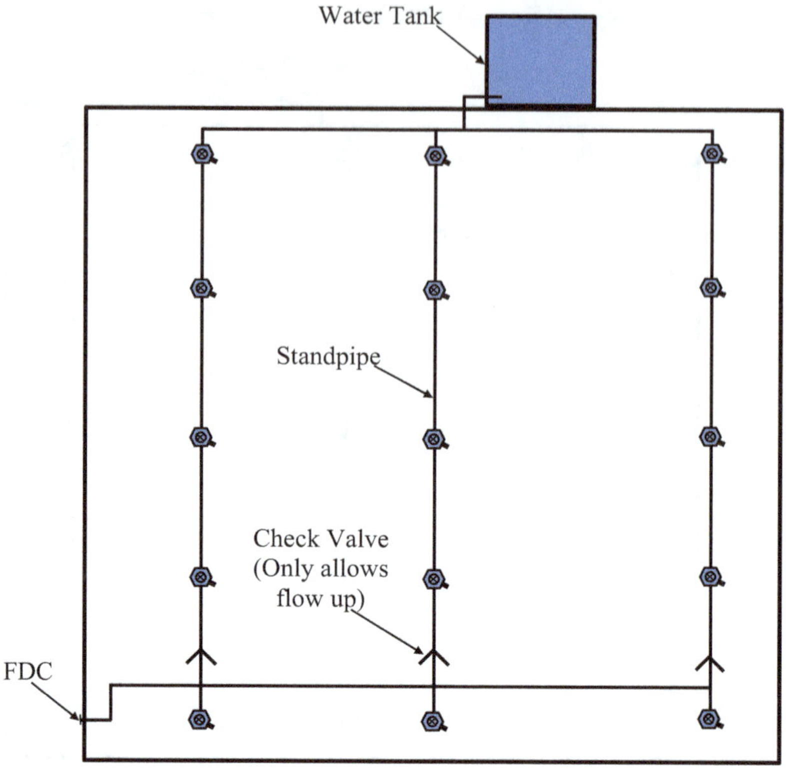

Fig. 5.5 Interconnection of standpipes at top and bottom with check valves

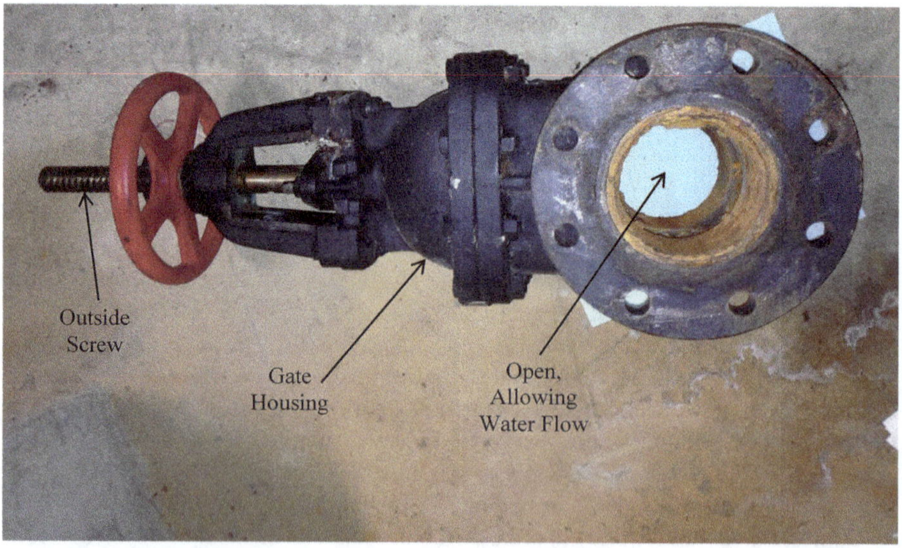

Fig. 5.6 OS&Y valve

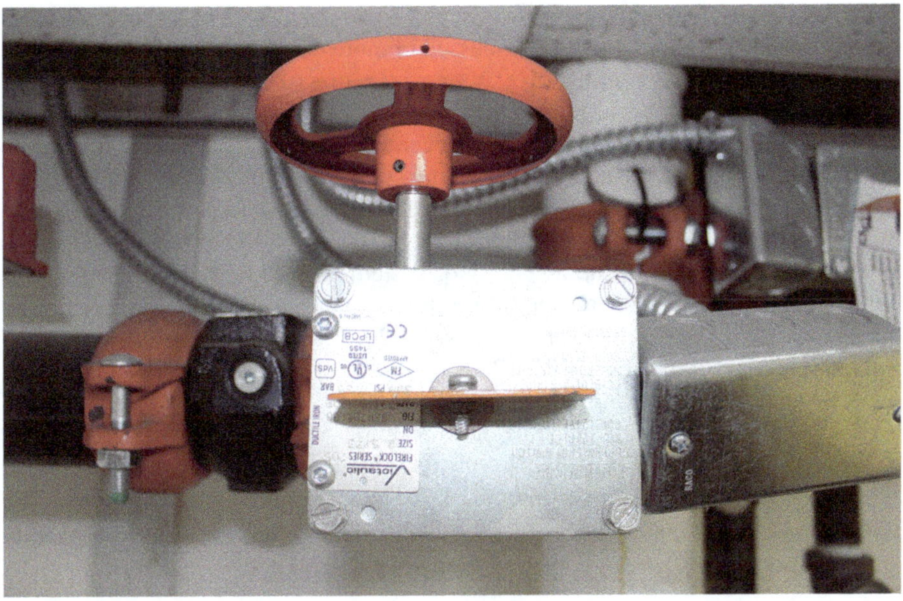

Fig. 5.7 Butterfly valve with paddle parallel to pipe indicating water can flow

Fig. 5.8 Butterfly valve with paddle perpendicular to pipe indicating water cannot flow

systems. The first two of these will be discussed in depth here. The last one (connections to sprinkler systems) will be discussed in more detail in Chap. 7 of this text.

NFPA 14 requires a control valve at the connection of each standpipe system to a water supply except for the connection to the fire department connection (sections 6.3.1.1 and 6.3.6.1.1). The purpose of this control valve is to cut off the main flow of water to the system, basically isolating the system from the supply in case repair of the main piping is required. Many water supplies for standpipe systems incorporate backflow preventers, pumps and/or tanks. Each of these devices has a control valve built in as a part of the assembly or system. A backflow preventer has two control valves as a part of its listed assembly. A pump has a control valve on the discharge side of the device required by NFPA 20. A tank has a control valve in the discharge pipe required by NFPA 22. Each of these control valves counts as the control valve required by NFPA 14.

Each of the individual standpipes in a standpipe system is also required to have its own control valve as an isolation valve so that the standpipe can be separated from the rest of the system (section 6.3.2). If repairs are necessary that require the standpipe to be drained, the control valve can be closed and the rest of the system can stay in service. Figure 5.9 shows a standpipe system with three risers and three

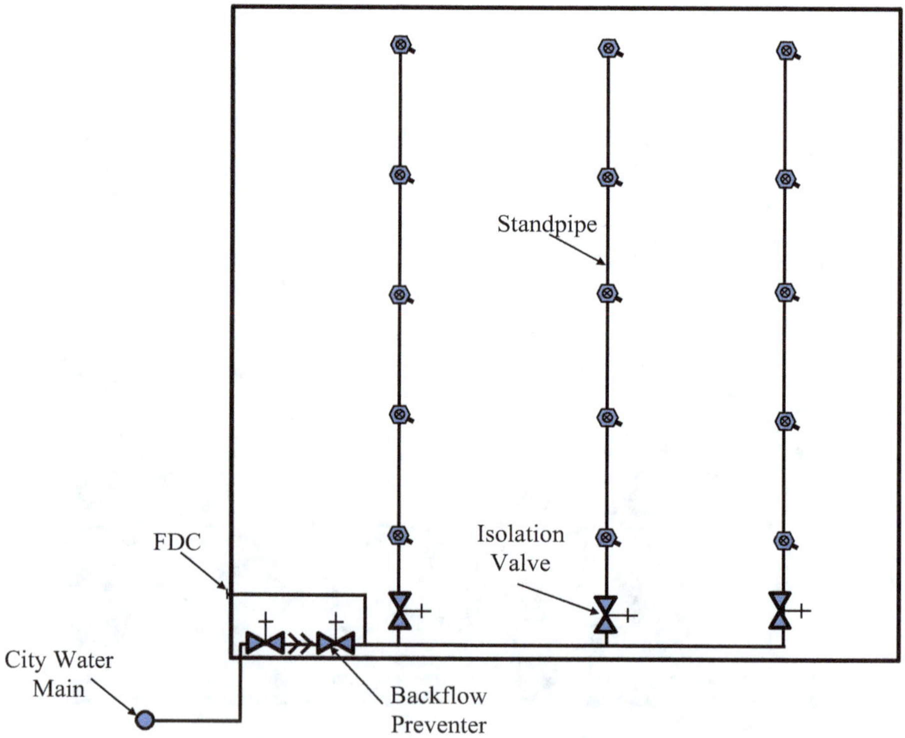

Fig. 5.9 Standpipe system with control valves on water supply and an isolation valve on each standpipe

control valves isolating each standpipe as well as a backflow preventer serving as the control valve from the water utility (city water main).

Since manual dry standpipe systems do not have a water supply, there is no requirement for a control valve to be installed at the water supply connection (section 6.3.1.4). But the requirement for each standpipe to have an isolation valve is still applicable to all manual dry standpipe systems. There have been so many questions on this subject over the years, that the committee recently added "including manual dry standpipes" to section 6.3.2 so that everyone would know that the valves are required.

A control valve is also required for branch lines that are more than 40 ft long. A branch line is a pipe that connects a remote hose connection to a standpipe as shown in Fig. 5.10. Branch lines can be installed on just a single floor as shown in Fig. 5.10 when the arrangement of walls and partitions just on that floor exceed the maximum permitted travel distance, or they can be installed on many floors if it was deemed more efficient than installing an additional standpipe. When the branch line exceeds 40 ft in length, there is a higher probability that the pipe may need to be isolated for repair. The purpose of the control valve is to allow the branch line to be isolated without having to take the whole standpipe out of service during the repair time. There is nothing magical about the length of 40 ft. The committee wanted a control

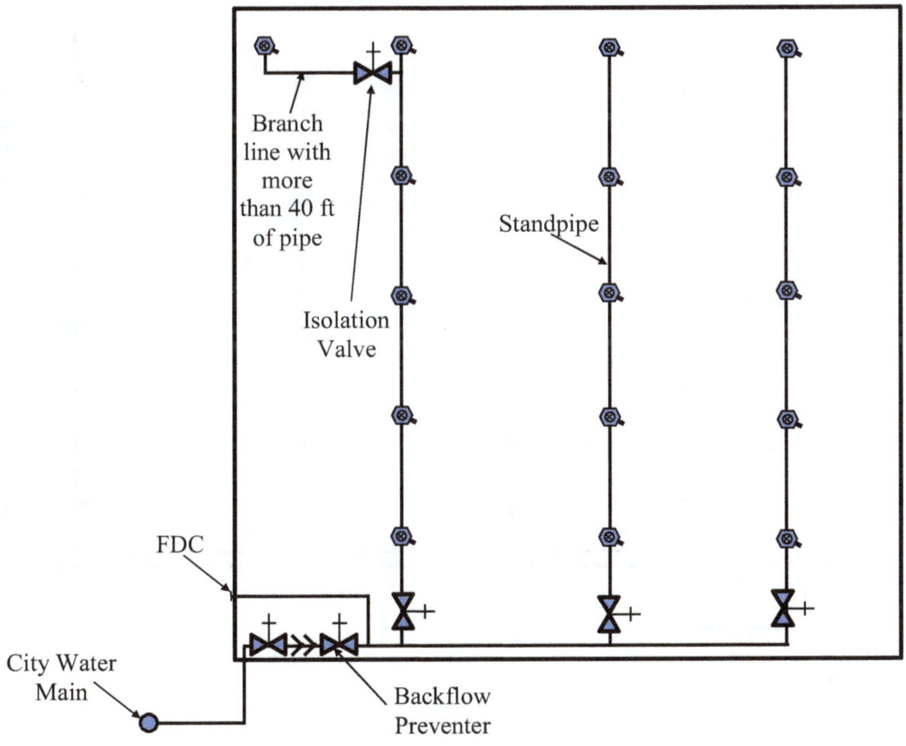

Fig. 5.10 Branch lined requiring control valve for isolation

valve on "long" branch lines, but using the term "long" in a standard is too difficult to enforce, so the committee came up with the requirement for 40 ft as a measurable requirement.

The requirement discussed earlier for the valve to be indicating does have an exception, but only when the authority having jurisdiction approves the non-indicating valve (see section 4.5.1.3). A typical use of a non-indicating valve is an underground gate valve connected to a road box that is operated with a t-wrench. This type of valve is a common replacement for a post indicating valve, which allows the system to be shut down from outside the building. Figure 5.11 shows the underground gate valve, road box, and t-wrench. The t-wrench would normally not be kept at the valve, but the owner should declare the location of the t-wrench and make sure that it is kept there so that it can be used in times of emergency.

All of the valves discussed here except the underground gate valve connected to the road box need to be supervised in the open position so that they will work during

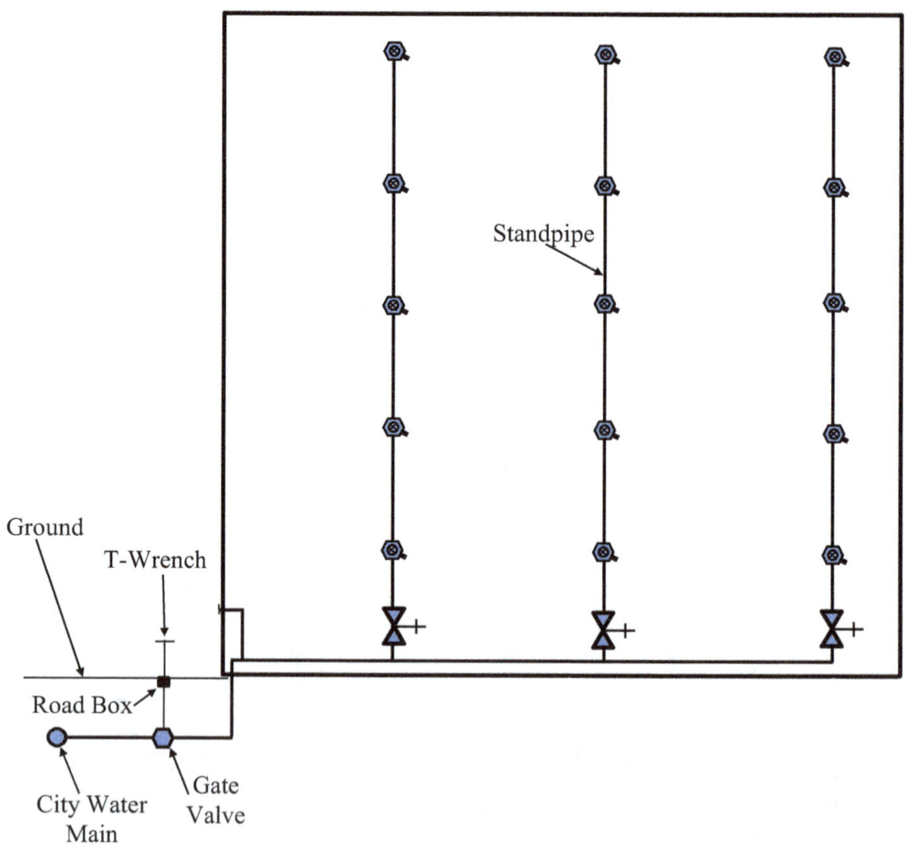

Fig. 5.11 Underground gate valve, road box and T-wrench

Table 5.3 Valve supervision

Type of supervision	Method of supervision
Electrical	Central station, proprietary or remote signaling service
	Local signaling service with an audible signal at a constantly attended location
Mechanical	Locking the valves in the open position
	Sealing the valve with a device that breaks if the valve is turned. This option is only valid if the seal is inspected on a weekly basis to insure that the seal is intact

a fire without someone having to find the valve and open it. NFPA 14 allows four different methods of supervision in section 6.3.7.1. Table 5.3 summarizes the acceptable types of valve supervision.

The first two options for supervision in Table 5.3 are electrical in nature. A switch is connected to the valve so that when the valve is closed, a signal is sent to a constantly attended location. The person monitoring the location is required to call the building owner when this signal comes in to let them know that someone is tampering with the sprinkler system. This type of signal is not treated as an alarm, so the fire department is not called. But it is important that the building owner determine what is causing the signal quickly so that the system is kept operational.

The third and fourth option in Table 5.3 are considered mechanical means of supervision. If someone decides to break the lock or seal, the owner may not know for some time. The use of a seal is typically done with a plastic tab or a lead wire wrapped around a working part of the valve that has to be broken if the valve is going to be closed. The sealed valve is required to be in a fenced enclosure that is only accessible by the building owner. The term "fenced" is rather loosely interpreted to include locked rooms and other locations only under the control of the building owner.

While NFPA 14 allows these four options for supervising the control valves in the correct (open) position, many building codes and fire codes do not have such flexibility. It is common to find that building codes and fire codes require electronic supervision, making only the first two options in Table 5.3 viable options in these cases. The codes base the requirement for electronic supervision on the belief that this dramatically improves the odds that the valves will be maintained open. There is some evidence from a few studies done by insurance companies that this is in fact correct.

Control valves are required to have a sign on them letting the reader know what system or portion of a system they control. Sign requirements are found in section 6.3.8 of NFPA 14. All signs are required to have red letters at least 2.5 in. in height on a white background. If the valves are in locked rooms, the sign needs to be located on the door to the room so that people know where the valve is located.

Check Valves

In addition to the check valves discussed above that need to be installed where standpipes are fed from both above and below, there are a number of other locations where check valves are required in standpipe systems. A check valve is required by NFPA 14 (section 6.3.1.1) at each connection to a water supply, including the fire department connection (FDC). This is different from the control valve rule. A check valve is needed at the FDC because the firefighters need to be able to disconnect the hose from the FDC and not have all of the water in the standpipe system drain onto them. The purpose of the check valve on the other water supplies that are not FDC's is to prevent water from one water supply from going back to another water supply. The check valves that are installed in the discharge piping of fire pumps and tanks serve to comply with this requirement. Backflow preventers are considered check valves for the purposes of this rule, so if a backflow preventer is installed as protection for a public water main, no additional check valves are required. Figure 5.12 shows a check valve installed in both the FDC and the connection to the private water main.

In Fig. 5.12, the check valve is shown a short distance to the right of the exterior wall with a drain in between the check valve and the wall. This is a common instal-

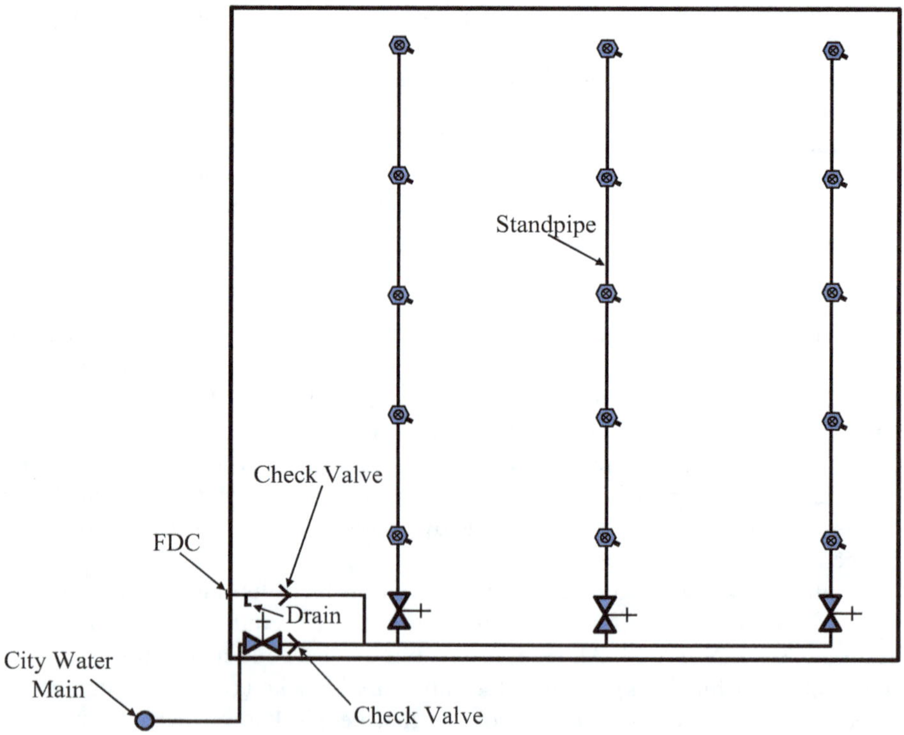

Fig. 5.12 Check valves in FDC and water main connections

lation in areas subject to freezing so that the system does not get damaged. The check valve holds the water in the standpipe system and the drain allows all of the water between the check valve and the wall to drain out. The type of drain typically installed here is a "ball drip" with a plastic ball slightly smaller than the diameter of the pipe it sits in. The ball floats when water is in the pipe and allows water to drip out a hole in the bottom of the pipe. When all of the water leaves this section of pipe, the ball settles down into the hole, closing it off.

Although there is no requirement for how far the check valve in an FDC needs to be away from the wall to prevent freezing, most designers try to keep the check valve at least 4 ft from the wall. The actual distance that would be needed would depend on the expected low temperature outside and the temperature that the building is heated to on the inside. Section 6.4.2.1 of NFPA 14 requires the check valve to be close to the inlets, in an accessible location, but not in a location subject to freezing. Rather than doing the complex convective and conductive heat transfer involved to figure out exactly where this will be, most people use the general guidance of 4 ft from the wall and make sure that the check valve can be accessed for inspection and maintenance.

In addition to the check valve, Fig. 5.12 shows a control valve on the pipe from the water supply to the standpipe system. The control valve is shown between the check valve and the water supply. This is not required by NFPA 14, but makes sense from an inspection and maintenance perspective. With the control valve in this position, the control valve can be closed and the check valve can be opened to make sure that the clapper inside is capable of swinging freely and preventing backward water movement. This internal inspection is required by NFPA 25 once every five years. If the control valve was installed between the check valve and the standpipe system, you would have to go back to the water main and shut it down in order to open up the check valve and perform the internal inspection every fifth year, which would be extremely difficult and might lead to people ignoring NFPA 25, which would be a poor design.

Check valves are also required at each connection to a sprinkler system from a standpipe system. There are a number of reasons for these check valves, which will be discussed in further detail in Chap. 7 of this text.

For dry standpipe systems that have automatic air maintenance systems, a check valve is required in the air delivery piping if the air maintenance system serves more than one fire protection system. This is to prevent air or water from traveling between systems through the air maintenance piping. See section 5.2.1.4.8 of NFPA 14 for more information on supplying air to a dry standpipe system.

Waterflow Alarms

Section 5.6 of NFPA 14 requires waterflow alarms on all standpipe systems except manual dry systems. The waterflow alarm needs to be installed in a location that would sense water flowing to any hose connection on the system. This could be anywhere between the water supply and the first (closest) hose connection.

For wet standpipe systems (automatic or manual), a paddle-type flow switch is acceptable and probably the flow switch of choice. For dry standpipe systems (automatic, semiautomatic, and manual), a paddle-type flow switch would not be allowed because there is a concern that the turbulence of the water pushing the air at the water/air interface might be rough enough to pull the plastic paddle off of its stem, which would potentially cause an obstruction in the piping downstream. Instead, on dry standpipe systems, pressure switches are used that either sense the drop in water pressure that occurs when water begins to flow in the system or that sense the difference between water pressure and air pressure in a system. Many dry standpipe systems use differential dry-pipe valves where low air pressures hold back high water pressures. The pressure switch can sense the higher water pressure when the valve trips and the water pushes out the air.

Downstream of the waterflow alarm needs to be a mechanism that can be opened to test the waterflow alarm. Since the standpipe system is designed to create flow between 100 and 250 gpm, this connection is not required to be a small 1 in. connection like you find on sprinkler systems. Instead, a hose connection can be used as long as the water can flow from the hose connection to a drain safely.

The waterflow alarm is not required by NFPA 14 to go to a constantly attended location. The waterflow alarm can just make a noise on the outside of the building. The purpose of the waterflow alarm is to let people know that water is flowing in the building and the hope is that someone would notice the alarm making noise outside of the building.

Many building and fire codes require that the waterflow alarm be monitored at a constantly attended location. Most of the time, when a standpipe system is being used, it is being used by a person, so it is not necessary to send a signal to a constantly attended location. But in those situations where someone is using the standpipe incorrectly (for a purpose that is not fire protection) or for situations where there is a leak, it is helpful to have the signal sent to a constantly attended location so that someone does something about the flow of water immediately. This type of monitoring service can potentially save a significant amount of money in water damage, so some codes require the signal to be monitored. If the control valves are being monitored by electronic tamper switches at a constantly attended location, then adding the cost of monitoring the waterflow switches is a minimal additional cost.

Drains

Standpipe systems have the potential to require several drains. The first type of drain, which is required on all standpipe systems, is a "main drain", which is a connection on the water supply piping between the main control valve for the system and the standpipes that permits the standpipe system to be drained and also allows for a test of the water supply and the main control valve. Section 7.11.2.1 of NFPA 14 requires a main drain to be installed on every standpipe system. Table 5.4

Table 5.4 Main drain sizing

Standpipe pipe size	Minimum drain size
Up to 2 in.	¾ in. or larger
2.5 in., 3 in. or 3.5 in.	1.25 in. or larger
4 in. or larger	2 in. or larger

summarizes the requirements for the minimum size of the main drain. The drain needs to be capable of being fully opened and needs to be capable of handling the flow created from this connection being fully opened. Rather than install a separate connection as a main drain, the designer is allowed to use the lowest hose connection in the system as a main drain as long as the connection can be opened fully with the drain being able to safely handle the flow, and as long as the authority having jurisdiction approves the installation.

Portions of standpipe systems that are trapped and will not drain back to the main drain need to be set up with auxiliary drains. The designer has three basic options for auxiliary drain arrangement. The three options are:

1. An auxiliary drain in accordance with NFPA 13, which has special allowances for portions of piping that are less than 5 gallons in size. The designer is allowed to just use an easily removable piping joint, like a flexible coupling, as an auxiliary drain. The 5 gallon limit comes from the idea that a person should be able to carry a 5 gallon bucket to the connection, take it apart and drain the system into the bucket, then carry the bucket outside to dump it. NFPA 13 does have some other allowances and the reader is encouraged to review section 8.16.2.5 of NFPA 13 for a more complete discussion of auxiliary drains.
2. A drain connection sized in accordance with Table 5.4.
3. A hose connection that will allow the water to be drained out of the trapped portion of piping to a drain that can safely handle the flow.

It is important to make sure that all of the water is eliminated from dry standpipe systems (automatic, semiautomatic, or manual) so that the water does not cause excess corrosion when air is added to the system (or in the case of manual systems, allowed to enter after the use of the standpipe is finished). In addition to making sure that the system can be drained, the designer needs to make sure that all portions of the piping system in all dry pipe standpipe systems is pitched (sections 6.1.3 and 6.1.4 of NFPA 14). The piping needs to be pitched at least ¼ in. for every 10 ft of pipe in non-refrigerated systems. In refrigerated systems (constantly kept at cold conditions) the pitch is required to be ½ in. per 10 ft.

The pitch discussed here is a vertical distance over a horizontal distance of 10 ft. So a pitch of ¼ in. for every 10 ft means that the pipe has to drop only ¼ of an inch vertically for every 10 ft of horizontal run of pipe. So, for a 200 ft horizontal run of pipe, the vertical drop would only have to be 5 in., which is not a difficult requirement to meet, but dramatically improves the long-term performance of the system by getting the water out.

Fig. 5.13 Typical drum drip at a low point in a dry standpipe

For all automatic or semiautomatic dry standpipe systems, there is a concern about water from condensation building up in the system over time. These types of systems have some compressed air in them under pressure. As temperatures change in the system, moisture in the air condenses out and settles into the bottom of the pipe, collecting at the low points in the system. If sitting water in the system freezes, the water will expand and has the potential to break the pipe. You might not find out about this break until the next time that the standpipe is used, and then the break can be catastrophic. In order to make sure that any collected condensate or stuck water eventually gets out of the system, special types of drains, called "drum drips" are installed at low points in the automatic and semiautomatic systems. A typical drum drip consists of two control valves on a vertical section of pipe with a larger diameter section of pipe installed in between as shown in Fig. 5.13. The drum drip is operated like an air lock so that water can be drained without losing air pressure in the system.

The proper way to operate a drum drip is to keep the lower valve closed and the upper valve open as shown in Fig. 5.13. The water collects in the larger diameter vertical section of pipe. When it is time to check the drum drip, the upper valve is closed and then the lower valve is opened. The plug is then removed and the drum drip is observed for water drainage. If water does drain out, wait until the water is done draining, then close the lower valve and open the upper valve. Listen for the sound of more water filling the device. Then close the upper valve and open the lower valve and see if more water comes out. Repeat this operation until all water has drained from the system.

For standpipe systems with pressure regulating valves at the hose connections, a drain riser is required so that the pressure regulating valve to be tested. The only way to properly test the pressure regulating valve is to flow 100 gpm (for 1.5 in. connections) or 250 gpm (for 2.5 in. connections) through the valve. There needs to be somewhere for this water to go. Running hose down the stairs in a tall building is not an appealing option. The drain riser makes the most sense for testing these devices. In the 1990s (before drain risers were required) a fire in Philadelphia brought out how important it is to test pressure regulating valves. Without a drain riser, building owners were not performing the tests, even though they could have done it with running hose down stairs. The standpipe committee decided to take the decision out of the hands of designers and force drain risers in all systems that have pressure reducing devices.

The size of the drain riser depends on the size of the pressure regulating valve. If the valve is a 1.5 in. valve, the drain riser needs to be at least 2 in. in size. If the valve is a 2.5 in. valve, the drain riser needs to be at least 3 in. in size. If the valve is some other size larger than 2.5 in., then the drain riser needs to be at least as large as the outlet of the valve.

The drain riser needs to have connections at least as large as the outlet of the pressure regulating valve installed at every other floor level. This means that the most hose that would need to be used during a flow test of the pressure regulating valves would be less than 50 ft, so a single length of 50 ft hose could be brought up in the building to conduct the tests.

The drain riser needs to terminate at a place low down in the building that can handle the flow from the test. Each pressure regulating valve is only required to be tested individually, so the most flow that would be required is 250 gpm from a 2.5 in. pressure regulating valve or 100 gpm from a 1.5 in. pressure regulating valve.

Even though the fire department will not be using the drain riser during a fire in the building, the connections on the drain riser need to approved by the local fire department and are generally required to match the threads that the local fire department uses (section 7.11.1.4 in NFPA 14). The thought behind this requirement is that the local threads used by the fire department will be installed on the outlet of the pressure regulating valve and during a test, hose needs to be run from the pressure regulating valve to the drain connection. The hose should have the same threaded connection at both ends unless the building owner wants to try and buy a hose with the local fire department threads on one end and the standard threads on the other end. This special type of hose might be too confusing to use since it could not be turned around during testing and it would be difficult to keep straight which end was the correct thread.

When drain risers are not in use, they are empty, with just air at atmospheric pressure inside. When the drain riser is being used during a test, water enters at the location of the connection and flows down the pipe towards the bottom. As the water flows, a vacuum develops in the drain riser above the location of the test connection. If there is no opening in the top of the drain riser, water only leaves the drain at the bottom in gulps. In order to improve the flow in the drain riser, some method of relieving the vacuum above the test connection would be helpful. Although NFPA

14 does not mention relief of the vacuum in the drain riser, contractors have developed the following methods for providing this relief:

1. An open pipe at the top of the riser with a check valve facing down the drain pipe. When the water starts flowing during the test, the vacuum causes the check valve to open, allowing water to enter at the top of the drain riser and relieve the vacuum. Rather than leave the pipe completely open at the top, an elbow or two can be used so that the opening is facing down. The advantage of this type of installation is that it will work automatically and the person performing the testing does not need to remember to do anything to relieve the vacuum.
2. An open pipe at the top of the riser with a control valve near the top. When it is time to run a test using the drain riser, someone needs to go open the valve. Rather than leave the pipe completely open at the top, an elbow or two can be used so that the opening is facing down. The advantage to this type of installation is that the drain riser is closed off when there is no testing being performed. The disadvantage to this arrangement is that someone needs to remember to open the valve before using the drain riser and they need to shut the valve when the testing is over.

Section 6.3.8.2 of NFPA 14 requires that all drain connections be identified with signs so that users can use them correctly for inspection, testing and maintenance of the system.

Fire Department Connections

The fire department needs some mechanism for pumping water into standpipe systems. This connection is more important on standpipe systems than any other fire protection system because the standpipe system is for the fire department's use. The name given to this connection is the "fire department connection" or FDC.

The rules for installing FDC's are scattered throughout NFPA 14 in sections 4.8, 6.4 and 7.12. In addition to these requirements, there are a number of common sense items that the designer will need to consider before deciding where to put the fire department connections. This text will cover both the requirements of NFPA 14 and the common sense items.

FDC's are required on all Class I and Class III standpipe systems (see section 7.12.1 of NFPA 14). Some standpipe systems are divided into different vertical zones in order to control the pressure. See Chap. 8 of this text for more information on why a system might be divided into different vertical zones. For these standpipe systems that have multiple vertical zones, an FDC is required for each zone in the system unless the zone is so high above the street level where the fire department will be pumping into the FDC that it is not reasonable to believe that the FDC is worthwhile. Exactly how high this might be depends on the fire trucks owned by the local fire department and the pressure these trucks can create.

Fig. 5.14 A fire department connection with two 2.5 in. inlet connections

Fire trucks can be set up to pump in series and create a great deal of pressure, so it all boils down to the standard operating procedures of the fire department how high they think they can produce a usable pressure. In some cities, the fire department has declared that they can reasonably get water 200 ft above the street level. In other cities, the fire department has declared that they can reasonably get water 350 ft above the street level. So, checking with the local fire department is critical in the design of standpipe systems for tall buildings. Sections 7.9.3 and 7.12.1.1 of NFPA 14 work together to state that FDC's can be eliminated from the upper zone of very tall buildings if there is a water supply internal to the building that feeds this high zone of the standpipe system.

Fire department connections are required to be listed by a testing laboratory. There are two types of fire department connections that have been listed. The first, and most common, is the combination of 2.5 in. inlet connections. Section 4.8.2 of NFPA 14 requires at least two 2.5 in. inlet connections, as shown in Fig. 5.14, however section 7.12.3 requires at least one 2.5 in. inlet connections for every 250 gpm of flow demand. So, a standpipe system with a 1000 gpm flow demand would need four inlet connections as shown in the FDC in Fig. 5.15. Since these connections are for the fire department, the threads on the inlets need to match the hose threads used by the local fire department. Since almost every type of building with a standpipe system has at least two exit stairwells, which creates a demand of at least 750 gpm, it is rare to find a fire department connection for a standpipe system with less than three 2.5 in. inlets. The buildings with two 2.5 in. inlets fall into four categories:

1. Those that were constructed before NFPA 14 required one inlet for each 250 gpm
2. Those that have a single exit stairwell (such as the special provision for a single exit stairwell for multi-family residential structures in the Life Safety Code.

Fig. 5.15 A fire department connection for a standpipe system with a demand of 1000 gpm and four inlet connections

3. Those where multiple FDC's are installed in a system. For example, a system with a 1000 gpm flow demand would need four 2.5 in. inlets, which could be provided using two FDC's, each on different sides of the building, each having only two 2.5 in. inlets.
4. Those where the AHJ has permitted the two 2.5 in. inlets typical to sprinkler systems.

The second type of FDC uses large diameter connections around 5 in. in diameter as shown in Fig. 5.16. Rather than make this connection with regular threads, this type of connection just requires that tabs be lined up with slots and rotated one-quarter of a turn. These connections were invented for connecting fire hose by a man named Carl August Guido Storz in 1882, so they are frequently referred to as "Storz connections". Today, they are manufactured by a bunch of different companies and are more frequently referred to under the generic term of "large diameter threadless couplings". At one time, the use of these devices on standpipe systems was problematic because there were no listed products, but several companies have now achieved a listing for their large diameter threadless couplings, so as long as these listed products are used and are installed in accordance with their listing, the installation can comply with NFPA 14.

Regardless of which type of fire department connection is used, the inlet needs to be protected by caps that can be easily removed by the fire department. The caps are important because they prevent people from sticking objects into the fire department connections. The caps in Fig. 5.14 are plastic breakaway caps that can be easily popped off by the firefighters if they need to connect their hose. The caps in Fig. 5.15 are heavy metal. Even though they are connected to the FDC by a chain, many

Fig. 5.16 Large diameter threadless coupling used as fire department connection for a standpipe system

building owners find that thieves steel them and turn them into scrap yards for money. To prevent this, many building owners switch to the easily removable plastic caps.

For automatic wet and manual wet standpipe systems, the fire department connection needs to be connected to the system on the system side of the main system control valve as shown in Fig. 5.9. For systems with fire pumps, the fire department connection needs to be tied into the system on the discharge side of the fire pump, on the standpipe system side of the pump's discharge control valve. The fire department connection is not permitted to be connected on the suction side of the fire pump because the fire department will be pumping the water in at significant pressure already from their fire trucks, they don't need the pressure to be boosted further by the fire pump for the standpipe system and such additional boost might overpressurize the system and cause damage due to the high pressure.

For an automatic dry standpipe system, the FDC needs to be tied into the system between check valve for the water supply and the dry pipe valve as shown in Fig. 5.17. The placement of this connection is important because the connection is made where there is water in the system. If the connection was on the other side of the dry pipe valve, the connection would be made where there is air in the piping, which could easily slip past the check valve in the FDC and cause the dry pipe valve to trip from the drop in air pressure. Also, the FDC needs to connect to the system downstream of the check valve so that when the fire department pumps into the system, the water does not just flow back to the city water main. Instead the water needs to be directed towards the standpipe hose outlets.

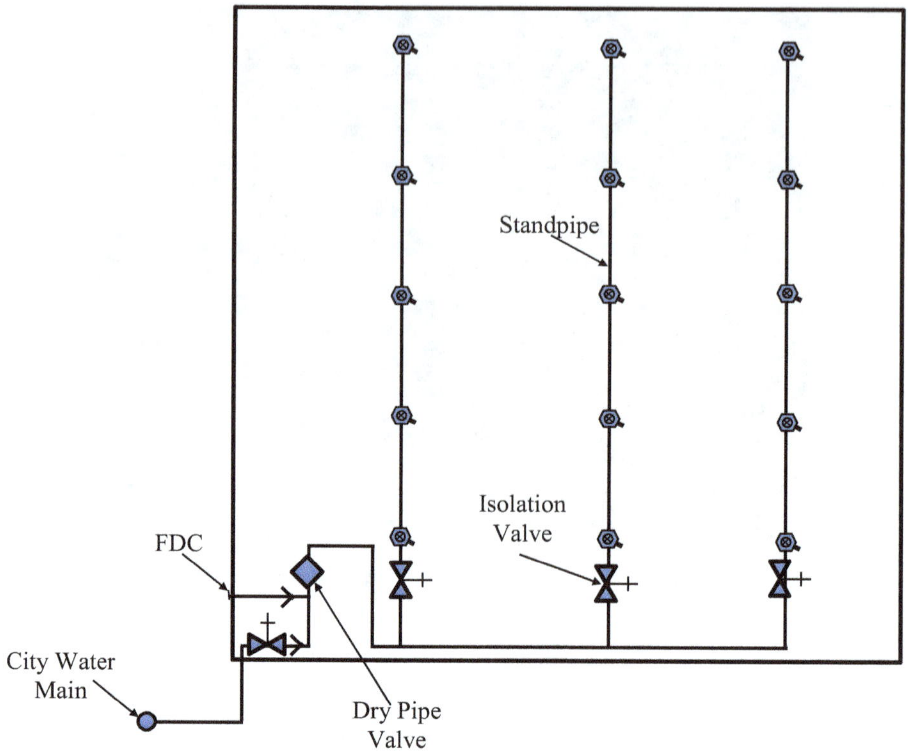

Fig. 5.17 Fire department connection for automatic dry standpipe system

For a semiautomatic dry standpipe system, the FDC needs to be connected downstream (on the standpipe hose connection side) of the preaction or deluge valve that is used to allow the water into the system, as shown in Fig. 5.18. In this case, the air will normally be in the piping to which the FDC is connected, which may allow some air to slip past the check valve in the FDC, but the loss of air will not cause the deluge or preaction valve to trip. The deluge or preaction valve will only trip if the pull station or button at the hose connection is activated, so the small loss of supervisory air in the standpipe system is not a concern.

For high-rise buildings (defined as a building with a floor level 75 ft above the lowest level of fire department vehicle access), each zone of a standpipe system is required to have at least two FDC's. The reason behind this requirement is that these buildings tend to be larger and the fire department might have to approach the building from different sides. Rather than take the time to run hose around the building, the fire department can start pumping into the connection faster if there is more than one FDC. Ironically, the standpipe system in a high-rise building should be automatic, so the fire department should not need the FDC if everything works correctly, but the requirements in NFPA 14 seem to be set up assuming that everything will not work correctly. There is no requirement for how far apart the FDC's need to be

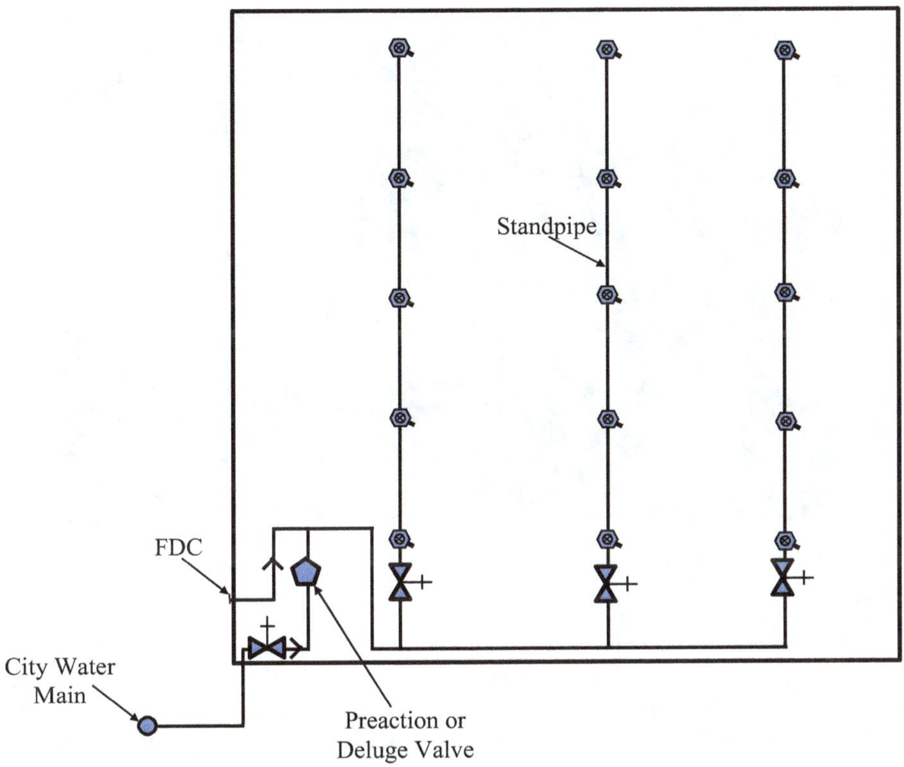

Fig. 5.18 Fire department connection for semiautomatic dry standpipe system

in a high-rise building, but the assumption is that the two FDC's for each zone are located in different places around the perimeter of the building that are each accessible to the fire department.

Section 6.4.5 of NFPA 14 requires that the FDC be visible and recognizable from the street, which can be extremely difficult given the design of the building, access to the building and location of objects around the building. For example, Fig. 5.19 shows a building where there are transformers permanently placed between the street and the FDC and the building is uphill from the street. It would be difficult to find the FDC on this building from the road. In these cases, a sign can be very helpful. The red and white sign on the building in Fig. 5.19 helps firefighters find the FDC on this building, even if it cannot be seen due to the building being uphill and the location of the transformers.

When deciding on where the FDC is going to be located on the outside of a building, the designer needs to take into account the accessibility of the FDC to the fire department apparatus, the location of the water supply that the fire department will be using, the potential for obstructions to the FDC like bushes, parked vehicles, or fences, and the use of the means of egress from the building by the occupants of the building. Each of these items will be discussed here in a bit more detail.

Fig. 5.19 FDC is difficult to see due to position of building uphill and transformer location

The first consideration is the accessibility of the fire department vehicles. If there is only one side of the building that faces the road, the FDC should be on that side. If there are multiple sides of the building that face a road, which one is the fire department more likely to use? Is there one that would be safer than the other to park a fire truck? Is there one that is less likely to cause traffic problems if a fire truck were to be parked there? The answers to these questions will help drive the decision as to where the put the FDC.

Parking lots and fire lanes around buildings can help firefighters gain access to all sides of a building, which increase the options for where an FDC can be located. Ultimately, it will be up to the fire department to determine where they would most likely want to set up operations and pump into and FDC, so it is a good idea to check with the local fire department before coming to a final decision on where the FDC will go. Having a suggestion or two in mind would help the fire department in the decision making, so it is best not to go into that discussion with no thoughts in mind.

The fire department is going to need to take water from some source in order to pump into the FDC. In urban and suburban areas with public water supplies, the location of the fire hydrants will help determine where the FDC needs to go. NFPA 14 requires that the FDC be within 100 ft of a fire hydrant, but then goes on to allow greater distances if the authority having jurisdiction allows. The trouble with this requirement is that many standpipe systems are located in jurisdictions that don't have fire hydrants at all. In these jurisdictions, the standpipe designer needs to have special permission just to put the FDC anywhere considering the way that NFPA 14 is written. Figure 5.20 shows an excellent location of an FDC for a building uphill and set back considerable from the road. Rather than put the FDC on the outside wall of the building (far from the street and hydrant), the FDC was tastefully

Fig. 5.20 FDC away from building near fire hydrant

installed near the hydrant in a place where the fire department would have easy access with a fire truck.

In the situation where the building is in a jurisdiction with no fire hydrants (such as rural areas), planning for the location of the FDC is even more important. The water supply that the fire department will use for the standpipe system might be a permanent tank or buried cistern, but it is much more likely that it will be some sort of temporary tank set up at the time of the fire. Rural fire departments have portable tanks that can be set up and filled from tankers on a moment's notice. One or more pumpers can draft from these tanks into the standpipe system through the FDC. The tankers then shuttle water from predetermined filling points back to the building that is on fire and dump water into the portable tanks. If there are enough tankers to shuttle the water fast enough, they can keep up with the demand at the building that is on fire. All of this takes careful planning and execution including planning for where the temporary tanks will be set up and where the FDC needs to be close to these tank or cistern locations.

The designer needs to take into account the potential for obstructions to the FDC like bushes, parked vehicles, and fences before deciding where to locate the FDC. The designer can't completely insure that the owner won't plant bushes or build fences, but the designer can consider where such things might be likely and avoid such locations. The designer can also educate the owner to the importance of keeping the FDC unobstructed.

It is easier to plan the location of the FDC so that it is not obstructed by parked vehicles. In those situations where the FDC is on the same side of a building as a parking lot, the area directly between the location where the fire department will pull up needs to be blocked off from parking, designated as a fire lane, or should be

Fig. 5.21 FDC on a freestanding post away from the building to keep egress clear

planted with low growth flowers to make sure that the FDC is accessible to firefighters in an emergency.

Another concern regarding the location of the FDC is the use of the means of egress from the building by the occupants of the building during a fire. You do not want to place an FDC in a location where the firefighters will connect hose that blocks a means of egress from a building. Figure 5.21 shows an FDC near the means of egress for a building. The designer understood that if the FDC was placed on the wall of the building at this location, hose connected by the firefighters would block the doors, effecting how many people could exit during a fire. By placing the FDC on the freestanding post between the sidewalk and the road, the designer created a much safer situation with the FDC still accessible to the location where the fire truck will pull up, and the exit door for the building clear even when hoses are connected to the FDC.

Gauges

Section 5.5 of NFPA 14 requires gauges to be installed in a number of places. The gauges are not required to be listed, they only need to be approved by the authority having jurisdiction. The gauges are attached to the piping system with a ¼ in. connection that will allow the gauge to read the air or water pressure in the piping at that location. Gauges are required at the following locations:

1. At the connection to any water supply (fire pump, pressure tank, or water main).
2. At each main drain connection.

3. At the top of each standpipe.
4. On both the upstream and downstream side of a master pressure regulating assembly. For more information on a master pressure regulating assembly, see Chap. 8 of this text.
5. Above and below each alarm check valve, dry pipe valve, deluge valve, backflow preventer, or system riser check valve where such devices are present.

The pressure gauge at the water supply is typically built into the device serving as the water supply. The fire pump already has a gauge at the discharge flange of the pump required by NFPA 20, which serves the dual purpose of the gauge required by NFPA 14. Similarly, the pressure gauge on a pressure tank that is required by NFPA 22 also serves as the gauge required by NFPA 14. For a water main, the pressure gauge installed below the alarm check valve, riser check valve, dry pipe, preaction or deluge valve will serve this purpose as well.

The gauge at the main drain is important so that a main drain test can be performed. The main drain test consists of reading the pressure in the system before opening the drain, after opening the main drain fully, and then after closing the drain. Without the gauge, the main drain test can't be performed. The gauge location with respect to the main drain is important so that the readings are accurate. Figure 5.22 shows an acceptable placement of a main drain while Fig. 5.23 shows an unacceptable placement because the turbulence caused by the change of direction of the water flowing through the drain will lead to an incorrect reading on the gauge.

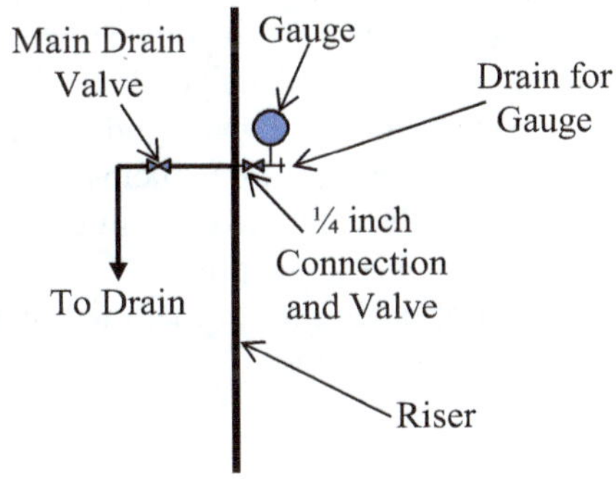

Fig. 5.22 Acceptable gauge placement for main drain

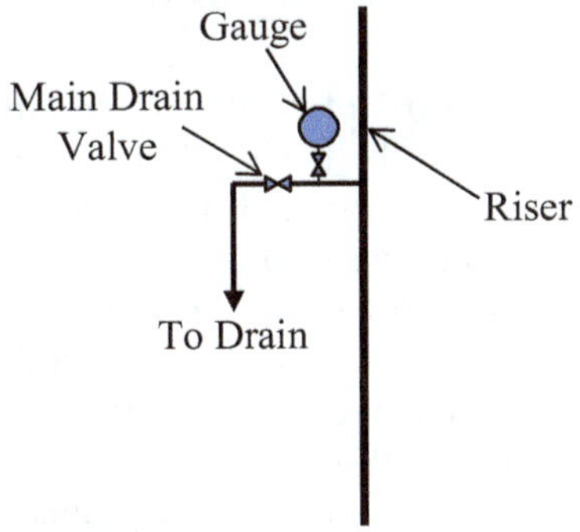

Fig. 5.23 Unacceptable gauge placement for main drain

Each connection for a gauge needs a control valve so that the gauge can be changed without shutting off the system. In addition, there needs to be a way to open up the tube and bleed the air off, letting the water fill the piping all the way to the gauge. In Fig. 5.22, the valve immediately to the right of the riser is the shutoff valve that allows the gauge to be changed without shutting down the system. To the right of the gauge is the cock that can be opened to bleed the air off and make sure the water can get all the way to the gauge. This cock can be opened to drain the water off after the control valve is closed as well.

The purpose of the gauge at the top of each standpipe is to verify the system pressure during normal system use and testing. For the situation where the standpipes are interconnected at the top of the system, as shown in Fig. 5.4, a single gauge at the top of the system is acceptable.

It is common to find pressure reducing valves at hose connections in order to keep the pressure to a reasonable level for firefighters. At these pressure reducing valves, a pressure gauge needs to be used during a test to determine the upstream water supply pressure during the test. However, this test only needs to be performed once every five years, and it seems to be a waste to install a gauge for a test that is only conducted so infrequently. Therefore, NFPA 14 does not require a gauge at every hose connection with a pressure reducing valve. But it does require a ¼ in. connection with a valve so that a gauge can be installed. When it is time to test the pressure reducing valves in a building, the testing contractor can bring one high quality calibrated gauge, use it at one location to test the pressure reducing valve, then move it to the next location to test the next pressure reducing valve.

It has been argued that you do not need a gauge on the upstream side of the pressure reducing valve since you have the gauge at the top of the riser and you could use the reading on that gauge (adjusted for elevation differences between the top of the riser and the pressure reducing valve) to determine the pressure at the upstream side of the valve. But the committee responsible for NFPA 14 wanted the gauge as close as possible to the pressure reducing valve so that a person could perform the test without having to do math to calculate the pressure at the pressure reducing valve.

Pipe: Underground

Section 6.2 of NFPA 14 requires the underground piping leading to a standpipe system to be designed and installed in accordance with NFPA 24. Note that the word "underground" in this context is pipe that is buried in soil. Pipe that is installed within a basement or garage below the surface of the ground that is exposed to the air is considered aboveground, not underground pipe.

NFPA 24 permits several different kinds of pipe to be used for underground service including lined ductile iron, cement, copper, brass, and several different kinds of plastic. While lined ductile iron pipe is rare to see aboveground, it is a very common selection for underground pipe. The concerns about potential mechanical damage are much less of an issue with ductile iron pipe installed below ground and covered with soil. Copper tube is rare underground because of the cost given that the pipe is going to be a relatively large size. While plastic pipe is typically not used aboveground in standpipe systems due to concerns over performance in a fire, there is no concern about exposure to fire underground, so this type of pipe is very popular in this application as well. The determination about which type of pipe to use will be made based on how corrosive the soil is and the material that the installation contractor likes to work with.

The lined ductile iron, cement, copper, brass, or plastic pipe that is used for underground service is not required to be listed as long as it is manufactured in accordance with one of the standards in Table 10.1.1.1 of NFPA 24 and joined in accordance with one of the standards listed in Table 10.2.1.1 of NFPA 24. Other types of pipe and tube are permitted to be used, but only if they are tested and listed by a laboratory specifically for underground service for fire protection systems. Likewise, the types of pipe listed in Table 10.1.1.1 can be joined by methods different from the fittings mentioned in Table 10.2.1.1 of NFPA 24 as long as the joining technique is tested and listed by a laboratory for underground service for fire protection systems.

One of the types of pipe that is not allowed underground in most applications is steel. While steel is a very popular choice aboveground, it typically reacts chemically with the soil and corrodes over short periods of time. Wrapping and coating the pipe tends to extend the lifespan of the pipe, but not enough for the committee responsible for NFPA 24 to feel comfortable in allowing the use of steel for most fire protection system use. In older editions of NFPA 24, a section used to permit the use of steel pipe as long as it was listed for underground service, but no steel pipe was ever able to achieve such listing from a laboratory, so the section was removed and is not included in the 2016 edition.

There is one permitted use of steel pipe underground by section 10.1.1.3 of NFPA 24 for fire protection systems. The piping from the hose connections on the FDC to the check valve at the connection to the standpipe system inside the building is permitted to be steel as long as the pipe is internally galvanized and externally coated and wrapped. Figure 5.24 shows the underground pipe permitted to be steel in red. Note that the water supply and other components discussed in this chapter have not been included in the diagram for simplicity. When the FDC is installed on the wall of the building, the concern about steel does not apply to the piping because the piping will not be run underground. This section of NFPA 24 is used when the FDC is installed on a post or in a remote location and the pipe needs to be run underground to get to the standpipe system. The FDC's shown in Figs. 5.20 and 5.21 are examples where underground pipe is needed to connect the FDC to the standpipe system. In these cases, internally galvanized and externally wrapped and coated pipe would have been permitted. Of course, line ductile iron, copper, or plastic (contained in Table 10.1.1 of NFPA 24) would have been permitted as well.

When the underground pipe gets into the building, if the type of pipe is not permitted to be used aboveground, then a transition must be made to the aboveground pipe within 24 in. of the entrance to the building. The 24 in. dimension is a somewhat

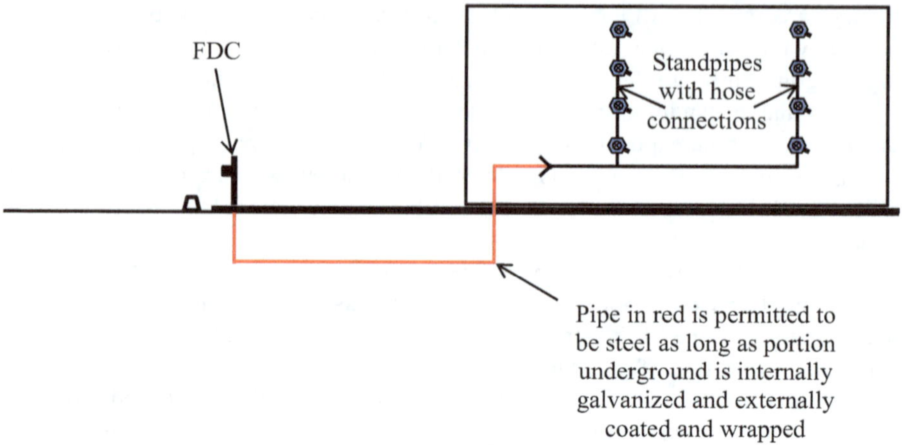

Fig. 5.24 Permitted use of steel pipe for FDC

arbitrary allowance that needs to be permitted because the installers of the pipe need some space in order to work with their hands to make the connection for the aboveground pipe. The committee deliberated for some time on the issue and decided that 24 in. was a reasonable amount. Note that if the underground pipe does extend 24 in. above the floor, some protection of the pipe might be needed because most types of underground pipe are more susceptible to mechanical damage than aboveground pipe. The protection might be a protective collar installed around the pipe or some construction in front of the pipe to prevent it from being struck from an angle in which it is likely to have an object come towards the pipe.

Special Considerations for Dry Systems

Dry standpipe systems are installed in situations where all of the piping cannot be installed in locations where the temperature cannot be reliably maintained at 40 °F or higher. Note that this is a bit above the freezing temperature of 32 °F because a small safety margin is important in fire protection systems. When a dry standpipe system is installed, the piping with air (or some other gas as discussed below) can be exposed to temperatures below 40 °F, but the piping from the water supply to the dry pipe, preaction, or deluge valve still needs to be maintained at 40 °F or higher.

Each of the different types of dry standpipe systems (automatic, semiautomatic, and manual) have special design and installation considerations. The concerns center around providing air to the systems (automatic and semiautomatic) and making sure that the delivery of water to the remote portions of the systems does not take too long. Each of the different types of systems will be discussed separately to make it clear which requirements apply to each different type of system.

Considerations for Automatic Dry Systems

The way that NFPA 14 controls the time that it takes to get water out to the most remote portion of the system is to limit the volume of the piping. The idea is that if the volume is limited, the total distance from the dry-pipe valve to the most remote hose connection will be relatively short and the water should be able to fill the system relatively quickly. If the system is kept to a volume of 750 gallons or less, then there is no other requirement for water delivery time (see section 5.2.1.2.1 of NFPA 14). In order to calculate the total volume of the system, the designer needs to know the internal volume of the pipe. One way to calculate this would be to calculate the internal cross sectional area of the pipe (in square inches since you know the diameter of the pipe in inches from Table 5.1 earlier in this Chapter and then convert that to square feet by dividing by 144. Then multiply by the length of the pipe in feet to get cubic feet. Finally, you would have to convert from cubic feet to gallons by multiplying by 7.48.

Luckily for us, the math has been done for the most common two types of steel pipe (Schedule 10 and Schedule 40) and the results are in Table A.5.2.1.2.1 of NFPA 14. Table 5.5 below has some of the internal volumes of the more common pipe sizes for Class I and Class III standpipe systems. The way to use Table 5.5 is to multiply the value in the table for the right type and nominal size of pipe by the length of the pipe in feet to get the total volume in that section of pipe in gallons. To determine the total volume of the system, the volume of all of the sections of pipe from the dry-pipe valve to the end of each portion of the system would have to be added up.

For example, consider an automatic dry standpipe system for a 5 story parking garage with three 4 in. standpipes that are each 40 ft long. If the system has 250 ft of 6 in. interconnection piping from the dry-pipe valve out to the most remote standpipe, and if all of the piping is Schedule 40 steel, the volume of the system piping can be calculated as follows:

Single Standpipe: $0.661 \times 40 = 26.44$ gallons
Three Standpipes $= 3 \times 26.44 = 79.32$ gallons
Interconnection Piping $= 1.501 \times 250 = 375.25$ gallons
Total Piping Volume $= 375.25 + 79.32 = 454.57$ gallons

Table 5.5 Internal volume for pipes 1 ft in length

Nominal pipe diameter (in.)	Volume of Schedule 10 pipe 1 ft in length (gallons)	Volume of Schedule 30 or 40 pipe 1 ft in length
2½	0.283	0.249 gallons (Schedule 40)
4	0.740	0.661 gallons (Schedule 40)
6	1.649	1.501 gallons (Schedule 40)
8	2.776	2.66 gallons (Schedule 30)
10	4.387	4.195 gallons (Schedule 30)

Note that in the example above, the total volume is less than 750 gallons, so no water delivery time is required, but keeping the total system volume under 750 gallons is not always going to be possible. So, NFPA 14 provides the designer with an option. If the water supply can deliver water to the most remote hose connection in 3 min or less, the system can be any volume larger than 750 gallons that the designer wants. Unfortunately, it's not so easy to calculate water delivery time to the most remote hose connection. A designer has two options when it comes to relying on the water delivery time:

1. Building the system and testing it, hoping that it passes the test and delivers water to the most remote outlet in 3 min or less (with experience and knowledge of the water supply and air supply, this can be done with a minimal amount of prayer).
2. Using the Tyco computer program for calculating water delivery time in a dry-pipe system from the layout plans (not that this author is pushing this particular company or program, but at the time of this publication, Tyco is the only company to offer such a calculation program).

The time requirement of 3 min is somewhat arbitrary, but it represents the amount of time that the committee thought would be reasonable for firefighters to wait for water after they had connected to the hose outlet in the system and opened the valve. Note that this is longer than the 1 min allowed by NFPA 13 for sprinkler systems, which is understandable because the firefighters should be in a protected location and the purpose of obtaining water is different.

NFPA 14 allows the measurement of time for water delivery to occur from the moment that a full 2.5 in. hose valve is opened. The air should evacuate from such a large opening fairly quickly. In order to help speed up the tripping of the dry-pipe valve and to speed up the delivery of the water in the dry-pipe system, NFPA 14 allows the installation of a quick opening device on the system. A quick opening device is either an accelerator that speeds up the tripping of the dry-pipe valve or an exhauster that helps get the air out of the system. They are rare in standpipe systems due to the size of the outlet from which the air is escaping, but they are more common in sprinkler systems when the air is trying to evacuate through a ½ in. orifice of a sprinkler.

The amount of time that it will take to get the air out of the dry-pipe system is a function of how high the air pressure is that has been stored in the system. Most dry-pipe valves that are used for standpipe systems have a differential design, meaning that the surface area of the valve that is in contact with the water is smaller than the surface area of the valve in contact with the air. This type of design results in a small amount of air pressure being able to hold back a larger amount of water pressure.

For example, a dry-pipe valve with a 6 to 1 differential has six times the amount of surface area in the valve in contact with the air than it does in contact with the water. If the expected high water pressure in the system is 120 psi, then the air pressure in the system could be as low as 20 psi and still hold back the water ($20 \times 6 = 120$). Most manufacturers specify an additional safety margin of 20 psi to take care of pressure surges and other fluctuations in the water supply, so it would not be uncommon to see air pressure of 40 psi in a system where the maximum expected water supply pressure of 120 psi was expected.

It should be noted that it is important for dry-pipe system performance that extra air pressure not be kept in the system. Any extra air in the system has to come out before the water can push through to the hose connection when it is opened. Extra air in the system slows down the delivery of water to the hose connection. People responsible for inspection, testing and maintenance of dry standpipe systems need to be kept aware of the air pressure expected by the designer, otherwise they may fill the system with too much air after working on the system, which will slow down the response of the system. The worst mistake that maintenance personnel can make is to put 120 psi of air pressure in a system that they think will have 120 psi of water pressure. It will take a great deal of time for the differential valve to trip with all of that extra air pressure in the system.

The dry-pipe valve for the standpipe system needs to be located in a lighted and heated room or enclosure (section 5.2.1.3.1 of NFPA 14) that protects the valve from mechanical damage as well as freezing. Heat tracing (electrical wires that heat up from electrical resistance and keep pipe from freezing) is not allowed to be used in order to meet this rule (section 5.2.1.3.2.3 of NFPA 14). The source of heating needs to be some other permanent source such as an approved electrical or gas heater.

Some dry-pipe valves (typically those with low differentials) are difficult to trip if water collects above the clapper due to the weight of the water. This water can collect from condensate from the air or if the system was put back in service before all of the water was drained out. Such valves are required to have an automatic drain or a high water level signal that lets the maintenance personnel know that water needs to be drained out of the system. The automatic drain or high level water signal can also help system maintenance personnel deal with dry-pipe valves that can be reset without opening them up, so they are permitted to be installed in that application as well.

Automatic dry standpipe systems are required to have air (or some other gas) pressure within the piping at all times except when the system is being used to fight a fire or when the system is being tested. The air (or gas) pressure in the system keeps the dry-pipe valve closed. It is not considered good practice to fill the system with water during the warm months because the system is not designed to be kept full of water.

Air has been the gas of choice in dry fire protection systems for over 100 years. It is inexpensive and easy to compress with relatively inexpensive equipment readily available from multiple suppliers. However, air does pose an increased risk of corrosion to steel pipe, especially steel pipe that has had water in it and is still moist from that water. In recent years, people have been switching from air to nitrogen in many dry-pipe systems. Nitrogen is also relatively inexpensive to obtain. Air is about 78 % nitrogen, so nitrogen generators can be set up that will take the nitrogen out of the air and put it into the standpipe system. While nitrogen generators are slightly more expensive than air compressors, they do help in the prevention of corrosion and can pay for themselves in the long-term life of a standpipe system. NFPA 14 allows the use of air, nitrogen or any other approved gas for keeping the dry-pipe valve closed (section 5.2.1.4.1).

If the gas used in the system is going to be air, then the air compressor or shop air system needs to be connected to the standpipe system at all times to make up for small losses in air pressure (due to leaks or changes in temperature) and prevent the dry-pipe valve from tripping due to these conditions (section 5.2.1.4.2.1 of NFPA 14). The compressor or shop air connection needs to be sized so that the total volume of the dry standpipe system can be filled to the proper air pressure within 30 min starting at atmospheric pressure (section 5.2.1.4.2.2 of NFPA 14). If the compressor is too small and takes too long to fill the system, the compressor may have to work too hard during the life of the system and may wear out prematurely, causing the system to fail and fill with water during cold weather when the air pressure drops due to the change in temperature. This is exactly when you do not want water in the system.

It is possible for a building or structure to have more than one automatic dry standpipe system. In this case, a single compressor is permitted to serve multiple automatic dry standpipe systems and the compressor is permitted to be sized for filling the largest single system by itself. There is no need to fill multiple standpipe systems at the same time.

NFPA 14 also requires a relieve valve to be installed in the air fill connection to prevent the system from putting too much air in the standpipe system (section 5.2.1.4.4). The relieve valve needs to be set at 10 psi in excess of the system air pressure that is supposed to be held in the system. So, for the system discussed before with the 6 to 1 differential valve and the maximum expected water pressure of 120 psi, the system would normally be set for 40 psi (20 psi for the air pressure to hold the valve closed an another 20 psi safety margin). This would mean that the relief valve should be set for 50 psi so that the standpipe system could not be overfilled with air pressure beyond 50 psi.

Considerations for Semiautomatic Dry Systems

Since semiautomatic dry standpipe systems are not held closed by air pressure, they have different design and installation considerations than automatic dry systems. The semiautomatic dry systems are held closed by a preaction or deluge valve. A separate electrical signal has to come from a remote control activation device (typically a button or manual pull station) installed near each hose connection. NFPA 14 requires the remote control activation device to be within 3 ft of each hose connection on the system (section 5.2.3.1). This remote control activation device needs to be installed in accordance with NFPA 72 since it is an initiating device for a fire protection system.

The preaction or deluge valve being used for the semiautomatic dry standpipe system also needs to have a mechanism for tripping the valve that is not using one of the remote control activation devices (see section 5.2.3.3 of NFPA 14). Typically this is a hydraulic or mechanical means of tripping the valve. The manufacturers of preaction and deluge valves provide this mechanism as a part of the auxiliary devices sold with the valve. The purpose of this alternate way to trip the valve is so that a person can be sent to the valve room to manually trip the valve in case the electronic means provided by the remote control activation devices is not working.

The preaction or deluge valve for the standpipe system needs to be located in a lighted and heated room or enclosure (section 5.2.3.5.2 of NFPA 14) that protects the valve from mechanical damage as well as freezing. Heat tracing (electrical wires that heat up from electrical resistance and keep pipe from freezing) is not allowed to be used in order to meet this rule (section 5.2.3.5.2.3 of NFPA 14). The source of heating needs to be some other permanent source such as an approved electrical or gas heater.

Semiautomatic dry standpipe systems are very similar to preaction sprinkler systems. As such, the designer is allowed to select a single interlock, double interlock, or non-interlocking system (see section 5.2.3.6 of NFPA 14). However, the complexity of the double interlock and the non-interlocking system tend to decrease the reliability of the system and most designers opt for the single interlock option. The different types of systems are defined as follows:

Single Interlock: The preaction or deluge valve only trips (allowing water into the system) if the remote control activation device is operated (the manual pull station is pulled or the button is pressed).

Double Interlock: The preaction or deluge valve will only trip (allowing water into the system) when both the remote control activation device is operated and a loss of air pressure is sensed (typically when the hose valve is opened).

Non-interlocking: The preaction or deluge valve will trip (allowing water into the system) if the remote control activation device is operated or if a loss of air pressure is sensed.

The non-interlocking system is seen as something of a failsafe device that allows the standpipe system to get water even if the activation devices are not working correctly. But given the fact that there will be a manual way of activating the preaction or deluge valve in the valve room for a single interlock system, the non-interlocking system seems to be a waste of money that could allow the valve to trip accidently and fill the system with water during cold weather. The difficulties associated with non-interlocking systems typically outweigh the advantages and few of these systems are designed or installed.

Double interlock systems were designed for situations like freezer storage warehouses where the accidental tripping of a single interlock system due to a malfunction in the initiating devices might have dire consequences. Once the water freezes in the pipes, you can't wait for the pipe to warm up to get the ice out; instead, you have to disassemble the pipe and take it out of the freezer to get the ice out, which is an extremely costly problem. There are few standpipe systems in freezer storage warehouses, but if they are installed, the double interlock system is the best option.

In circumstances like parking garages, single interlock systems are all that is really needed if a semiautomatic system is going to be installed. In the rare case of a malfunction of an initiating device, if the system is accidentally filled with water, it can be quickly drained. In the extremely rare circumstance where the water turned to ice in the system before it can be drained, the maintenance personnel can wait until the next warm day to drain the system and perform any necessary repairs. The system may be out of service until then, and impairment procedures (discussed in more detail in Chap. 14 of this text) should be implemented until the system is put back in service. Still, this is better than taking each piece of the system outside to thaw and drain.

Semiautomatic dry standpipe systems generally have a small amount of air put into the system for maintenance purposes. If the air pressure drops, maintenance personnel know that there is a leak somewhere in the system and can fix it before the system is filled with water and the leak causes water damage. Assuming that the system is a single interlock system, it is not dependent on air pressure to work in a fire protection situation, so the air supply is not required to be permanent. The air can be added by a temporary connection and then monitored through frequent inspection of the air pressure gauge required above the preaction or deluge valve.

For double interlock systems, the air pressure is more important. Still, the system is not going to trip unless the remote control activation device is operated, so the air maintenance system is permitted to be temporary if the designer chooses. The designer is allowed to install a permanent air delivery system, but if the system is protecting a space constantly subject to freezing, the special rules of NFPA 13 for preaction systems in such spaces should be followed including the requirement for dual air lines so that when one line freezes up, the other can be put in service.

For non-interlocking systems, the air pressure is extremely important. A loss of air pressure will cause the valve to trip and the system to fill with water. NFPA 14 does not contain any specific requirements for the air pressure system in non-interlocking situations, but since these systems act like automatic dry standpipe systems, it would be a good design practice for the rules of automatic systems to be followed.

As discussed under the portion of this Chapter for automatic dry standpipe systems, some designers prefer to use nitrogen instead of air to improve the corrosion performance of standpipe systems. Such use in semiautomatic systems is completely acceptable and this text has used the term "air" for convenience. If the designer wants to use nitrogen in a semiautomatic system, the same concepts apply as discussed for air.

Considerations for Manual Dry Systems

There are very few special considerations for manual dry standpipe systems that have not already been discussed in the earlier parts of this chapter. Manual dry pipe systems do not have any water supply, so there is no concern for a valve room or for any part of the system being protected from freezing. Manual dry systems are not required to be filled with air under pressure or any other type of gas, so none of those rules apply here either. The designer could decide to monitor the integrity of the piping with compressed air or some other type of gas, but this author has never seen that in 30 years of experience as a fire protection engineer.

Signs

NFPA 14 has a few special requirements for signs on all standpipe systems, wet or dry. The purposes of the signs are to provide information to the user (typically the firefighter) or the maintenance personnel. For each of the signs required by NFPA 14, the sign needs to be constructed of weatherproof metal or rigid plastic (section 4.10). The sign is also supposed to be "permanently marked". There is no definition of what "permanently marked" means, but most AHJ's will accept a process that presses the letters into a thin piece of metal causing the letters to be raised or causing an impression in the metal. Some AHJ's will also accept legible writing with a permanent marker on a piece of metal or plastic. It is best to check with the local AHJ before deciding what method you will use. The signs also need to be attached to the device or the building wall with corrosion resistant chains or fasteners (section 6.6). The following is a summary of the signs required along with a reference for where the requirement occurs in NFPA 14:

1. All main and sectional system control valves, including water supply control valves, need a sign indicating the portion of the system that is controlled by the valve (section 6.3.8.1). Where one of these valves is located in a closed room or concealed space, the location of the valve needs to be indicated by a sign in an approved location on the outside of the door or near the opening to the concealed space. (section 6.3.8.4)
2. All drain and test connection valves need signs indicating their purpose (section 6.3.8.2).
3. Sprinkler systems fed by more than one standpipe need to have a warning sign indicating the location of the additional valve that needs to be closed before working on the sprinkler system (section 6.3.8.3 of NFPA 14). See Chap. 7 of this text for more information on this installation situation.
4. Each fire department connection (FDC) needs a sign, with letters at least 1 in. in height that reads "STANDPIPE." If automatic sprinklers are also supplied by the fire department connection, the sign or combination of signs shall indicate both designated services. For manual systems, the sign needs to indicate that the system is manual and that it is either wet or dry (section 6.4.5.2). Note that some AHJ's will accept this information being cast into the valve is shown in Fig. 5.14 or etched into the plate around the FDC as shown in Fig. 5.15. Some AHJ's require an extra sign high up in the building as shown in Fig. 5.19 so that the FDC can be found from the road by the first responding firefighters.

5. A sign is required at the inlets of an FDC to indicate the pressure required to deliver the standpipe system demand (section 6.4.5.2.2). In the experience of this author, this requirement is rarely met. The standard operating procedures of most fire departments are to connect to the FDC and charge it to 150 psi. If the pressure demand at the standpipe is greater than 150 psi, then the fire department might want a sign letting them know that a higher pressure is necessary. Note that none of the figures in this chapter showing FDC's have a sign indicating the pressure that needs to be pumped in.
6. Where a fire department connection services multiple buildings, structures, or locations, a sign needs to indicate the buildings, structures, or locations that it serves (section 6.4.5.3).
7. Where a fire pump is provided, a sign needs to be located in the vicinity of the pump indicating the minimum pressure and flow required at the pump discharge flange to meet the system demand (section 6.7). This sign is extremely helpful in determining the pass/fail criteria for the annual pump test since the ultimate goal of the fire pump is to meet the demand of the fire protection system.
8. A hydraulic design information sign needs to be provided with the basis for the system design (section 6.8). This sign is required to be at the water supply control valve for automatic and semiautomatic systems. The AHJ will need to approve the location of the sign for manual wet or dry standpipes, although the connection to the water supply seems like a good idea for manual wet systems. The sign needs to include five pieces of information:

   (a) The location of the two most hydraulically remote hose connections or hose stations.
   (b) The design flow rate for the most remote two hose connections.
   (c) The residual inlet and outlet pressures for the remote two hose connections. Note that for systems without pressure reducing devices at the hose connections, the inlet and outlet pressures will be the same. But for systems with any kind of device that reduces the pressure (pressure regulating, control or restricting devices), the outlet pressure will be less than the inlet pressure.
   (d) The system flow and pressure demand at the system control valve or the discharge flange of the pump.
   (e) The system flow and pressure demand at each FDC. Note that this may be a different pressure demand than required in part (d) because the piping may be a different size or length going to the FDC. Also note that different FDC's might have different pressure requirements because they are located on different sides of a building with different lengths of pipe to the connections. See Fig. 5.25, which shows a Hydraulic Design Information Sign for a High Rise Building with Two FDC's.

Fig. 5.25 Hydraulic design information sign for a theoretical high rise building with two FDC's

Standpipe Hydraulic Design Information Sign

Location of two most hydraulically remote hose connections:	Stairwell C, 9th and 10th Floor
Design flow for the two most hydraulically remote hose connections:	250 gpm each
Design residual inlet and outlet pressures for the two most hydraulically remote hose connections:	100 psi inlet and outlet
System demand at water supply control valve or pump discharge flange:	1000 gpm at 161 psi
System demand at FDC's:	Main St: 1000 gpm at 158 psi 3rd Ave: 1000 gpm at 167 psi

Special IBC Rules

The International Building Code (IBC) contains some additional rules beyond those required by NFPA 14. These additional rules fall into two categories: cabinets and valve supervision. The rules can be found in sections 905.7 and 905.9 of the IBC. They are summarized here for the convenience of the reader.

Any standpipe equipment that is within a cabinet (such as a hose connection in a hallway or a hose station containing 100 ft of hose) needs to be located so that the cabinet is not blocked from use or obscured from view. This means that the cabinet door needs to be able to swing fully open so that the equipment can be properly used. The cabinets need to either have a clear panel so that a person can see what is inside or they need to be identified with a sign that has at least 2 in. high letters telling the reader what is in the cabinet. The letters on the sign need to have a contrasting color to the background. If the door is not large enough to accommodate a sign, an approved picture or symbol indicating what is in the cabinet is sufficient.

In general, cabinets are not allowed to be locked. However, there are a number of provisions on the IBC that allow locked cabinets. The first is where the door to the cabinet is clear or there is a vision panel that lets a person see what is in the cabinet. In this case, the cabinet is allowed to be locked if there is an approved frangible material that can be broken to allow access to the cabinet.

The second condition where locked cabinets are allowed is where the locking arrangement has been approved by the AHJ. This would typically be a situation where the equipment was not intended to be used by the general public and the people responsible for using the equipment would be the ones with the keys to unlock the cabinet. For example, in a building with a trained fire brigade, the cabinets to the Class II hose systems might be locked and the keys would only be given to the members of the fire brigade so that untrained people would not be tempted to try and fight a fire with the available hose.

The third situation where locked cabinets are allowed is in Group I-3 occupancies, which are jails or correctional facilities. In these cases, inmates are not provided with access to firefighting equipment, but guards or correctional officers might have keys, or access to places with keys, so that they can open the cabinets and use the equipment in case of emergency.

The concept of valve supervision was discussed earlier in this Chapter in the section on control valves. In that discussion, it was mentioned that NFPA 14 allows both mechanical and electrical supervision of control valves, but that some local codes might only allow the electrical supervision. The IBC is one of those codes that only allows electrical supervision. However, there are two exceptions to the requirement for electrical supervision:

1. For underground gate valves with roadway boxes provided by the water utility or the municipality, not supervision is required. So, not only is this an exception to the rule for electrical supervision, it is a complete exemption from the requirements for any supervision. If this valve gets accidentally closed, the main drain test required by NFPA 25 periodically should pick this up.
2. In buildings not equipped with a fire alarm system, valves are permitted to be locked (provided with mechanical supervision) in the appropriate (open) position as long as the lock is inspected periodically. See Chap. 14 of this text for more information about the inspection requirements.

Test Yourself

5.1. List three materials that are permitted for aboveground pipe:

(a) ____________________
(b) ____________________
(c) ____________________

5.2. Which of the following types of steel pipe and joining techniques would not be allowed by NFPA 14?

(a) Unlisted Schedule 10 pipe joined with threaded fittings
(b) Unlisted Schedule 40 pipe joined with threaded fittings
(c) Unlisted Schedule 10 pipe joined with roll grooves and plain end fittings
(d) Unlisted Schedule 40 pipe joined with cut grooves and plain end fittings

5.3. In a standpipe system with all 4 in. pipe using cast iron fittings where the maximum expected working pressure of the system is 200 psi, what type of fittings will be required to be used?

(a) Lightweight pattern fittings
(b) Standard weight pattern fittings
(c) Heavy weight pattern fittings
(d) Extra heavy weight pattern fittings rated up to 300 psi

5.4. In a standpipe system with all 4 in. pipe using malleable iron fittings where the maximum expected working pressure of the system is 200 psi, what type of fittings will be required to be used?

(a) Lightweight pattern fittings
(b) Standard weight pattern fittings
(c) Heavy weight pattern fittings
(d) Extra heavy weight pattern fittings rated up to 300 psi

5.5. In a Class I manual wet standpipe system, a 50 ft section of 2½ in. pipe leading to a single remote hose connection cannot be positioned to drain towards the main drain and will not drain out of the hose connection. Which of the following drain situations is required?

(a) A fitting that can easily be taken apart at a low point
(b) An auxiliary drain at a low point piped to a location that can handle the flow
(c) A drum drip drain that can be operated like an air lock so that the air does not leave the system

5.6. For the Class I automatic dry standpipe system shown in Fig. 5.26, what is the total volume of the system and what is the required water delivery time to the remote hose connection (Node A)? Assume that all pipe is Schedule 40.

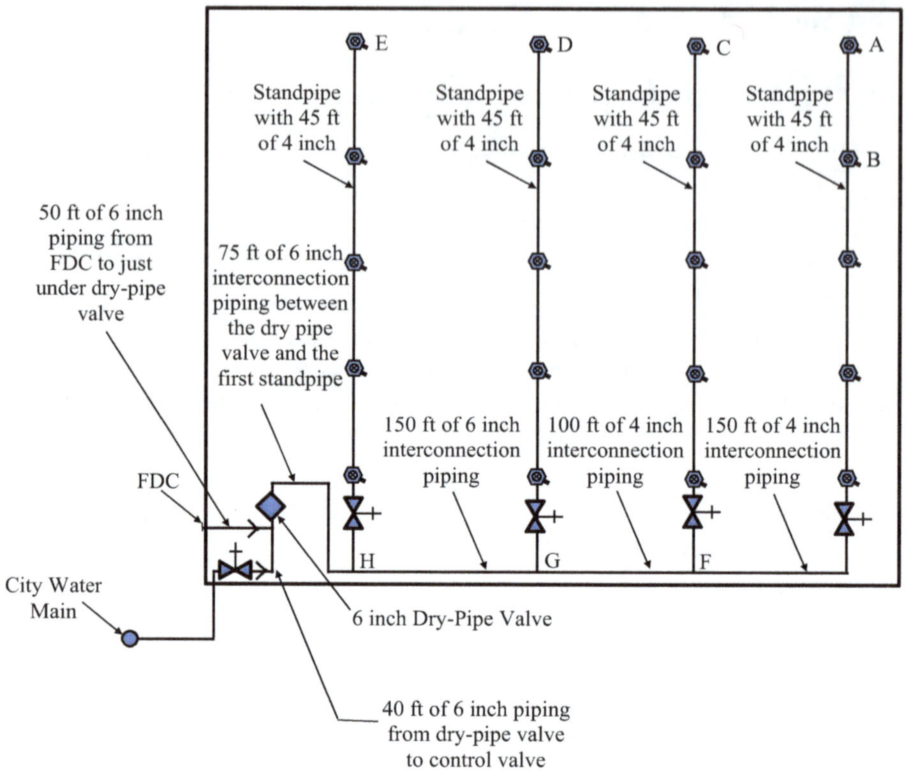

Fig. 5.26 Class I automatic dry standpipe system for question 5.6

Chapter 6
Hydraulic Calculation of Standpipe Systems

The whole standpipe system exists to provide water for firefighting purposes. In the design of a standpipe system, the designer needs to make sure that there will be a water supply that can provide a sufficient flow of water at a sufficient pressure for a sufficient period of time. This water supply could be a fixed supply dedicated to the standpipe system, or it could be a portable supply brought by the fire department. The calculations necessary to insure that the water flow, pressure and duration are going to be there when they are needed are called hydraulic calculations.

Another reason for performing hydraulic calculations is to determine the pipe sizes that will be used and whether these sizes are appropriate. If cost were not an issue, standpipe systems would all use very large piping (10 or 12 in.) and hydraulic calculations would be simple to perform in one or two steps. But cost is an issue with all fire protection systems and it is not realistic to believe that every building owner will want to install 10 or 12 in. pipe for a whole standpipe system. Larger pipe costs more to purchase and the labor costs associated with installing it go up as the pipe size increases as well. So, sizing pipe becomes a balancing act, trying to minimize the cost by keeping the pipe size down, while still being able to provide the flow and pressure that the system needs. Hydraulic calculations become critical to proving that the pipes are not too small and that the flow and pressure can be delivered to the remote portions of the building.

NFPA 14 has some minimum requirements for the size of the pipe. The rule that applies to most Class I and Class III standpipe system piping is that the minimum size is a nominal 4 in. Pipe that is part of a partial sprinkler system that does not meet the requirements of NFPA 13 or NFPA 13R, needs to be a minimum of 6 in. nominal size. For branch lines, which are pipes going to a single Class I or Class III outlet, pipe is allowed to be a minimum 2.5 in. nominal size. There is no specific minimum size for Class II standpipe system piping, but whatever size is selected needs to be proven by hydraulic calculations.

This Chapter will cover the basic concepts of hydraulic calculations beginning with a quick primer on terminology and concepts. Then, the Chapter will cover the

K.E. Isman, *Standpipe Systems for Fire Protection*,
DOI 10.1007/978-3-319-47750-3_6

requirements of NFPA 14 and how to make sure that the water supply can provide all of the demands. The much more complex issue of pressure control at the hose connections will be covered in Chap. 8.

Basic Terminology and Concepts

There are three basic terms that people performing hydraulic calculations needs to understand: flow, pressure, and duration. Here is a quick definition of each of these three terms:

Flow: The quantity of water that passes a given point over a specific period of time.

Pressure: The energy stored in the water. The energy is responsible for moving the water, creating flow, and getting the water to overcome the fire.

Duration: The period of time that the water needs to last.

The flow is generally measured in units of gallons per minute (gpm), although it could be measured in terms of cubic feet per second, cubic feet per minute (cfm), or liters per minute (if you like the metric system). Water absorbs the heat from a fire. The larger the fire, the higher the quantity of water that is needed to fight the fire. Flow becomes the critical variable for firefighters to make sure that they will be able to extinguish a fire.

One might expect that flow would be represented in formulas with the variable "F", but the variable "F" has been so associated with force in physics equations that the people developing hydraulic equations decided that the use of the variable "F" for flow might cause confusion. So, the variable "Q" is used for flow, giving recognition to the concept that the flow is related to the quantity of water.

Pressure, which is represented by the variable "P" in most hydraulic equations, is an expression of energy in terms of force over a unit of area. The most common units of force are pounds per square inch (psi). Other units of pressure include pounds per square foot (psf) and bar (associated with the barometer that measures atmospheric pressure).

There are many different pressures associated with the flow of water in a pipeline. This text will concentrate on one pressure, the total gauge pressure of the water. This type of hydraulic analysis, which ignores the velocity pressure (energy lost to actually get water moving), is generally conservative and is common in most fire protection systems. This type of analysis is also simpler and easier to perform, so most people prefer this method of performing hydraulic calculations. More precise (and more challenging) calculations can be performed by taking the velocity pressure into account, but the reader will need to reference a more advanced text on hydraulics to master this type of calculation.

Pressure can be created in two ways: elevation and mechanical work. When water is elevated over a specific location, it gains energy due to gravity relative to that location. Close to the surface of the earth, the pressure created by elevating the

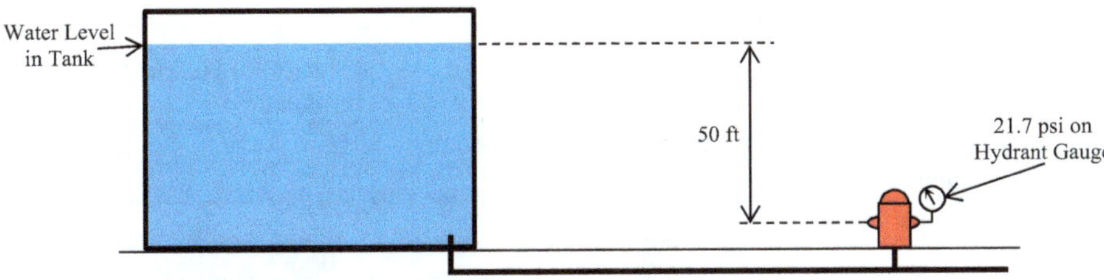

Fig. 6.1 Elevation of water above a fire hydrant creates pressure at that hydrant

water is a constant created by the earth's gravitational field. On other planets (or the moon), the pressure created by the elevation of water would be different. On earth, water gains 0.433 psi for every foot above a point where it is elevated. As Fig. 6.1 shows, water elevated 50 ft above a fire hydrant creates a pressure at that hydrant of 21.7 psi (50 × 0.433 = 21.7).

As water flows through a piping system, the water loses, or gains, pressure as the elevation changes. Since the elevation changes occur frequently in standpipe systems, the designer of the system must take these changes into account in order to figure out how much pressure is needed from the water supply. As the water goes down in elevation, the water gains pressure. As the water goes up in elevation, the water loses energy. Since gravity is a constant all over the surface of the earth, the path that the water takes has no effect on the pressure created by gravity. All that matters in the calculation of the water pressure due to elevation is the starting elevation of the calculation and the ending elevation.

As water flows through pipe and rubs against the walls, the water loses energy, which means that the pressure goes down. Just how much energy the water loses depends on the flow of water in the pipe, the internal diameter of the pipe and just how smooth the walls of the pipe are. At the same time that the water is losing energy from rubbing against the pipe wall, it is also losing energy when it changes direction in the piping. Every time that the water flows through a tee or elbow, the change in direction creates turbulence in the flow stream, which causes the water to lose energy. All of these losses are lumped together into a single term called "friction loss".

Around the year 1900, two civil engineers from the University of Michigan, Allen Hazen and Gardner Stewart Williams, developed a formula for calculating the friction loss in pipe. Their formula, while more than 100 years old, is still used in fire protection today because it is both simple and conservative. The Hazen-Williams formula is as follows (with the variables discussed below the formula):

$$P_L = \frac{4.52Q^{1.85}}{C^{1.85}d^{4.87}}$$

P_L = The friction loss in a single one foot section of pipe. In order to get the total friction loss, the length of the pipe needs to be multiplied by P_L. Unlike elevation differences, P_L is path dependent, meaning that it is very important to keep track of

Table 6.1 Standardized C-factors for aged pipe

Type of pipe	C-factor for Hazen-Williams formula
Steel in an automatic, semiautomatic or manual dry system	100
Steel in an automatic or manual wet system	120
Cement lined ductile iron	140
Copper	150
Plastic	150

the length of pipe and fittings between points when calculating the friction loss of the water between those points.

Q = The flow of water in the pipe in gallons per minute.

C = The relative roughness (or smoothness) of the pipe's wall. The larger the number, the smoother the pipe. Table 6.1 shows the various C-factors that need to be used when calculating standpipe systems. Note that these numbers do not represent the actual smoothness of new pipe. Instead, these numbers represent the smoothness of aged pipe after years of service, taking into account potential corrosion that might occur within the pipe. Table 6.1 contains many different types of pipe for comparison purposes. Not all of the types of pipe shown in Table 6.1 are allowed to be used aboveground in standpipe systems, but they are all allowed to be used either aboveground or underground.

d = The actual inside diameter of the pipe. Note that this is different for each type of pipe. See Table 5.1 (back in Chap. 5) for the inside diameters of typical sizes of steel pipe used in standpipe systems.

The following is an example of how to use the Hazen-Williams formula to calculate the friction loss of 500 gpm flowing through 200 ft of 4 in. Schedule 40 steel pipe in a manual wet standpipe system:

$$P_L = \frac{4.52Q^{1.85}}{C^{1.85}d^{4.87}} = \frac{4.52(500)^{1.85}}{(120)^{1.85}(4.026)^{4.87}} = 0.072$$

Total Friction Loss = 200 ft × 0.072 psi/ft = 14.4 psi

Hydraulic calculations are then performed by starting at the most remote portion of the system and figuring out what the flow and pressure demands are at this location. Then working back towards the water supply and incorporating the elevation losses and friction losses as the water flows between points (called nodes) and adding additional flows where needed. Once back at the water supply, the total flow and pressure demands are compared to what is available from the water supply. If the water supply can provide what is needed, the supply is considered adequate. If the supply cannot provide what is needed, the water supply will need to be augmented.

If the water supply is deficient in pressure, then a pump can be added to the system to increase the available pressure. If the water supply is deficient in flow, a pump cannot solve the problem by itself. A fire pump does not produce water, therefore it cannot increase the available flow of the water supply. If the water supply is deficient in flow, then a tank or other water supply needs to be added to the system. A pump can help to get the water from this new supply to the standpipe system, but just adding a pump alone will not correct the problem of not having enough flow.

As mentioned before, when water flows through tees, elbows, and valves, the water loses energy (pressure) due to the change in direction. Just how much energy the water loses depends on how abrupt the change in direction is within the tee, elbow or valve. This loss in pressure needs to be accounted for in the hydraulic calculations.

There are many ways that the loss of pressure from a tee, elbow or valve could be accounted for. NFPA 14 presents a standardized way of accounting for these losses in such a way that everyone in the fire protection industry can follow the calculations and make sure that the proper losses were counted. This method is called the "Equivalent Pipe" method. Basically, each type of fitting and valve is given a certain amount of pipe (in ft) that creates the same amount of friction loss as the fitting or valve. This "Equivalent Length" is then added to the pipe in the hydraulic calculation step to account for the friction loss in that fitting or valve.

For example, assume that there is an elbow on the end of the 200 ft of 4 in. pipe that we calculated earlier in this chapter that had 500 gpm of water flowing through it. We calculated that there was 0.072 psi/ft of friction loss in the pipe and we multiplied that by the 200 ft of pipe to get the total friction loss in that portion of the system. But with an elbow at the end, the friction loss changes a bit. According to NFPA 14, a 4 in. elbow on a piece of steel pipe in a manual wet standpipe system has an equivalent length of 10 ft (assuming that it is a standard turn elbow and not a long turn elbow). This does not mean that the elbow is physically 10 ft long. Instead, this means that the water flowing through the elbow loses the same amount of pressure as the same flow of water through 10 ft of straight 4 in. Schedule 40 pipe in a manual wet standpipe system. So, the friction loss through the pipe and elbow will be equal to the friction loss through 210 ft of pipe, which we can calculate as 15.1 psi ($210 \times 0.072 = 15.1$).

Knowing the equivalent length of each different type of fitting and valve becomes important to being able to perform hydraulic calculations. NFPA 14 provides the user with equivalent lengths for eight different types of valves and fittings over 14 different nominal sizes of pipe in Table 8.3.1.3. This table however is only valid to be used when the standpipe system is Schedule 40 steel pipe and the system is a manual wet or automatic wet standpipe (C-factor is 120). For other types of systems using other types of pipe, there are two choices for how to obtain the equivalent length of the fitting or valve:

1. The manufacturer can provide an equivalent length for a fitting or a valve.
2. The value in Table 8.3.1.3 can be used if two correction factors are applied. One correction for the C-factor and one correction for the internal pipe diameter.

The correction for the C-factor is straightforward. If a type of pipe is being used that has a different C-factor from 120, then the fitting's equivalent length needs to be multiplied by one of the following:

- 0.713 if you are using pipe with a C-factor of 100.
- 1.33 if you are using pipe with a C-factor of 140.
- 1.51 if you are using a pipe with a C-factor of 150.

The correction factor for the different pipe diameters is a bit more complicated. The correction factor needs to be calculated by taking the internal diameter of the pipe you are using, dividing by the internal diameter of Schedule 40 pipe, and then raising that quotient to the 4.87 power. The equation for calculating the correction factor is as follows:

$$Correction_{Diameter} = \left(\frac{d_{pipe}}{d_{Schedule\,40}} \right)^{4.87}$$

The following is an example of the correction factors applied to an automatic dry standpipe system with 4 in. Schedule 10 steel pipe (C = 100). The first correction is for the C-factor, which is 0.713. The second correction needs to be calculated using the equation above:

$$Correction_{Diameter} = \left(\frac{d_{pipe}}{d_{Schedule\,40}} \right)^{4.87} = \left(\frac{4.26}{4.026} \right)^{4.87} = 1.32$$

The correction factors need to be multiplied together. So, for 4 in. Schedule 10 steel pipe being used in an automatic dry standpipe system, the complete correction for fittings and valves is 0.94 (0.713 × 1.32 = 0.94). This means that all of the values for 4 in. pipe in Table 8.3.1.3 of NFPA 14 can be multiplied by 0.94 and applied to Schedule 10 steel pipe used in an automatic dry standpipe system. A standard turn elbow with an equivalent length of 10 ft in the table becomes 9.4 ft when used in an automatic dry standpipe with Schedule 10 steel pipe. A tee with an equivalent length of 20 ft in the table becomes 18.8 ft when used in an automatic dry standpipe with Schedule 10 steel pipe.

Calculation of Class I and Class III Systems

The calculation of Class I and Class III systems is the same. NFPA 14 does not consider the hydraulic demands different for these two types of systems. For simplicity, this Chapter will refer to Class I systems, but in all cases, the calculation of a Class III system would utilize the same rules and the same techniques.

For all fire protection systems, the process of performing the hydraulic calculations is basically the same. You start from the most remote point and work back to the water supply. Since most fire protection systems have one water supply, you only need to perform one set of hydraulic calculations. The difference with standpipe systems is that there is a high probability that multiple hydraulic calculations will have to be performed because there are multiple water supplies. Most fire protection systems do not consider the fire department connection (FDC) to be a water supply for which hydraulic calculations are necessary, but for a standpipe system, the FDC is critical to system performance, so hydraulic calculations are necessary for each FDC (see section 7.7.1 in NFPA 14). For a system with an automatic water supply and two FDC's (a common arrangement for a high-rise building) three sets of hydraulic calculations will need to be performed: one for the water supply and one for each FDC).

The calculations discussed here will be for standpipes that are vertical in nature. A standpipe meeting the definition of a horizontal standpipe (where pipes serve two or more hose connections on the same level) will need to be calculated using a slightly different technique than described here. Requirements for horizontal standpipes will be discussed in Chap. 12 of this text.

In the fire protection industry, the term "demand" is used to describe the needs of the fire protection system. So, when a person uses the term "flow demand" they are talking about how much flow the system needs in order to be able to fight a fire. When a person uses the term "pressure demand" they are talking about how much pressure the system needs at a specific point in order to work properly. The term "system demand" is typically used to describe the combination of the flow and pressure needed at the water supply. The pressure and flow demands will be discussed here separately and then will be combined to show how the calculation is performed with all of the information put together.

Flow Demand

For typical (vertical) standpipes, NFPA 14 requires the system to be capable of delivering 250 gpm from the most remote hose connection on the most remote standpipe plus 250 gpm from the second most remote outlet on the most remote standpipe. To this 500 gpm, an additional 250 gpm needs to be added from the most remote outlet on each of the other standpipes until a maximum flow is reached (see section 7.10.1.1 of NFPA 14). The maximum flow in unsprinklered buildings is 1250 gpm (note that any building that is not sprinklered in accordance with NFPA 13 is considered an "unsprinklered building" for the application of this rule). The maximum flow in a sprinklered building (sprinklered in accordance with NFPA 13) is 1000 gpm (see section 7.10.1.1.5 of NFPA 14). If all of the standpipes have flow accounted for before the maximum flow is achieved, then the system demand is the lower flow, not the maximum.

For example, in a building with two standpipes (whether it is sprinklered or not), the flow demand would be 750 gpm (500 gpm from the most remote two outlets on the first standpipe and 250 gpm from the next standpipe). Since there are only two standpipes in this system, no additional flow is required.

As a second example, consider a sprinklered building with five standpipes. The flow demand would be 1000 gpm (500 gpm from the two most remote outlets on the first standpipe, 250 from the most remote outlet on the second standpipe and 250 gpm from the most remote outlet on the third standpipe). Once the 1000 gpm maximum is reached, there is no need to include additional flow even though there is no flow from the fourth or fifth standpipe.

This explanation has been a bit wordy and it is possible that not everyone has been able to follow it. A graphic explanation might be better for some people. Figure 6.2 shows a standpipe system with two standpipes in a sprinklered building. The flow demand is 750 gpm with 250 from each of the two most remote outlets on the most remote standpipe (Nodes A and B) and the most remote outlet on the second standpipe (Node C).

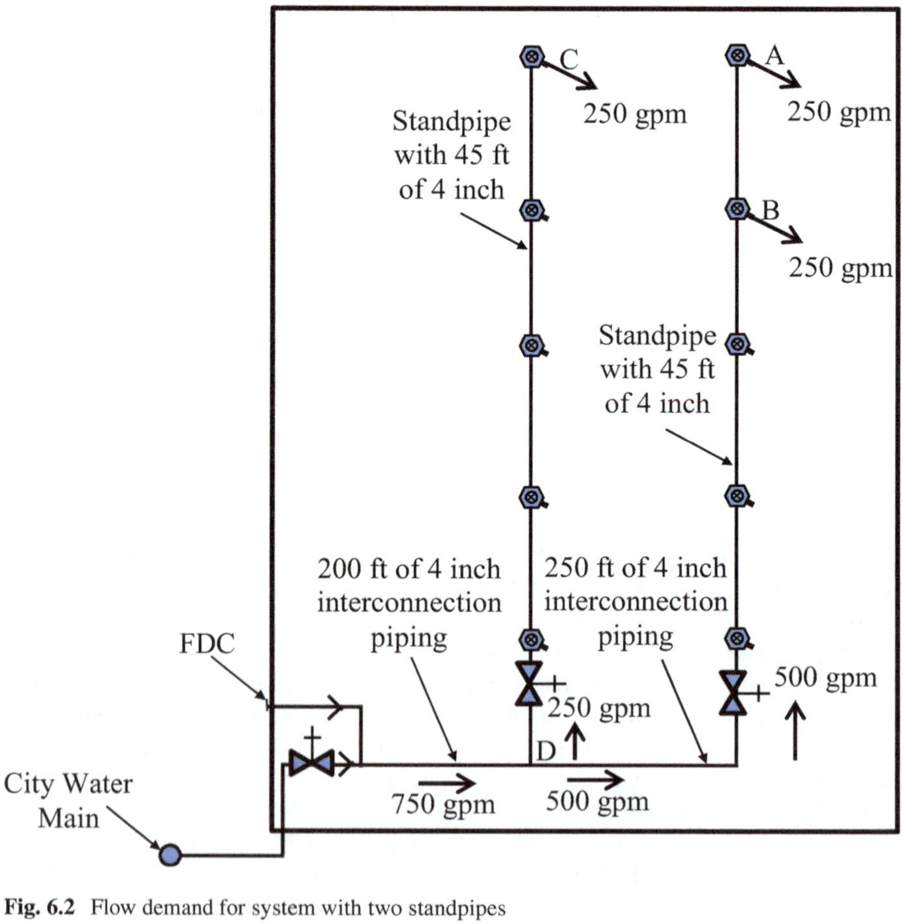

Fig. 6.2 Flow demand for system with two standpipes

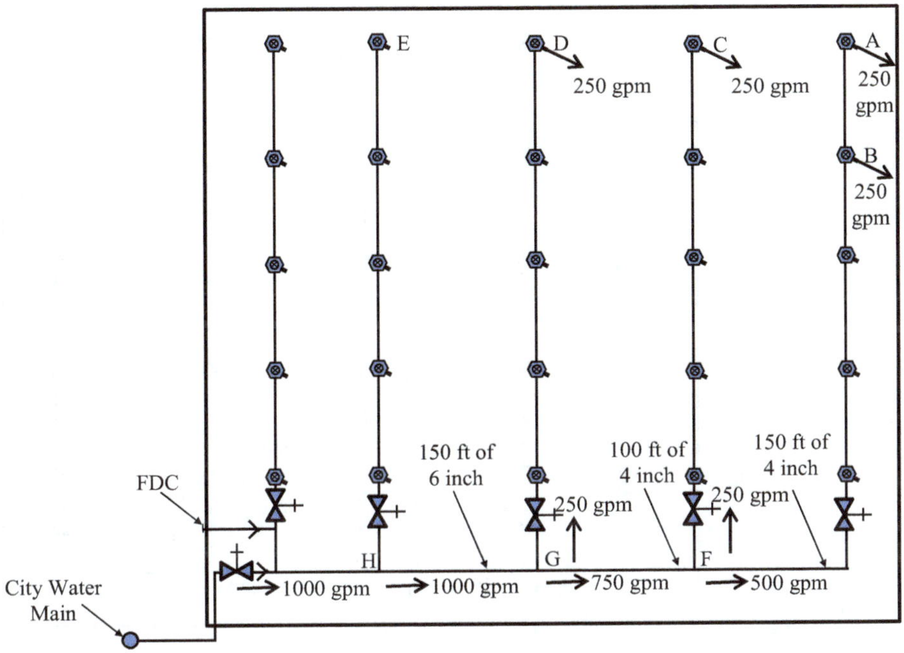

Fig. 6.3 Flow demand for sprinklered building with five standpipes

Figure 6.3 shows the flow demand for the sprinklered building with five standpipes. A flow of 250 gpm is required from Nodes A, B, C, and D. Once that flow is achieved, the maximum demand of NFPA 14 has been achieved (1000 gpm) and no additional flow is required from any of the other standpipe outlets.

The explanation for the flow demand is that the designer needs to plan for a fire occurring in the most difficult place to get water (the most remote portion of the building from the water supply), which is typically the highest floor. The firefighters will put two hose lines in service (one on the fire floor and one from the floor below to protect the first group of firefighters) and start to fight the fire. In the meantime, another group of firefighters go to the top floor in another portion of the building, and advance a hose towards the fire. Still another group of firefighters then go to the top floor in another portion of the building and start advancing a hose towards the fire. At this point, in a sprinklered building, there should not be any additional hose connections close to the fire. The fire should be controlled by the sprinklers and should not ever get that large. In an unsprinklered building, one additional hose line is planned, but by the time that five hose lines are in service, the firefighters should be able to control the fire.

In a building that is more than 80,000 ft^2 in area, the rules change a bit for the second standpipe. In such buildings, the firefighters might want to take a second hose from the floor below the fire to protect the firefighters coming from the second standpipe, so the flow demand is increased to 500 gpm from the second standpipe as well as the first. The maximum flow demands stay the same for sprinklered

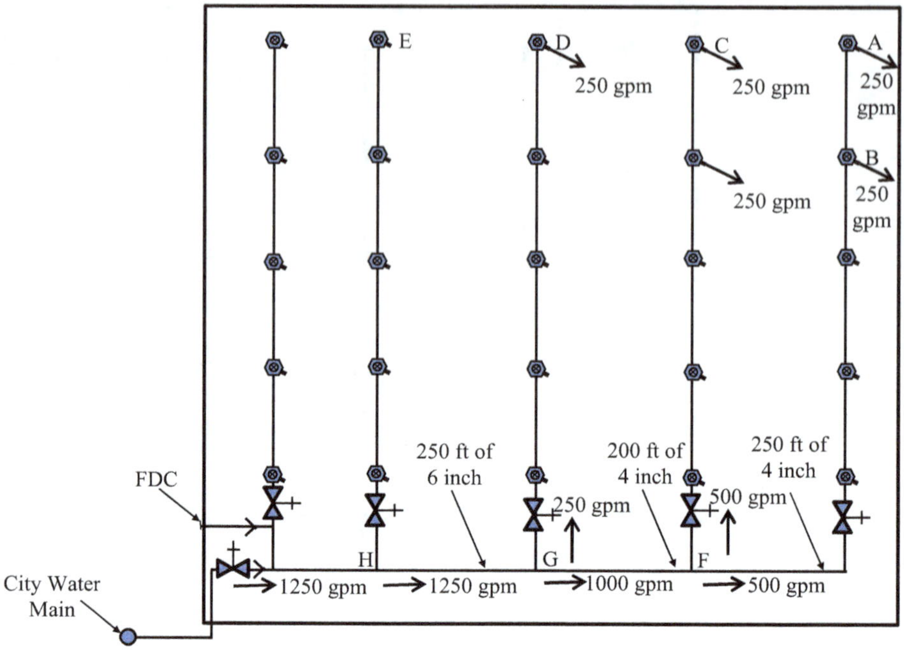

Fig. 6.4 Flow demand for an unsprinklered building that is more than 80,000 ft^2 in area

buildings (1000 gpm) and unsprinklered buildings (1250 gpm). Figure 6.4 shows the flow demand for a standpipe system in an unsprinklered building that is more than 80,000 ft^2.

The examples so far have been buildings where all of the standpipes have gone up to the same floor level of the building. But it is possible that a building can be constructed with different numbers of floors in different parts of the building. In this case, the standpipes will be installed in the exit stairs and there will be different heights to each standpipe. For these situations, NFPA 14 requires separate hydraulic calculations to be run for different portions of the building (see section 7.10.1.2.1.1). For example, consider the sprinklered building situation shown in Fig. 6.5. Two sets of hydraulic calculations need to be provided for this standpipe system. The first needs to prove that the taller portion of the building can get 750 gpm using Nodes A, B and C in the two risers in that portion of the building. The second needs to prove that 1000 gpm can be provided using Nodes D, E, F and G in that portion of the building. The calculations are performed separately and are not required to be added together. Each calculation is assuming a fire in a different portion of the building.

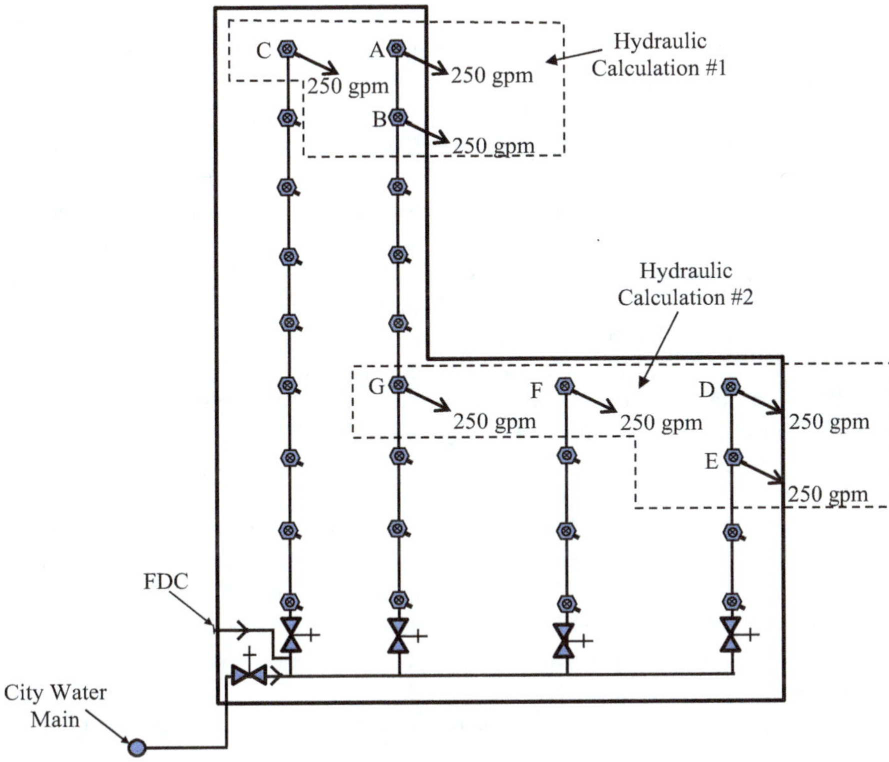

Fig. 6.5 Building with high-rise portion and low-rise portion

Pressure Demand

NFPA 14 now requires that a minimum pressure of 100 psi be available at the most remote outlet. This minimum pressure represents something of a compromise between two different groups of fire protection professionals that has taken more than 20 years to work out, so it is possible that there is a different requirement for an existing standpipe system that you might be working on. It is also possible that the local fire department has passed a requirement for some other minimum pressure based on the hoses and nozzles they want to use when fighting a fire, so check with the local fire department before performing hydraulic calculations on a standpipe system.

Going back more than 20 years ago, NFPA 14 required a minimum pressure of 65 psi at the hose connections. This value was calculated based on the fact that the type of nozzles that fire departments were typically using needed a water pressure of 50 psi at the nozzle and that there was typically about 15 psi of friction loss in the hose between the hose connection and the nozzle. The simple math of 15 + 50 led to the requirement of 65 psi needed at the hose connection.

There were a number of fire departments that were not happy with the 65 psi minimum pressure. These fire departments had made the switch to a different type of nozzle that needed more pressure. In order to take advantage of the full range of options in using these new nozzles, firefighters needed a minimum pressure of 100 psi at the nozzle. Accounting for friction loss in the hose, these firefighters would need 115 psi at the hose outlet in order to get the pressure at the nozzle for these devices to work completely. But designers quickly found that starting hydraulic calculations at 115 psi at the most remote portion of the system drove up the system demand too high at locations closer to the water supply, making it very difficult to keep the pressure below acceptable maximum numbers. So, a compromise was reached in the 1994 that made the base requirement 100 psi for standpipe systems, but allowed local fire departments to reduce the requirement to 65 psi if they used those nozzles that only needed 50 psi at the base of the nozzle.

For many years, this situation of having a 65 psi or 100 psi requirement for the hose connections on the standpipe system was the way that NFPA 14 was written. But more recently, the committee started to consider the fact that the fire departments currently using the nozzles that only need 50 psi at the base might want to switch to the newer nozzles in the future that need more pressure. The committee wanted the standpipe systems that are being installed today to be able to handle this newer technology. In 2003, NFPA 14 was changed to require the 100 psi minimum at the hose connection regardless of the type of nozzle currently used by the fire department.

The 100 psi requirement is not a perfect one for the most remote hose connection. With 15 psi of friction loss between the hose connection and the nozzle, the water will arrive at about 85 psi. At this pressure, water will flow through the nozzle, but all of the features of the nozzle may not be usable. Those people who argued in favor of the 100 psi requirement point out that the 100 psi is a worse-case situation that only occurs when the fire occurs simultaneously with the water supply being at its lowest point and the fire department using all of the flow demand of the system simultaneously, and then only if the fire occurs at the most remote portion of the building. The likelihood of all of these things happening simultaneously is small enough that it makes sense to keep the requirement at 100 psi rather than dealing with all of the problems that would occur throughout the system if the minimum is raised to 115 psi. For more information on the effect of pressure throughout systems, see Chap. 8 of this text.

Duration Demand

Section 9.2 of NFPA 14 requires the water supply to last for a duration of at least 30 min. For automatic and semiautomatic standpipe systems where the water supply is the public water main, that supply is considered to meet the 30 min duration demand since these water supplies can handle a much greater demand. If the water supply is going to rely on a tank, then the tank will need to be sized based on the system demand flow multiplied by 30 to achieve the number of gallons that need to be stored.

For manual standpipe systems, the fire department will need to determine how they are going to put water into the fire department connection (FDC). This might mean the installation of a tank, cistern, pond or reservoir near the FDC that the fire trucks can pull water out of. Without a water supply near the FDC, the fire department will have to plan on using a portable tank and use shuttles of tank trucks to fill the tank and replace the water used by the firefighters as they fight the fire.

For sprinklered buildings, it is common that the flow demand be 1000 gpm. If this is the case, then the size of the tank that would be needed for a standpipe system would be 30,000 gallons (30 min × 1000 gpm = 30,000 gallons). If the building owner does not have the room for a 30,000 gallon tank, then it is possible to install a slightly smaller tank and rely on the refill rate into the tank to provide some of the duration demand (as long as the standpipe system is not for a super high-rise building). See Chap. 10 of this text for more information on the use of tanks in very tall high-rise buildings.

If you are going to rely on the refill rate of the tank to handle some of the duration, it is critical to figure out how quickly the tank can be refilled from whatever source the water is coming from. For example, if the refill rate of 100 gpm can be guaranteed, then over a 30 min period of time, the refill of the tank would account for 3000 gallons (100 × 30 = 3000). In this situation, the tank could be constructed to hold 27,000 gallons with the other 3000 gallons being provided through the refill mechanism of the tank.

Putting It All Together

For Class I (and Class III) standpipe systems, the flow demand of 250 gpm at each of the 2.5 in. hose connections needs to be available at a pressure of at least 100 psi. In order to make sure that this occurs, a series of calculations needs to be performed starting at the most remote node in the system and working back towards the water supply. In order to keep track of the numbers, a standardized work sheet has been developed. While NFPA 14 does not mandate the use of a standardized worksheet, NFPA 13 does. This text will use the standardized worksheet required by NFPA 13 and shown in Fig. 6.6. Whether the calculation is done by hand or by computer, the output can be put into a worksheet that looks like this so that everyone can find the information they need to make sure that the calculations were done correctly.

The first column in the hydraulic calculation worksheet is the "Node" column. Each step in the calculation procedure has a starting node and an ending node. Each row of the worksheet is divided into a top half and a bottom half. The top half of the row applies to the node at the beginning of the step and the bottom half of the row applies to the node at the end of the step. The first column is where the user of the worksheet lets the user declare the beginning and the ending of each step.

Node 1	Elev 1 (ft)	K-Factor	Flow added – this step (q)	Nominal ID	Fittings–quantity and length	L ft	C Factor	total (P_t)	Notes
Node 2	Elev 2 (ft)		Total flow (Q)	Actual ID		F ft	P_f per foot (psi)	elev (P_e)	
						T ft		frict (P_f)	

Fig. 6.6 Blank hydraulic calculation worksheet

The second column in the worksheet is the elevation column. The user puts the elevation of the beginning node at the top of the row and the elevation of the ending node at the bottom of the row. If the elevation changes between nodes, then the elevation pressure will need to be taken into account in the pressure summary.

The third column is the k-factor column. This is only for discharge devices out of fixed orifice devices like fire sprinklers. Nozzles from hoses on standpipe systems have the ability to vary their flow, so the k-factors are potentially infinite for these devices. Rather than worry about the variety of k-factors, the procedures for calculating standpipe systems will ignore this column.

The fourth column in the worksheet is the flow column. This is also split into two pieces of information in each step. The top of the row is for the small flow that is leaving the node at the beginning of the step. This flow is given the variable of a small case "q". The bottom of the row is for the total flow in the pieces of pipe between the beginning and end node for the step. This variable is given the variable of an upper case "Q". For the first step in the calculations, the small q and the big Q are typically the same. For the other steps in the calculation, there will either be no small q, or the small q will be less than the big Q.

The fifth column in the worksheet is the pipe size column. In the upper half of the row, the user inputs the nominal diameter of the pipe. In the lower half of the row, the user enters the actual inside diameter of the pipe being used. It is this actual inside diameter of the pipe that will be needed for the Hazen-Williams formula.

The sixth column of the worksheet is for the user to create a list of the fittings and valves in each step of the calculations. It is important to account for the friction loss through elbows, tees, valves and other devices. The user of the form lists the devices here in the step and then adds the equivalent length for these devices in the next column.

The seventh column of the worksheet is for the length of the pipe and fittings. This column is split into three parts. The first part (L) is where the user enters the actual length of pipe between the starting node and the ending node. In order for the calculations to be done correctly, the distance needs to be calculated from the center of the fitting representing the starting node to the center of the fitting representing the ending node following the center of the pipe along the route between nodes. The second part (F) is for the total equivalent length of all of the fittings listed in the previous column. The third part (T) is for the total of all of the pipe and equivalent lengths of the fittings (T = L + F).

The eighth column is for the Hazen-Williams formula. At the top of the row, the user enters the C-factor for the pipe in the step. The Hazen-Williams friction loss per foot is entered in the bottom portion of the row. It is customary to report this value to three digits to the right of the decimal point. This is the most accurate value that is considered reasonable.

The next to the last column (ninth) is the pressure summary and is broken into three parts in each step. The top part of the step (P_t) is used to show the total pressure demand at the beginning node of the step. The second part (P_e) is used to show the elevation pressure change between the starting and ending node in the calculation step. Note that this could be positive or negative depending on the direction of the elevation change. Since systems are calculated in the opposite direction of flow, the elevation loss is positive when the ending node is below the beginning node. The elevation loss is negative when the ending node is above the beginning node. The third part (P_f) is the friction loss between the beginning and ending node in the step. This is calculated by multiplying the Hazen-Williams friction loss per foot (from the previous column) by the total length of pipe (T from two columns back). Once the three values are entered in this column, they can be added up and entered into the first part of the next step as the total pressure demand at the next node.

The last column in the worksheet is the Notes column. There is nothing technically required to go into the Notes column, but a good designer will provide information here to help the reader understand the numbers entered into that row of the worksheet. Any breadcrumbs that the designer of the form can leave in this column will help a future reader follow the calculations and can improve the value of those calculations.

Each row of the hydraulic calculation worksheet is considered a step in the calculations. A different step will be necessary any time the flow, c-factor of the pipe or diameter of the pipe changes. These three variables are the critical variables

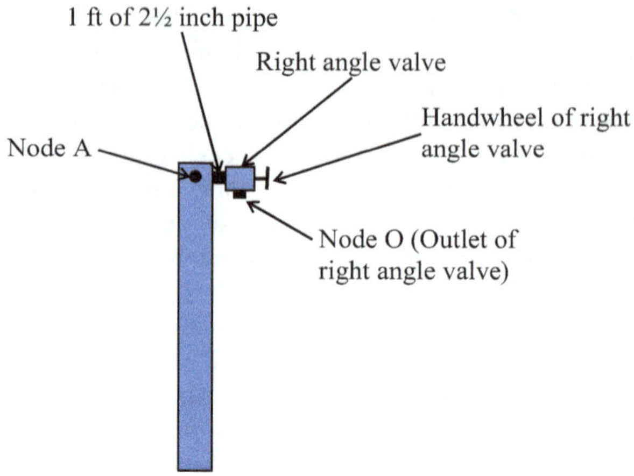

Fig. 6.7 Close-up of Class I hose connection

because they are the three variables of the Hazen-Williams formula, which needs to be used in each step. If any of these variables change, a different step is necessary so that the variables can be entered into the formula.

The following is a step-by-step procedure for the calculation of the extremely simple automatic wet standpipe system shown in Fig. 6.2 with just two standpipes. As the figure shows, there are three points in the system, Nodes A, B and C where water will leave the system at a flow of 250 gpm. However, in order to perform the calculation correctly, each node needs to be examined a little more closely. Figure 6.7 shows a blow-up of a hose connection where Node A is actually at the standpipe riser itself, but Node O is at the outlet of the hose connection, where the calculation needs to start. The 100 psi pressure needs to be available at the outlet of the hose connection, not back in the riser.

The first step in the calculation is to go from Node O to Node A with a flow of 250 gpm. The water needs to arrive at Node O at a pressure of 100 psi. The water is flowing through the 1 ft length of 2.5 in. pipe and the right angle valve. The first step is shown in the worksheet in Fig. 6.8. This calculation will assume that all of the piping in the system is Schedule 40 steel. Note that there is a very tiny difference in elevation between Node O and Node A, but such change is negligible and there is no reason to worry about it. The conservative approach is to just use the elevation of Node A, which is what this calculation does.

The first step of the calculation has determined that the pressure demand at Node A is 109.5 psi. The next step is to calculate down the standpipe to Node B. Note that there is no extra flow added in this step. The 250 gpm that is exiting the system at Node A is the same 250 gpm that is flowing up from Node B towards Node A. The second step in the calculation, from Node A to Node B is shown in Fig. 6.9.

Note that in the second step, there is an elevation change between Node A and Node B of 10 ft The elevation change can be seen in the second column of the calculation worksheet and the change in pressure is accounted for in the pressure summary (ninth) column. In this particular case, the change in elevation is the height of

Node 1	Elev 1 (ft)	K-Factor	Flow added this step (*q*)	Nominal ID	Fittings quantity and length	L ft	C Factor	total (P_t)	Notes
Node 2	Elev 2 (ft)		Total flow (*Q*)	Actual ID		F ft	P_f per foot (psi)	elev (P_e)	
						T ft		frict (P_f)	
O	45		250	2½	Right angle valve (31)	1	120	100	
						43	0.215	0	
A	45		250	2.469	Tee (12)	44		9.5	
A								109.5	

Fig. 6.8 First step of calculation from Node O to Node A

each story of the building since the hose connections are at the same location on each floor.

The third step of the calculation can go all the way down the most remote standpipe and across the interconnection piping to the base of the second standpipe (Node D). The calculations can go this far because there is no change in flow, C-factor or pipe size in this entire run of pipe, so there is no need to put any other nodes in the system or to make more steps out of the calculation. In this step, there will be 285 ft of pipe and one elbow. The flow in this step will be the 500 gpm that needs to get up the standpipe to Nodes A and B. The third step is shown in Fig. 6.10.

At Node D, the total pressure demand is 150.2 psi. From here back towards the water supply, the system will continue with 4 in. pipe. The friction loss here will be significant, and if the water supply can't handle such a loss, this might be a good section of pipe to increase to 6 in. But for now, let's see what would happen with 4 in. pipe. The flow in this section of pipe will be 750 gpm including the 250 gpm that needs to discharge from Node C and the 500 gpm that needs to discharge from Nodes A and B. There is no node designated at the end of the 200 ft of pipe shown

Node 1	Elev 1 (ft)	K-Factor	Flow added this step (q)	Nominal ID	Fittings quantity and length	L ft	C Factor	total (P_t)	Notes
Node 2	Elev 2 (ft)		Total flow (Q)	Actual ID		F ft	P_f per foot (psi)	elev (P_e)	
						T ft		frict (P_f)	
O	45		250	2½	Right angle valve (31)	1	120	100	
A	45		250	2.469	Tee (12)	43	0.215	0	
						44		9.5	
A	45		0	4	None	10	120	109.5	$P_e = 0.433 \times 10$
B	35		250	4.026		0	0.020	4.3	$P_e = 4.3$
						10		0.2	
B								114.0	

Fig. 6.9 Second step of calculation from Node A to Node B

in Fig. 6.2, but we will assume that this node will be at the transition between the underground pipe and the aboveground pipe just under the elbow to the left of the control valve and the check valve. For the sake of hydraulic calculations, we will label this Node T for the transition between the underground and aboveground pipe. Figure 6.11 shows the hydraulic calculation for this step.

The last step will be to calculate through the underground from Node T to Node U (where the underground pipe connects to the city water main). The length of the underground is not shown, but let's assume for the sake of this calculation that it is 75 ft plus the elbow where the underground pipe makes the turn from being vertical to horizontal and the tee where it connects to the city water main. Let's also assume for the sake of completing this calculation that the underground pipe is 6 in. Class 52 cement lined ductile iron, which has an internal diameter of 6.16 in. (see Table A.10.1.3 in NFPA 13). Figure 6.12 shows the final calculation and represents the whole hydraulic calculation for the standpipe system being fed through the city water main.

Node 1	Elev 1 (ft)	K-Factor	Flow added this step (*q*)	Nominal ID	Fittings quantity and length	L ft	C Factor	total (P_t)	Notes
Node 2	Elev 2 (ft)		Total flow (*Q*)	Actual ID		F ft	P_f per foot (psi)	elev (P_e)	
						T ft		frict (P_f)	
O	45		250	2½	Right angle valve (31)	1	120	100	
						43	0.215	0	
A	45		250	2.469	Tee (12)	44		9.5	
A	45		0	4	None	10	120	109.5	$P_e = 0.433 \times 10$
						0	0.020	4.3	$P_e = 4.3$
B	35		250	4.026		10		0.2	
B	35		250	4	Elbow	285	120	114.0	$P_e = 0.433 \times 35$
						10	0.072	15.1	$P_e = 15.1$
D	0		500	4.026		295		21.2	
D								150.2	

Fig. 6.10 Third step of calculation from Node B to Node D

It is worth noting that there is no elevation change shown on the last step even though the pipe runs from inside the building to several feet underground. The justification for this is that the pressure measurements made for the underground pipe are not made at the elevation of the underground pipe. Nobody digs down to the pipe underground and installs a gauge on the underground main in order to determine the water pressure available in this main. Instead, the pressure in the water main is measured at a fire hydrant above the ground, so the final step in a hydraulic calculation always considers the elevation change between the fire hydrant where the city water main was measured to the last node in the building. With no change shown in the calculations, we are showing that the last node in the building (Node T) is at the same elevation as the fire hydrant where the water main was measured.

The equivalent length of the fittings in the last step of Fig. 6.12 requires a bit more explanation. The underground pipe has an elbow and a tee, which would have

Node 1 / Node 2	Elev 1 (ft) / Elev 2 (ft)	K-Factor	Flow added this step (q) / Total flow (Q)	Nominal ID / Actual ID	Fittings quantity and length	L ft / F ft / T ft	C Factor / P_f per foot (psi)	total (P_t) / elev (P_e) / frict (P_f)	Notes
O	45		250	2½	Right angle valve (31)	1	120	100	
A	45		250	2.469	Tee (12)	43	0.215	0	
						44		9.5	
A	45		0	4	None	10	120	109.5	$P_e = 0.433 \times 10$
B	35		250	4.026		0	0.020	4.3	$P_e = 4.3$
						10		0.2	
B	35		250	4	Elbow	285	120	114.0	$P_e = 0.433 \times 35$
D	0		500	4.026		10	0.072	15.1	$P_e = 15.1$
						295		21.2	
D	0		250	4	Check Valve (22) Gate Valve (2) Elbow (10)	200	120	150.2	
T	0		750	4.026		34	0.152	0	
						234		35.6	
T								185.8	

Fig. 6.11 Fourth step of calculations from Node D to Node T

equivalent lengths of 14 ft and 30 ft if they were installed in Schedule 40 steel pipe. But the underground pipe is cement lined ductile iron, so two adjustments need to be made to the equivalent lengths: one for the C-factor of the pipe and the other for the different internal diameter. The C-factor adjustment is simply 1.33 (see Table 8.3.2.2 in NFPA 14) and the diameter adjustment is calculated as 1.08 $[(6.16/6.065)^{4.87} = 1.08]$. This makes the equivalent length of the elbow 20.1 ft ($14 \times 1.33 \times 1.08 = 20.1$) and the equivalent length of the tee is 43.1 ft ($30 \times 1.33 \times 1.08 = 43.1$). The total equivalent length of the fittings is then 63.2 ft ($20.1 + 43.1 = 63.2$).

As Fig. 6.12 shows, the water supply needs to create a pressure of 187.7 psi at a flow of 750 gpm in order to supply this standpipe system. This creates a slight problem since some of the equipment on the standpipe system might only be rated for 175 psi. In addition, the static pressure in the water supply will likely be higher than 187.7 psi, which might over-pressurize the standpipe system. So, it might be a good

Node 1 Node 2	Elev 1 (ft) Elev 2 (ft)	K-Factor	Flow added this step (q) Total flow (Q)	Nominal ID Actual ID	Fittings quantity and length	L ft F ft T ft	C Factor P_f per foot (psi)	total (P_t) elev (P_e) frict (P_f)	Notes
O A	45 45		250 250	2½ 2.469	Right angle valve (31) Tee (12)	1 43 44	120 0.215	100 0 9.5	
A B	45 35		0 250	4 4.026	None	10 0 10	120 0.020	109.5 4.3 0.2	$P_e = 0.433 \times 10$ $P_e = 4.3$
B D	35 0		250 500	4 4.026	Elbow	285 10 295	120 0.072	114.0 15.1 21.2	$P_e = 0.433 \times 35$ $P_e = 15.1$
D T	0 0		250 750	4 4.026	Check Valve (22) Gate Valve (2) Elbow (10)	200 34 234	120 0.152	150.2 0 35.6	
T U	0 0		0 750	6 6.16	Elbow (20.1) Tee (43.1)	75 63.2 138.2	140 0.014	185.8 0 1.9	Equivalent Length Adjustments C-factor = 1.33 $(6.16/6.065)^{4.87} = 1.08$
U								187.7	

Fig. 6.12 Final hydraulic calculation for system fed from city water main

idea to go back and select some 6 in. pipe in the building rather than using all 4 in. pipe. As discussed earlier, the section of pipe from Node D back to Node T could be changed to 6 in., which would likely keep the pressure well below 175 psi. See Chap. 8 of this text for more information on pressure control.

Figure 6.12 shows the hydraulic calculations on a standpipe system from the most remote hose connection to the city water main, but this is not the end of the hydraulic calculation requirements of NFPA 14. Another calculation needs to be made from the most remote hose connection to the FDC. The first part of these calculations would look the same as shown in Fig. 6.12, but would change at the end going to the FDC rather than the underground. See the information later in this chapter regarding calculation of manual systems for more information on calculating to the FDC.

Class II Systems

Class II standpipe systems are much easier to calculation than Class I (or Class III) systems. There is only one demand, to provide 100 gpm at 65 psi at the most remote 1.5 in. hose station. The requirement for 100 gpm as the flow demand can be found in section 7.10.2.1.1 of NFPA 14. The requirement for 65 psi as the pressure demand can be found in section 7.8.1 of NFPA 14.

If the pipe leading to the most remote hose station from the water supply is all the same size, the hydraulic calculation can be done in a single step. A flow of 100 gpm creates a pressure loss of 0.317 psi per ft in 1.5 in. Schedule 40 pipe (inside diameter of 1.61 in.). In a system with 200 ft of pipe (including equivalent lengths of fittings), this would be a friction loss of 63.4 psi ($200 \times 0.317 = 63.4$). Considering that there would also be elevation loss in the system as well, this 200 ft distance seems to be the effective limit for 1.5 in. pipe feeding Class II hose stations. If the pipe size is increased to 2 in., the friction loss drops to 0.094 psi per ft, which would allow about 600 ft of pipe and fittings before reaching a critical limit.

Calculations for Manual Systems

The hydraulic calculations for manual standpipe systems are the same as those for automatic systems except that they end at the FDC rather than the water supply. Once these calculations are complete, you compare the results of the demand to the ability of the fire department to determine whether the system pipe sizes are acceptable. This means that you need to have some knowledge of what the fire department is capable of providing.

The pumps on fire trucks are very similar in design to the pumps used in fixed fire protection systems, however the rating system for these pumps is very different. While pumps in fixed fire protection systems are allowed to be used at 150 % of their rated flow, the pumps on fire trucks are not permitted to be used this way (because the rating point is determined in a different manner). The ratings for the pumps on fire pumps are regulated by NFPA 1901, Standard for Automotive Fire Apparatus.

The rating for the pump on a fire truck is intended to be a maximum flow. The pump is required to achieve this maximum flow at a net pressure of 150 psi. The pump is then supposed to be designed to provide 75 % of the rated flow at a net pressure of at least 200 psi and 50 % of rated flow at a net pressure of at least 250 psi. For example, consider a fire truck with a pump rated at 1500 gpm. The pump is required to create a pressure of at least 200 psi at a flow of 1125 gpm and a pressure of at least 250 psi at a flow of 750 gpm. The pump curve formed by these design points is shown in Fig. 6.13.

Fire departments specify the pump size when the buy the fire truck. Popular sizes of the pumps on fire trucks are 750, 1000, 1250, and 1500 gpm. The 750 gpm sizes were fairly popular in the 1960s and 1970s, but it is this author's experience that in

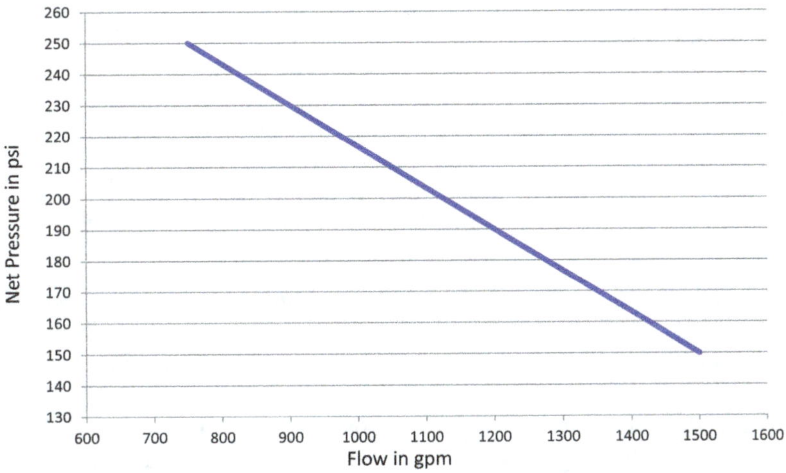

Fig. 6.13 Pump curve for fire truck with 1500 gpm rated pump

more modern times, the 1000 gpm and 1500 gpm pumps are much more popular. Before designing a manual standpipe system, check with the local fire department to determine what size pump will be typical on the trucks that they will use to respond to the building in question. Once you have that information, the following procedure should be used to determine the ability of the fire department to provide water to the FDC (this procedure is for a system with a fire hydrant fed from a municipal water supply near the hydrant):

1. Determine the residual pressure at the hydrant closest to the FDC using the standpipe system flow demand for the flow from the city water main. Call this pressure P_R.
2. Calculate the friction loss through the hose from the fire hydrant to truck using a flow of the standpipe system flow demand. Call this friction loss F_T.
3. Calculate friction loss through hose from truck to the FDC at a flow of the standpipe system flow demand. Call this friction loss F_{FDC}.
4. Calculate the elevation loss from the fire hydrant to the FDC. Call this pressure difference E_{FDC}.
5. Determine the suction pressure at the intake of the pump on the fire truck. Call this pressure P_S and calculate it using the formula $P_S = P_R - F_T - E_{FDC}$
6. Determine the discharge pressure at the exit of the pump on the fire truck. Call this pressure P_D and calculate it using the formula $P_D = P_S + P_N$ (P_N is the net pressure from the pump on the truck. You may need to draw a curve like the one in Fig. 6.13.)
7. Calculate the water supply pressure that will be provided to the FDC and call this pressure P_{FDC}. Calculate it using the formula $P_{FDC} = P_D - F_{FDC}$
8. Compare the water supply pressure (P_{FDC}) to the pressure demand from the hydraulic calculations. In order to be acceptable, the P_{FDC} must exceed standpipe system pressure demand.

Table 6.2 Hose constants (c) for calculating the friction loss

Type of hose	Hose constant (c)
1½ in. rubber lined	24
2½ in. rubber lined	2
3 in. with 2½ in. couplings	0.8
4 in.	0.2
4½ in.	0.1
5 in.	0.08

For systems where the fire hydrant will be pulling water from a standing water supply like a tank, pond, cistern or reservoir, the same procedure can be used as discussed above with the pressure in Step 1 as zero. The other steps should be the same.

The procedure discussed above relies on the user being able to calculate the friction loss of water flowing through fire hose. There have been debates on exactly what formula to use for this procedure. While research is being done on this subject at the time of publication of this text, the following formula should be adequate: $FL = cq^2L$ where the units are as follows:

FL = friction loss through the hose in psi

c = a constant that depends on the size of the hose and the quality of the lining (Table 6.2 shows the "c" values for a number of common sizes and types of fire hoses)

q = the flow of water in the hose in hundreds of gpm (so for a flow of 500 gpm, q = 5)

L = length of hose in hundreds of feet (so for a 100 ft section of hose, L = 1)

As an example, consider a Class I manual wet standpipe system where the demand of 1000 gpm at 165 psi has been calculated from the most remote hose connection to the FDC. The water supply at a fire hydrant 30 ft below the elevation of the FDC has a static pressure of 50 psi and a residual pressure of 20 psi at 2000 gpm, producing the water supply curve in Fig. 6.14. Will this be acceptable as a pressure demand at the FDC if the local fire department has only 1500 gpm rated pumps on their fire trucks?

In order to answer the question asked in the example, the pressure that the fire department can deliver to the FDC can be calculated and compared to the system demand pressure as follows:

1. Determine the residual pressure at the hydrant. $P_R = 42$ psi at a flow of 1000 gpm, See the city water main pressure in Fig. 6.14 at 1000 gpm.
2. Calculate the friction loss through the hose from the fire hydrant to truck. Assuming that the fire department uses a 5 in. hose, carries the hose in 100 ft lengths, and uses one length of hose from the hydrant to the truck, the formula is $F_T = 0.08(10)^2(1) = 8$ psi.
3. Calculate friction loss through hose from truck to the FDC at a flow of the standpipe system flow demand. Call this friction loss F_{FDC}. Assume that the firefighters lay two 100 ft lengths of rubber lined 2.5 in. hose from the truck to the FDC. Each

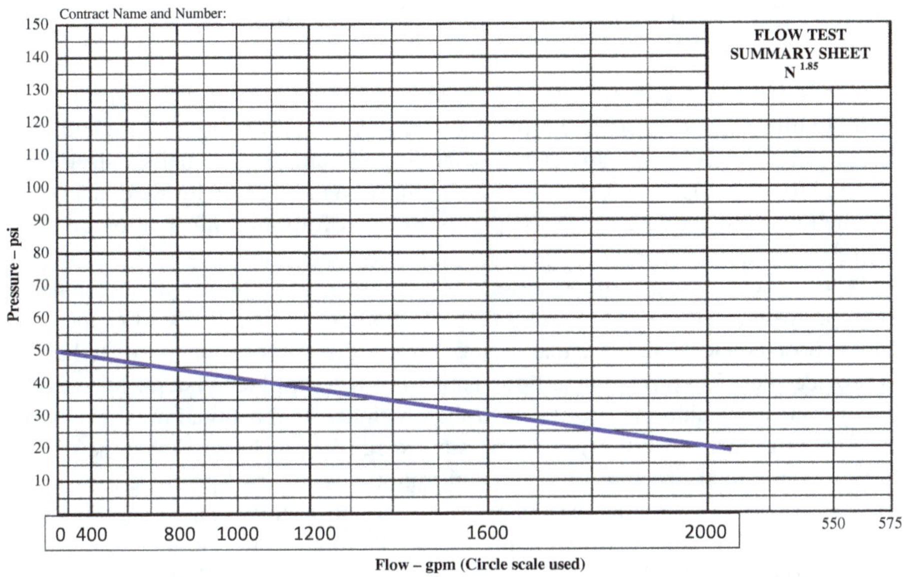

Fig. 6.14 Water supply curve for FDC example

hose will carry 500 gpm and the friction loss can be calculated as $F_{FDC} = 2(5)^2(1) = 50$ psi

4. Calculate the elevation loss from the fire hydrant to the FDC. Call this pressure difference $E_{FDC} = 0.433 \times 30 = 13$ psi.
5. Determine the suction pressure at the intake of the pump on the fire truck. Call this pressure P_S and calculate it using the formula $P_S = P_R - F_T - E_{FDC} = 42 - 8 - 13 = 21$ psi.
6. Determine the discharge pressure at the exit of the pump on the fire truck. Call this pressure P_D and calculate it using the formula $P_D = P_S + P_N = 21 + 218 = 23$ 9 psi (the net pressure of the pump was determined from Fig. 6.13 at a flow of 1000 gpm).
7. Calculate the water supply pressure that will be provided to the FDC and call this pressure P_{FDC}. Calculate it using the formula $P_{FDC} = P_D - F_{FDC} = 239 - 50 = 189$ psi.
8. Compare the water supply pressure (P_{FDC}) to the pressure demand from the hydraulic calculations. In order to be acceptable, the P_{FDC} must exceed standpipe system pressure demand. In this case, the fire truck can supply 189 psi to the FDC, which exceeds the 165 required to make the system work, so this is acceptable. The equipment on the FDC should be rated to handle 189 psi and the hose connections low down in the building may need pressure control valves. See Chap. 8 of this text for more information on pressure control in standpipe systems.

Special IBC Rules for Shopping Malls (See 905.3.3)

Section 905.3.3 of the International Building Code contains some special rules for mall buildings that do not have full standpipe systems. There are requirements for certain mall buildings to have a full standpipe system. Those malls that are not required to have a full standpipe system still have to have Class I hose connections installed in a few locations. See Chap. 4 of this text for more information on these partial standpipe systems.

When these partial standpipe systems are installed, a hydraulic calculation needs to be performed to make sure that a flow of 250 gpm can be provided to the most remote hose connection. If the mall is sprinklered, the water supply needs to be able to provide the 250 gpm plus the flow demand of the sprinkler system. In addition, the hydraulic calculation needs to show that the maximum residual pressure loss in the standpipe portion of the system does not exceed 50 psi. The way that the section is written, the total pressure loss due to elevation and the total pressure loss due to friction loss together cannot exceed 50 psi. If the most remote outlet is 20 ft above the FDC, then the elevation loss is 8.7 psi, so the total friction loss cannot exceed 41.3 psi so that the total residual pressure loss does not exceed 50 psi.

Hose Connections on Sprinkler Systems That Are Not Standpipes

Towards the end of Chap. 4 in this text, there was a discussion of hose stations that are extensions of the fire sprinkler system, not separate standpipe systems. These hose stations are common in storage warehouses and backstage areas of theaters. When these hose stations are installed, NFPA 13 (in Chaps. 11 and 12) requires that 50 gpm be added to the sprinkler system hydraulic calculations at the two most remote connections to the hose stations. There is not requirement for the hose station to operate at any specific pressure. The idea is to provide 100 gpm total of "inside hose stream demand" for these situations.

These systems are not considered standpipes and are not required to meet the rules for Class II standpipe systems. They are just considered a part of the fire sprinkler system.

Test Yourself

6.1. If a vertical piece of standpipe piping is 100 ft long and the pressure needs to be 110 psi at the top, what is the minimum pressure required at the bottom when no water is flowing?

6.2. Using the Hazen-Williams formula, calculate the friction loss of 250 gpm flowing through 10 ft of 4-in. Schedule 40 steel pipe in a wet pipe standpipe system.

6.3. Using the Hazen-Williams formula, calculate the friction loss of 500 gpm flowing through 90 ft of 4-in. Schedule 40 steel pipe in a wet pipe standpipe system.

6.4. What is the pressure demand at the bottom of the single vertical standpipe discussed in questions 6.1, 6.2, and 6.3 for the condition where 250 gpm is discharging from the most remote outlet on the standpipe (at the top) and 250 gpm is discharging from the second most remote outlet, 10 ft down from the top?

6.5. What is the equivalent length of a 90 tee fitting in 6 in. Schedule 10 pipe being used in an automatic wet standpipe system?

6.6. How would the answer to question 6.5 change if the standpipe system is a manual dry system?

6.7. What is the total friction loss of 1000 gpm flowing through 500 ft of 6 in. Schedule 10 pipe with five 90 tee fittings in a manual dry standpipe system?

6.8. The standpipe system shown in Fig. 6.15 is an automatic dry standpipe system protecting an unsprinklered open parking garage that is 90,000 ft^2 in area with 10 ft between floors. The five vertical standpipe risers are 4 in. Schedule 40 pipe connected to the horizontal interconnection piping (nodes P to L) through a 5 ft horizontal piece of 4 in. Schedule 40 pipe with a OS&Y control valve on it (Node L to Node K). The distance from Node A to Node K is 55 ft The pipe from Node M to Node L is 200 ft of 4 in. Schedule 40. The pipe from Node N to Node M is 200 ft of 6 in. Schedule 40. The pipe from Node O to Node N is 200 ft of 6 in. Schedule 40. The pipe from Node P to Node O is 200 ft of 8 in. Schedule 30. The pipe from Node P to the dry-pipe valve (measured to center) is 10 ft of 8 in. Schedule 30. The dry-pipe valve has an equivalent length of 15 ft with a C-factor of 120. The pipe from the dry-pipe valve (center) to the floor flange is 5 ft long 8 in. Schedule 30. The underground pipe is 8 in. Class 50 lined ductile iron (inside diameter 8.39 in.). From the floor flange to the city water main is 75 ft of actual pipe and the fire hydrant where the city main was tested is 8 ft below the floor flange. Answer the following questions:

(a) What is the demand (flow and pressure) of the standpipe system to the city water main?
(b) How many inlets are required at the fire department connection?
(c) If the available flow and pressure of the city water main, adjusted for reasonable seasonal and daily fluctuations, is shown in Fig. 6.16, will this system work without a fire pump?
(d) If the answer to question (c) is "no", what is the minimum net pressure necessary from the fire pump at the system demand flow to make the standpipe system work?

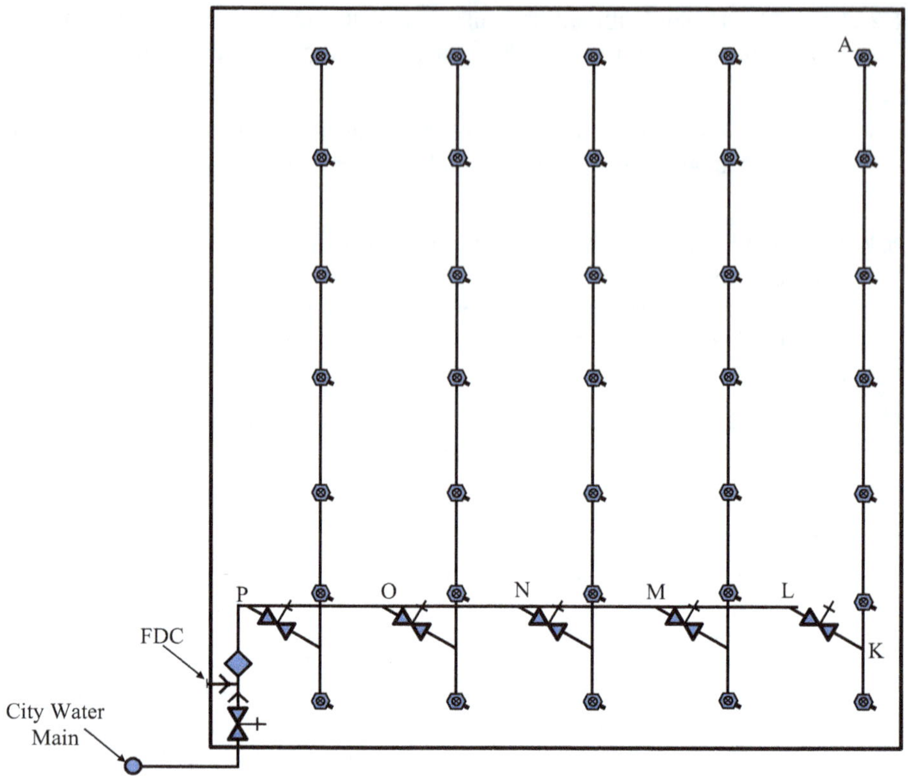

Fig. 6.15 Automatic dry standpipe system for question 6.8

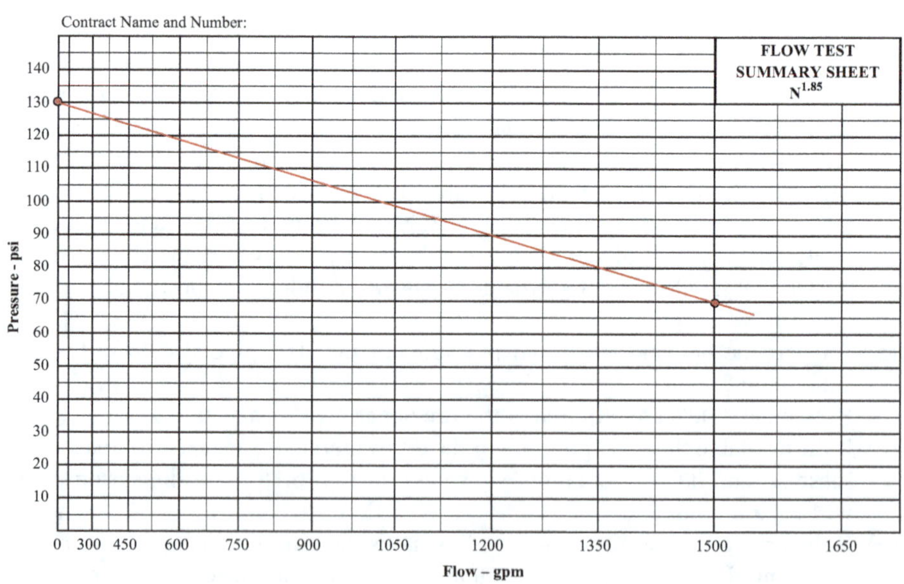

Fig. 6.16 Water supply for city water main in question 6.8

Chapter 7
Combined Sprinkler/Standpipe Systems

So far in this text, the topic of fire sprinkler systems and standpipe systems in the same building has been handled in a generic fashion. The rules for the installation and hydraulic calculation of the standpipe system depend on whether the building is sprinklered or not. But the combination of the two systems using some of the same piping has not yet been discussed. This chapter will delve into the rules of NFPA 13 and NFPA 14 regarding the combination of the sprinkler system and standpipe system for at least some of the piping in the building. See section 7.10.1.1.5 in NFPA 14 for more information on this requirement.

Both NFPA 13 and NFPA 14 allow the fire sprinkler system to be interconnected to the standpipe system, but neither document requires the piping systems or water supplies to be combined. Logic and economics usually end up leading to the conclusion that the water supply can serve both systems and that the piping can be shared, at least for some of the systems.

The old BOCA Basic Building Code (which in the last few years of its existence changed its name to the BOCA National Building Code) did have a requirement for a partial sprinkler system protecting a very small portion of a building (six or fewer sprinklers in any single fire area of the building) to be connected to the standpipe system in a building that had a standpipe system. When the BOCA code combined with the other model building codes to become the International Building Code (IBC), this rule continued into that code. Today, the same requirement exists in the IBC as section 903.3.8.3. For a full sprinkler system that meets NFPA 13 or NFPA 13R, there is no requirement for the sprinkler system or the standpipe system to be combined.

Layout of Pipes and Equipment

Even if there is no requirement to combine the sprinkler and standpipe systems, it is still a very common design concept. A single water supply in a building that handles both systems can save the building owner a significant amount of money. Combining

K.E. Isman, *Standpipe Systems for Fire Protection*,
DOI 10.1007/978-3-319-47750-3_7

the piping that goes up through the building can save building owners as well. Typically, one or more of the vertical standpipes in the building also serves as the pipe that brings water to the sprinkler system on each floor of the building.

Section 6.3.5.1 of NFPA 14 requires, "Each connection from a standpipe that is part of a combined system to a sprinkler system shall have an individual control valve and check valve." The combination of the control valve and check valve at the connection for the sprinkler system at the standpipe system is shown in the sketch in Fig. 7.1 and in the fire protection system in Fig. 7.2.

In Fig. 7.1, the first blue valve to the right of the standpipe is the main control valve for the fire sprinkler system. The blue valve to the right of the control valve is the check valve. The purpose of the control valve is to be able to close that valve, so that you can isolate the sprinkler system and drain it for repairs without having to drain the standpipe system or take the standpipe system out of service. The purpose of the check valve is to keep the pressure that gets created by a fire pump or occurs during high pressure situations from the water supply in the sprinkler system. Also, if the standpipe system has to be taken out of service and needs to be drained, the water can be taken out of the standpipe system without having to close all of the control valves on each sprinkler system on each floor of the building. Another reason for having the check valve on each sprinkler system at each connection from the standpipe system is to prevent all of the waterflow switches from going into alarm when the standpipe system is returned to service.

In Fig. 7.1, the red box to the right of the check valve is the waterflow switch. Downstream of that is the system pressure gauge. Immediately opposite of the pressure gauge is the main drain for the fire sprinkler system. This allows the system gauge to be used during the main drain test. The drain connection is usually 1.25 in. pipe as long as the sprinkler pipe is 2.5 or 3 in. pipe. However, the drain connection is also used to test the waterflow alarm, which is why this connection is downstream of the alarm. The 1.25 in. connection is too large to simulate the flow of a single

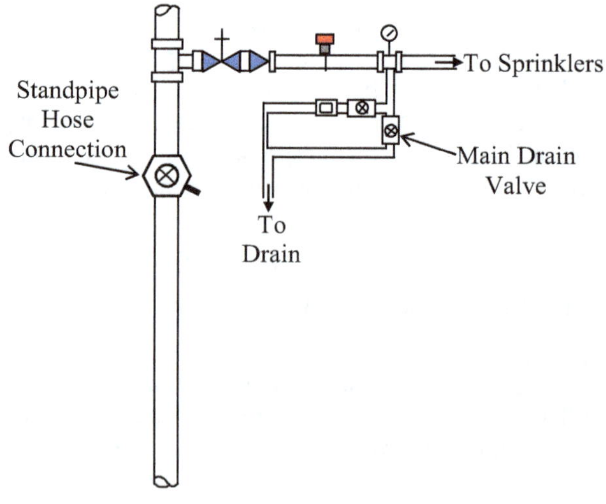

Fig. 7.1 Sketch of check valve and control valve at the sprinkler system connection to the standpipe system

Fig. 7.2 Combined sprinkler and standpipe system

sprinkler, which NFPA 13 requires for the alarm test line, so many installers put an alarm test bypass line around the drain valve with an orifice that necks down to the orifice of a typical sprinkler on the system. This bypass line has a valve to initiate the flow and a sight gauge so that you can see if water is actually flowing in the pipe.

The connections shown in Figs. 7.1 and 7.2 are typically installed on each individual floor of a building. Since the definition of a fire sprinkler system is a system with a control valve, check valve and a drain, this makes each of the fire sprinkler piping arrangements on each floor a separate fire sprinkler system. A multistory building piped in this manner then has many fire sprinkler systems in it, one on each floor, rather than a single system. This has serious consequences when it comes to periodic testing of systems. NFPA 25 requires internal inspection of sprinkler systems on a regular basis (once every five years). Since each system on each floor is a separate system, the internal inspections need to be done on each system, not just one of the systems in the building. NFPA 25 does allow alternate systems to be inspected in buildings with multiple systems so that each system is actually tested every 10 years.

The definition of a "sprinkler system" also includes the concept of a "system riser". Looking at the sprinkler system in Fig. 7.1, you may ask where the "system riser" for the sprinkler system is. The sprinkler system technically starts at the connection to the standpipe, and the piping from there to the sprinklers is horizontal not vertical. The answer is that even though the piping to the sprinklers that contains the control valve and the check valve is horizontal, it still meets the definition of a "system riser". Section 3.5.13 of NFPA 13 defines a "system riser" as, "The aboveground horizontal or vertical pipe between the water supply and the mains (cross or feed) that contains a control valve (either directly or within its supply pipe), a pressure gauge, a main drain, and a waterflow alarm device." Since the horizontal piping in Fig. 7.1 contains this equipment, it meets the definition of a "system riser".

In Fig. 7.2, there are actually two sprinkler systems connected to the standpipe at the same elevation. One connection is on the side of the standpipe closest to the photographer and the other is on the exact opposite side. This was done to feed two separate systems on the same floor because the building exceeded 52,000 ft^2 on this floor, so the building needed separate sprinkler systems. For the sprinkler system coming off the standpipe in the direction of the photographer, there is a butterfly control valve, which is the silver box with the red turn wheel on top, then there is a wafer check valve, which is black and difficult to distinguish from the pipe, but it is between the red couplings. Down the pipe from the check valve is the pressure gauge and opposite of the pressure gauge is the test and drain connection. The test connection is directly piped to the drain riser (which is behind the standpipe) and there should be an orifice in that connection that simulates the orifice of a sprinkler on the system (the smallest sprinkler on the system since this is a wet pipe sprinkler system). There is no separate drain connection, but a hose can be connected to the spigot that is shown if the system needs to be drained faster than what can be done through the test connection. Note that this arrangement does not allow main drain testing except through the hose spigot, which was evidently acceptable to the authority having jurisdiction in this case.

Variations Depending on Floor Landing of Standpipe Installation

In both Chaps. 1 and 4 of this text, there were short discussions of the decision that needs to be made regarding the installation of the standpipe and the hose connections on either the main floor landing in a stairwell or on the intermediate floor landing. There are pros and cons to locating the hose connections in either location. One of the considerations as to whether the hose connections would be better at the main floor landing or the intermediate floor landing is the sprinkler system connection. Sprinkler system connections when the hose outlets are on the main floor landing are much easier to make than if the hose outlets are on the intermediate floor landing.

To illustrate the problems when the hose connections are at the intermediate floor landings, consider Figs. 7.3 and 7.4. Figure 7.3 shows the hose connections at the intermediate floor landing for a stairwell where the stairs go all the way to the roof. Figure 7.4 shows the hose connections at the intermediate floor landing for a stairwell where the stairs only go up to the top floor of the building.

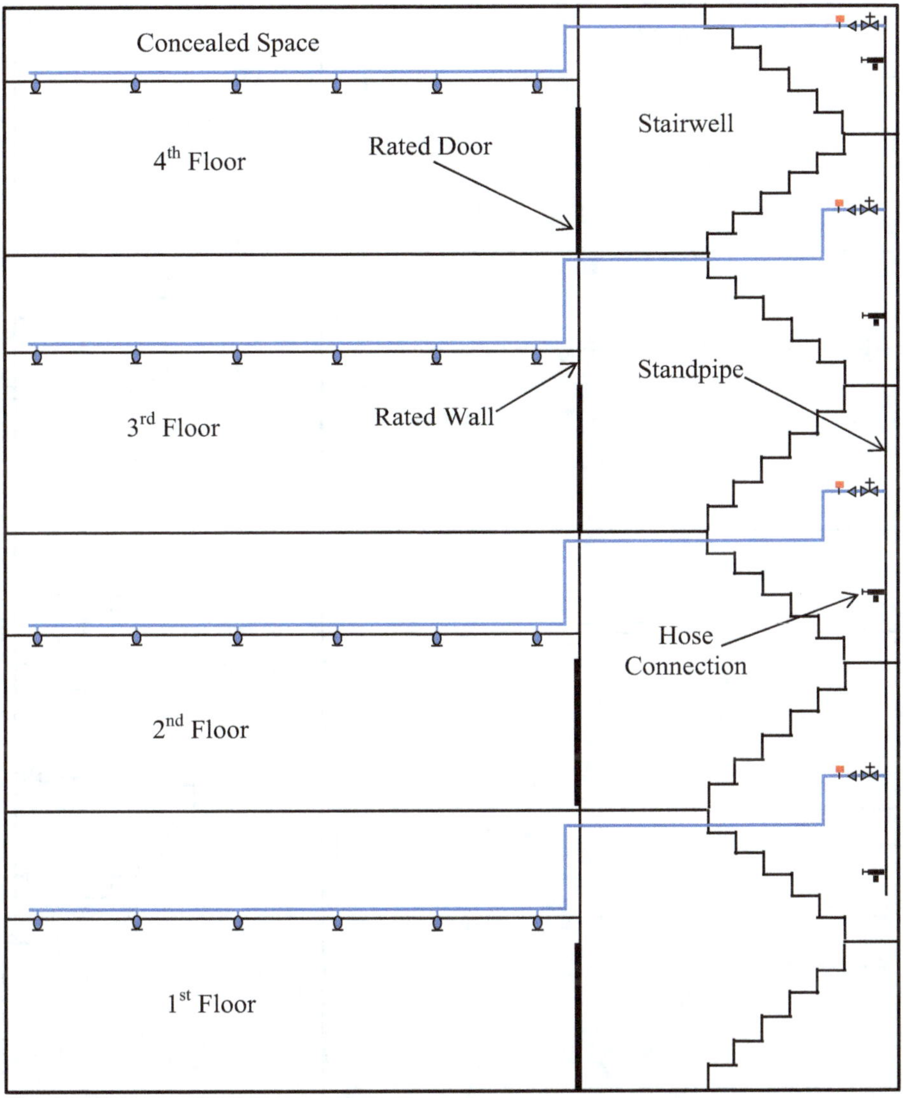

Fig. 7.3 Combined standpipe/sprinkler system with hose connections at intermediate landings in a stairwell that does not go to the roof

Consider Fig. 7.3 first. The sprinkler piping is shown as the blue lines with the control valve, check valve, and waterflow switch at the connection to the standpipe. The control valve, check valve and waterflow alarm need to be up and out of the way for a number of reasons. If these devices were installed lower down, people would bang their head on the pipe as they walk down the stairs, the devices would be more susceptible to damage, and the devices would be more susceptible to tampering. But it is difficult to install these devices up and out of the way. The

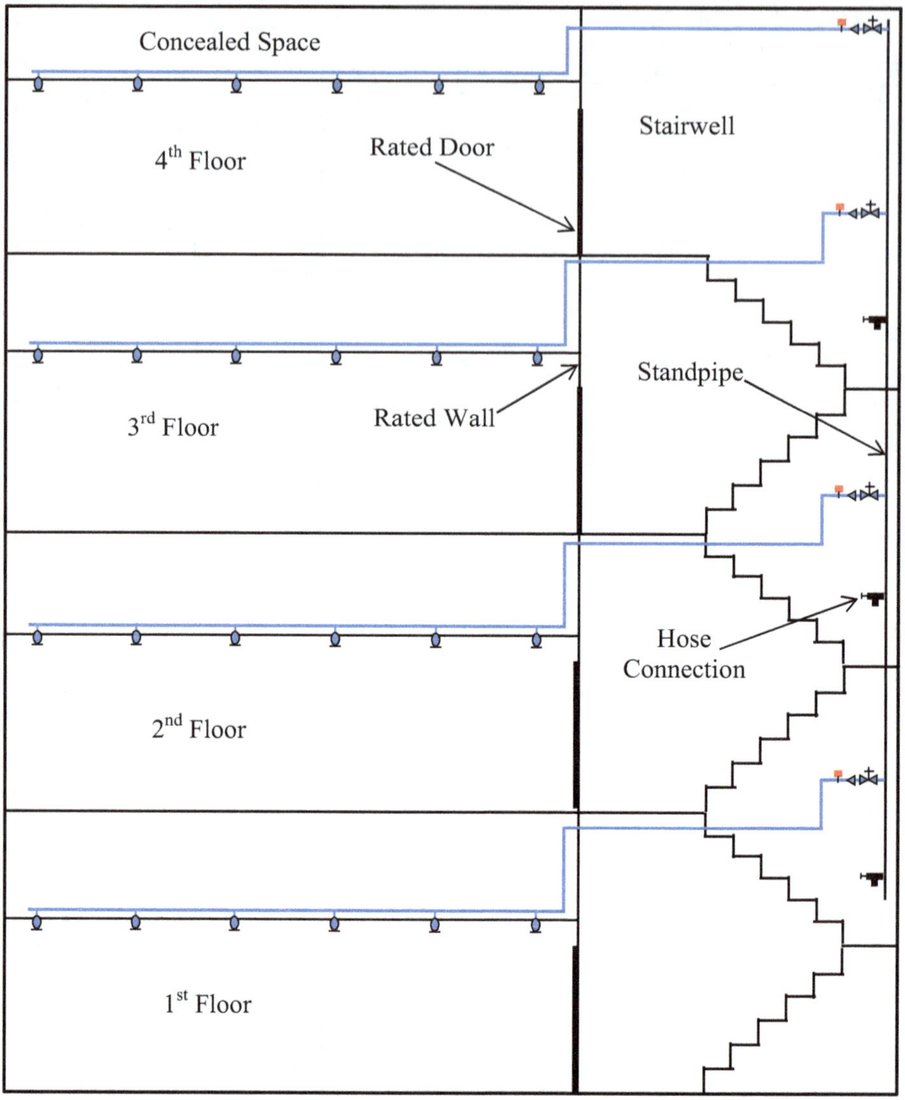

Fig. 7.4 Combined standpipe/sprinkler system with hose connections at intermediate landings in a stairwell that goes all the way to the roof

intermediate landings are smaller than the main floor landing and it is more difficult to get a safe ladder on these landings. Often, scaffolding needs to be constructed in the stairwell so that the fitters can tap into the standpipe for the sprinkler system outlet.

Once the standpipe has been tapped for the sprinkler system, it is difficult to get the piping across the stairwell to the sprinklered portion of the building. In Fig. 7.3, the piping is shown going across and down, but based on the configuration of the stairs, that might not be possible. Figure 7.3 has been simplified as a two-dimensional drawing, but in reality, it might take a great deal of extra pipe and fittings to get the water from the sprinkler connection at the riser to the sprinklers on each floor.

Once the sprinkler system is installed, testing and maintenance become an issue. Just as it was difficult to install the tap for the fire sprinkler system in Fig. 7.3, it would be difficult to get to the equipment for testing and maintenance. Immediately downstream of the waterflow alarm needs to be a test and drain combination as shown in Fig. 7.1. This includes a sight glass so that you can see water flowing in the drain piping when you run a waterflow alarm test. But if the sight glass and control valve are well above the intermediate landing, and the intermediate landing is too small to safely set up a ladder, it will not be possible to conduct the test or see water flowing in the sight glass. A more important concern would be shutting off the flow of water in the system after a fire or if a pipe breaks or leaks. If the intermediate landing does not allow the safe operation of a ladder, extra water damage might occur before a person can safely get up and close the control valve.

Looking at Fig. 7.4, the problem gets worse if the stairs do not go all the way to the roof. The control valve for the sprinklers on the top floor (the fourth floor in the figure) is 15 ft above the intermediate landing between the third and fourth floor. This is extremely difficult to reach from an installation perspective and would be very hard to reach for maintenance and testing purposes. Section 8.16.1.1.1.1 of NFPA 13 requires that the control valve be in an accessible location. A spot 15 above a small floor service would be difficult to defend as "accessible".

Figure 7.5 shows the situation where a standpipe is located so that the hose connections are at the main floor landings, making for a much easier sprinkler system installation. The location of the stairs does not interfere with the piping from the standpipe to the sprinklers. In Fig. 7.5, the control valve, check valve and waterflow switch are at an unusual angle, but that is just so that you can see them in the picture. In reality, they would be installed horizontally off of the standpipe and then a simple connection would be made to the sprinkler piping on the other side of the stairwell as shown in Fig. 7.2.

The arrangement in Fig. 7.5 is easier to install because the main floor landing is typically larger than the intermediate landing, making it easier to work on a ladder safely. Not only will the installation be easier, but testing and maintenance will be easier as well. A tall person can reach the control valve from the main floor landing directly or by standing on a very small step ladder. The sight glass can be installed at a location where someone can see into it either directly standing on the landing or on a small step ladder. Since there is less piping and fewer fittings to get the water from the standpipe to the sprinklers, the installation is less expensive. Finally, it does not matter whether the stairs go all the way to the roof or not since all of the main floor landings are served with a hose connection.

Ultimately, the design professional will need to choose whether the standpipe hose connections will be at the main floor landing or the intermediate floor landing, in consultation with the fire department and the building officials. This decision will need to be made taking into consideration local codes, fire department preferences, fire department operating procedures, and input from the fire sprinkler contractor. If

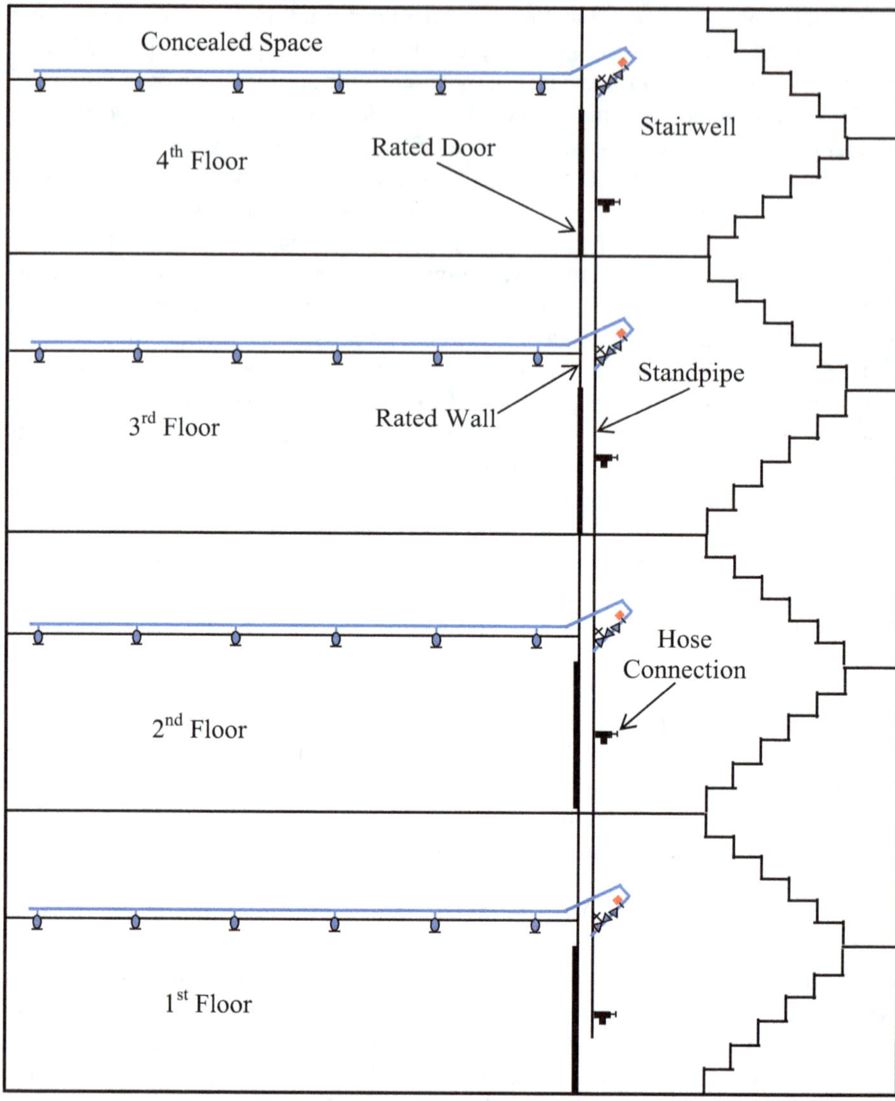

Fig. 7.5 Combined Standpipe/Sprinkler System with Hose Connections at Main Floor Landings

the decision is made to put the standpipe system at the intermediate landings, then it is entirely possible that the sprinkler contractor might choose to run separate risers through the building for the fire sprinkler system rather than use the standpipes. The separate sprinkler riser might be a better option that can save the building owner money. Even if separate sprinkler and standpipe piping are installed, the sprinkler system can still be connected to the same water supply as the standpipe system.

Sprinkler Systems Fed by Multiple Standpipes

So far in this text, we have been discussing fire sprinkler systems with one connection to a standpipe. There are hydraulic and reliability advantages to having a sprinkler system fed from two risers. Hydraulically, with two supply pipes sharing the water delivery, each pipe will cause less friction loss, which reduces the total pressure demand of the sprinkler system. From a reliability standpoint, the sprinkler system can be kept in service, even when one of the standpipes needs to be taken out of service by closing the control valve between the sprinkler system and the out of service standpipe, but keeping the control valve to the other standpipe open.

Figure 7.6 shows a standpipe system feeding a sprinkler system through two standpipes. Note that the control valve, check valve and waterflow alarm are

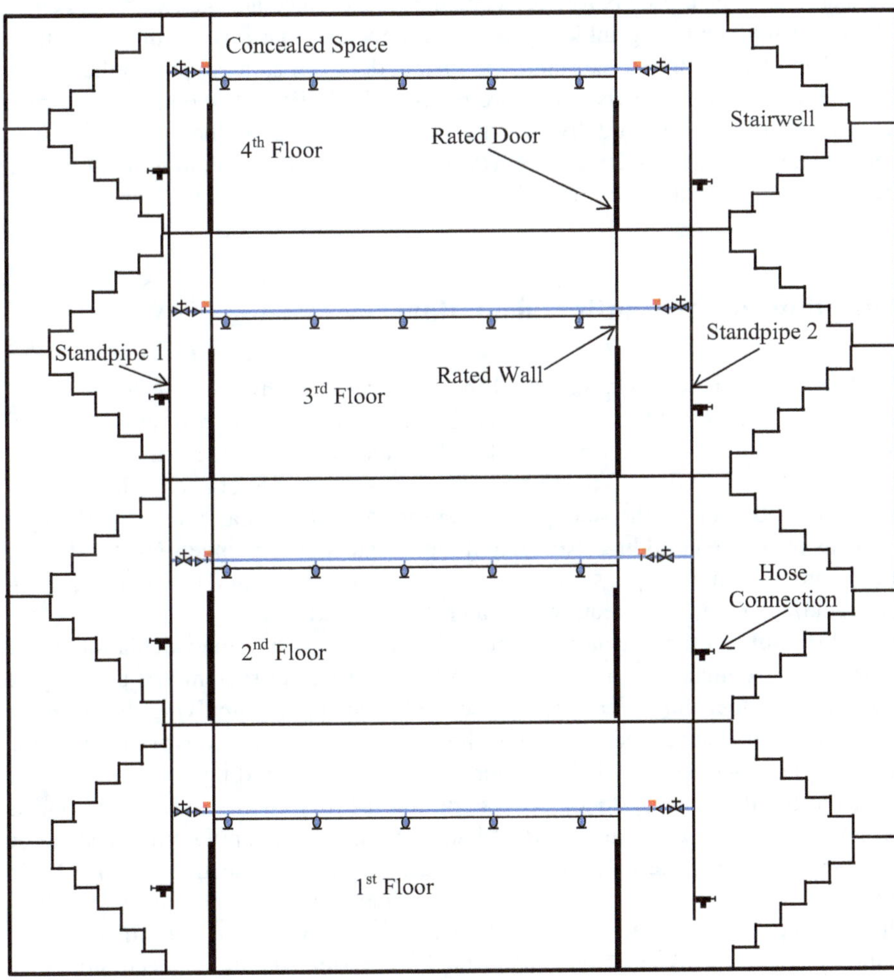

Fig. 7.6 Sprinkler system fed from two standpipes

installed on each connection. The control valve and check valve need to be installed on each connection due to the way that NFPA 14 is written because it specifically requires the check valve and control valve on each connection. The waterflow alarm is important because a single sprinkler that opens near one of the connections might not create enough flow to operate a waterflow alarm at the other end if only one waterflow alarm was installed.

If a sprinkler system is connected to multiple standpipe risers, it becomes important to identify that for anyone that is going to need to know how to shut down the system. If a fire occurs, or the system needs to be shut down and drained, the person responsible needs to know that there is more than one valve that needs to be found and closed. If only one of the valves is closed, and the system is taken apart, significant water damage can occur. Section 6.6.4.3.1 of NFPA 13 requires that the sign installed on one of the control valves needs to explain where the other control valve(s) are that need to be closed before the system can be drained and safely worked on.

Note that when the sprinkler system is designed so that it is getting water from more than one standpipe, the standpipe system does not gain any hydraulic advantage. Due to the check valves that are required by NFPA 14, the sprinkler piping does not share in the water delivery to the standpipe system, so the standpipe system needs to be calculated as a tree system (with only one path for the water to flow from the water supply to the open hoses).

Variations in Hydraulic Calculations

In Chap. 6 of this text, we discussed the basic rules for hydraulic calculations of standpipe systems and the effect that having a sprinkler system had on those calculations. Basically, for a Class I or Class III standpipe system, having a sprinkler system that meets NFPA 13 in the building allows the user to only calculate a maximum of 1000 gpm for the standpipe system instead of the maximum of 1250 gpm in an unsprinklered building. Note that if the sprinkler system meets NFPA 13R, the 1250 gpm maximum applies due to the very real possibility of a fire in the unsprinklered attic, which might require a greater flow to fight.

But the subject that has not yet been addressed regarding the calculation of the water supply demand for a building with both a sprinkler system and a standpipe system is whether the water supply needs to be capable of supplying the sprinkler demand and the standpipe demand at the same time. The answer to this question depends on how extensive the sprinkler system is in the building.

If the building has a standpipe system and a fire sprinkler system that meets NFPA 13, both NFPA 13 and NFPA 14 say that the water supply only needs to be capable of handling the single most demanding system. See section 7.10.1.3.1.1 in NFPA 14 for the exact language of this requirement. Most of the time, this will be the standpipe system with a demand of 1000 gpm. A light hazard or ordinary hazard building with a sprinkler system and a standpipe system should not need more than a total of 1000 gpm, even with the two systems operating simultaneously. While the

1000 gpm is calculated through the standpipe system piping from four hose connections, the reality is that the 1000 gpm is really for three hoses and the sprinklers. The fire department should not need more than three hoses (750 gpm) for fighting a fire in a fully sprinklered building and the other 250 gpm should be plenty for the sprinkler system.

One of the frequently asked questions regarding buildings with fire sprinkler and standpipe systems, is whether the inside hose stream demand needs to be calculated when sizing the fire sprinkler piping at the Class I or Class III hose connections. The answer to that question has changed over the years. Older editions of NFPA 13 implied in the annex that it was appropriate to add 50 gpm to the sprinkler demand at each of the two most remote Class I or Class III outlets when calculating the sprinkler demand. The most recent edition of NFPA 13 (2016) specifically says that this addition of inside hose demand is not necessary. The reality is that it should not make much of a difference to the total sprinkler system demand since the standpipe needs to be sized for a greater flow anyway.

If the building is sprinklered in accordance with NFPA 13R, the same rules apply as for NFPA 13 regarding the addition of the sprinkler demand and the standpipe demand (section 7.10.1.3.1.1 of NFPA 14); however, the maximum flow increases to 1250 gpm instead of 1000 gpm. So, the sprinkler system demand or the standpipe system demand needs to be provided by the water supply, whichever is greater. Most of the time, this will be the standpipe demand, which is 1250 gpm (assuming there are enough standpipes in the system to get to this number).

If the building just has a partial sprinkler system, then the rules are very different. Section 7.10.1.3.2 of NFPA 14 requires that the sprinkler demand be added to the standpipe demand, with some maximum numbers. If the sprinkler demand is for a light hazard occupancy, the maximum amount that you add for the sprinkler system is 150 gpm. If the sprinkler demand is for an ordinary hazard occupancy, then the maximum amount you would add would be 500 gpm. These sprinkler demands would be added to the standpipe system demand, which could be as high as 1250 gpm (if the building has enough standpipes), so the water supply might have to be capable of providing a flow of as much as 1750 gpm (1250 for the standpipe and 500 for the ordinary hazard sprinkler system).

Sprinkler Systems with Standby Hose Stations

In both Chaps. 4 and 6 of this text, the concept of hose stations directly connected to the sprinkler system, which are not considered standpipe systems, was discussed. When these hose stations are added to the sprinkler system, they are treated as inside hose stations and 50 gpm needs to be added to the sprinkler demand at the most remote two connections to the hose stations.

The flow demand for the inside hose stations takes the place of some of the total hose stream demand. Table 11.2.3.1.2 of NFPA 13 requires a total of 100 gpm of hose stream demand for light hazard, 250 gpm for ordinary hazard and 500 gpm for extra hazard. If there are two or more hose stations in a building, the inside hose stream demand will be 100 gpm and the outside hose stream demand becomes 0 gpm in a light hazard occupancy, 150 gpm in an ordinary hazard occupancy and 400 gpm in the extra hazard occupancy.

For storage occupancies, the total inside and outside hose stream demand depends on what is being stored, how it is being stored, and the type of sprinklers used to protect the storage. For storage warehouses with two or more inside hose stations and 100 gpm of inside hose stream demand at those hose stations, the outside hose stream demand will be 100 gpm less than the total hose stream demand.

Test Yourself

Answer the following questions to reinforce the concepts discussed in this chapter.

7.1. List at least two reasons why putting Class I standpipe hose connections at the main floor landing would be advantageous.

7.2. List at least two disadvantages of putting Class I standpipe hose connections at the main floor landing.

7.3. List at least two advantages to putting Class I standpipe hose connections at the intermediate floor landing.

7.4. List at least two disadvantages to putting the Class I standpipe hose connections at the intermediate floor landing.

7.5. List at least two advantages to supplying a sprinkler system from two standpipes.

7.6. List at least one disadvantage to supplying a sprinkler system from two standpipes.

7.7. What equipment does NFPA 14 require on each standpipe connection to a sprinkler system?

7.8. A building has a fire sprinkler system designed and installed in accordance with NFPA 13 and a standpipe system with four Class I standpipes. If the sprinkler system flow demand is 200 gpm, what is the total flow demand required for the fire protection in this building?

7.9. A building has a fire sprinkler system designed and installed in accordance with NFPA 13 and a standpipe system with two Class I standpipes. If the sprinkler system flow demand is 200 gpm, what is the total flow demand required for the fire protection in this building?

7.10. A building has a fire sprinkler system designed and installed in accordance with NFPA 13R and a standpipe system with four Class I standpipes. If the sprinkler system flow demand is 60 gpm, what is the total flow demand required for the fire protection in this building?

7.11. A building has a fire sprinkler system designed and installed in accordance with NFPA 13R and a standpipe system with six Class I standpipes. If the sprinkler system flow demand is 60 gpm, what is the total flow demand required for the fire protection in this building?

7.12. A building has a standpipe system with four Class I standpipes and six sprinklers in an ordinary hazard area connected to the standpipe system. If the sprinkler system flow demand is 180 gpm, what is the total flow demand required for the fire protection in this building?

Chapter 8
Pressure Control in Standpipe Systems

Previous chapters in this text have discussed the concept of water pressure, minimum pressure requirements, and the calculations involving the changes in pressure that occur when water flows through pipes. Pressure differences occur due to changes in elevation and due to friction loss in the piping. By adding up the pressure losses between the place where the water starts and the place where the water ends up, you can determine the pressure demand at the water supply.

In this chapter, we will discuss the maximum pressure allowed in the system at different points and the ways in which you can control the pressure in order to make sure that you do not over-pressurize a system. The next chapter in this text will focus on fire pumps, which provide pressure to standpipe systems. It is difficult to discuss controlling the pressure in standpipe systems without getting into fire pumps, so they will also be discussed in this chapter. If the reader is not familiar with fire pumps, it may be advantageous to go back and forth between this chapter and the next chapter if questions arise about fire pump performance.

Minimum Pressure

For a Class I or Class III standpipe, the minimum pressure that is required at the most remote hose connection is 100 psi. There may be some local fire departments that have adopted some higher minimum pressure and there also may be a fire department that allows a lower minimum pressure (some used to allow Class I and Class III systems to have a minimum pressure of 65 psi), but as far as the NFPA and model codes are concerned, the minimum pressure in the system has to be 100 psi when the system is being used. While a system is standing by (ready to use, but not yet in use), the water pressure in the system may be much lower than 100 psi as long as a pump starts or the fire department connects a pump when it is time to use the system.

K.E. Isman, *Standpipe Systems for Fire Protection*,
DOI 10.1007/978-3-319-47750-3_8

For a Class II standpipe system, the minimum pressure required at the remote hose connection or hose station is 65 psi. The reasons that this pressure is lower than the Class I or Class III system is that the flow is expected to be lower, so you won't need as much pressure, and the reaction force that is caused by the pressure is more of a concern for lesser trained individuals that will probably be using the Class II systems. Since these people are not as well trained as the people using Class I or Class III systems, they should not have as much pressure at the outlet.

Maximum Pressure

NFPA 14 has a great deal to say about the maximum water pressure that will be allowed in the system. There are two concerns with high pressure. First, there is a concern that the pressure might exceed the rating for which the components can be used. If this happens, the components might blow apart from the inside, causing damage to people and property in the area while also taking the fire protection system out of service, potentially during a time of emergency when the system is needed.

The second concern with high pressure is the reaction force caused by water flowing out of the nozzle at the end of the hose. The higher the pressure, the higher the reaction force and the stronger the person needs to be that is holding on to the nozzle. If the pressure gets too high, the person holding the nozzle might lose control and let go. The reaction force would then cause the hose to fly around uncontrollably, which could injure or kill a person.

In order to prevent all of these problems, NFPA 14 has a series of requirements regarding the maximum allowable water pressure at different points in the system. At some points, the maximum will be expressed in terms of the static, or nonflowing, condition. This will be the situation where no water is flowing, but the piping is filled with water under pressure. Other requirements in NFPA 14 refer to the maximum residual pressure at a point in the system, which is the pressure when water is flowing at that point. The residual pressure is the energy left over in the water after the energy that caused the water to start moving has been taken away. Therefore, the residual pressure will always be lower than the static pressure at the same location.

For Class I and Class III systems, NFPA 14 states that the maximum static pressure at the 2.5 in. outlets is 175 psi (see section 7.2.3.2). In a situation where a pressure higher than 175 psi is going to be provided by the water supply, a device will need to be installed that reduces both the static and residual pressure. Such situations are common enough that there are manufacturers that make right angle hose connections with pressure reducing valves built in.

In order to determine where the pressure might be over 175 psi, the design professional starts with hydraulic calculations. Starting at the highest and farthest hose connection with a pressure demand of 100 psi at a flow of 250 gpm, the elevation loss and friction loss are added (in addition to the flow at the next hose connection) working back to the water supply. Once the pressure demand gets above 175 psi, some sort of pressure control will be necessary. But this is not the only situation where pressure control will be necessary. Even if the pressure demand is below 175 psi, the water supply might provide more pressure than the demand, in which case the pressure needs to be controlled at the outlet as well. This is one of the situations where water supplies that deliver a great deal more water at a higher pressure than the demand can cause a problem for the fire protection system.

Consider a single riser on a standpipe system with a 4 in. pipe that is 120 ft vertically from the top hose connection to the bottom hose connection. With a demand of 100 gpm at the top of the riser, the demand at the lowest hose connection would be about 169 psi, which is just under the limit of 175 psi. While it looks like the lowest hose connection would not need some sort of pressure control, the pressure would exceed 175 psi if the water supply exceeded the demand by only 6 psi. Taking into account variations in water supply, static pressures being greater than residual pressures, and churn pressures of pumps being greater than the flowing pressure, it is likely that the bottom hose connection, as well as additional hose connections on the standpipe will need pressure control.

For Class II hose stations, the maximum residual pressure allowed by NFPA 14 is 100 psi at the connection from the standpipe to the hose. Since it is the residual pressure that is being regulated, the type of device that controls the pressure only has to create friction loss during flowing conditions. Interestingly, Class II systems do not have any specific static pressure control or maximum. The person that grabs the hose and starts using it might have to deal with a very high initial pressure until the water starts flowing.

The pressure limitations discussed so far have been concerned with the outlet where the hose is connected. In addition to this concern, NFPA 14 also limits the pressure in the system in two other ways. First, in any portion of the system that has hose connections attached, the maximum pressure allowed is 350 psi (see Section 7.2.1). The portion of the system that is of concern here is the piping between the water supply and the hose connections. The limit of 350 psi is in place because there is a concern that the pressure control valves that are used at the hose connection might fail in the open position, which would expose firefighters to high pressure and nozzles that are difficult or impossible to control. By limiting the pressure anywhere in the system that has hose connections to 350 psi, NFPA 14 is limiting the maximum pressure to which firefighters could be exposed.

The 350 psi limit does not apply to express risers. An express riser is a pipe that does not have hose connections on it, but takes water from the water supply directly up into the building to a higher zone. Express risers are common in situations with fire pumps installed in series, which will be discussed in more detail in Chap. 9 of this text.

The second mechanism that NFAP 14 uses to control the maximum pressure is the rating on the system components. Like all NFPA standards, NFPA 14 limits the use of equipment to the pressure for which it is rated, so while there is no official maximum pressure for express risers, Section 7.2.2 limits the pressure in express risers to the pressure limit in the listing of the pipe, valve or fitting. There are some very high pressure situations where there may not be any equipment listed for fire standpipe system use. In these situations, the authority having jurisdiction is allowed to approve other equipment that is pressure rated for other types of systems, but the authority needs to be careful regarding the safety factors involved in the pressure rating. Some plumbing systems are pressure tested with burst pressure safety factors less than that used for sprinkler systems, so it might not be safe to just assume any plumbing pipe might be acceptable.

Devices that Control and Restrict Pressure

NFPA 14 describes several different devices that can be used to control or restrict pressure. All of these devices fall into the broad category called "Pressure-Regulating Devices", which are defined by Section 3.3.13 as any type of device designed to reduce, regulate, control or restrict water pressure. Figure 8.1 shows the entire family of devices that reduce or restrict the pressure in one way or another and how they relate to each other. NFPA 14 divides these devices into two separate categories; devices that reduce both the static and residual pressure and devices that only reduce the residual pressure. Each category has its purpose as described earlier in this chapter where Class I and Class III standpipe systems need devices that can reduce both the static and residual pressures while Class II standpipe systems are allowed to use devices that only reduce the residual pressure.

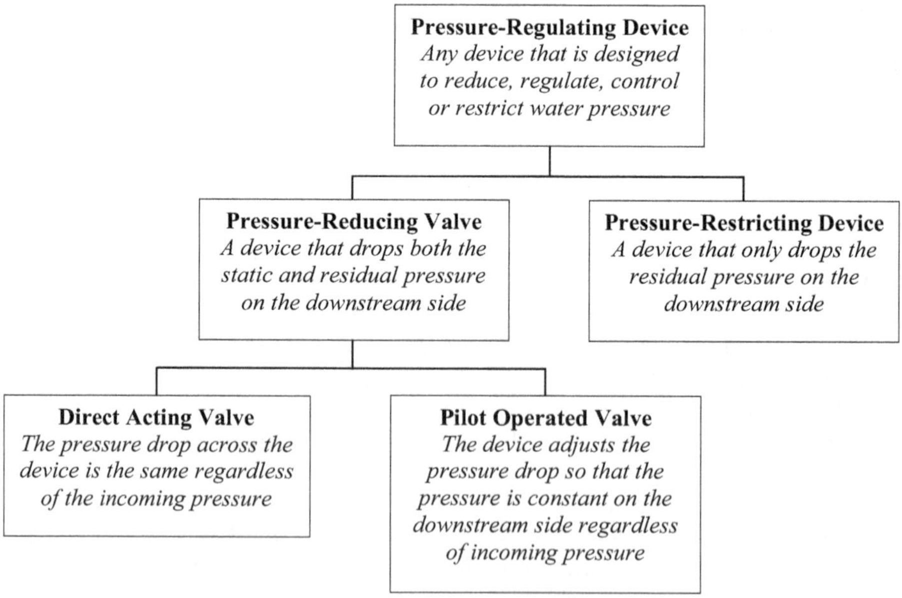

Fig. 8.1 Explanation of terminology for devices that reduce or restrict pressure

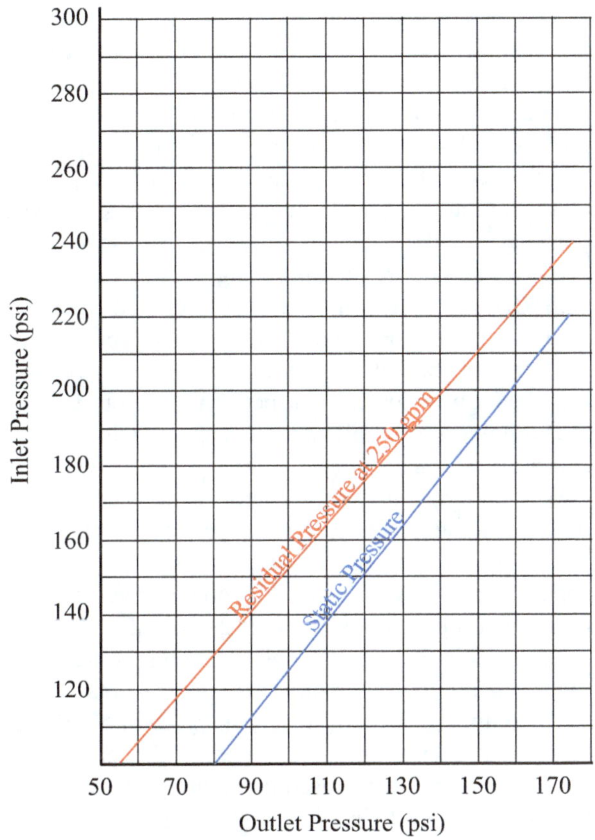

Fig. 8.2 Static and residual pressure reduction from a specific model direct acting pressure reducing valve

The devices that reduce both the static and residual pressure are called, "Pressure-Reducing Valves". These devices can be further subdivided into two categories: "Direct Acting" and "Pilot Operated." The direct acting pressure-reducing valves take a certain amount of pressure away from the water as it moves through the valve. This amount of pressure reduction is dependent on the design of the valve and will follow a linear reduction in pressure based on the inlet pressure with a lower pressure reduction at a low inlet pressure and a higher pressure reduction at higher inlet pressures. Figure 8.2 shows the performance of a specific direct acting pressure reducing valve. There are two situations shown in Fig. 8.2. The first, represented by the red line, is the pressure reduction that occurs when a flow of 250 gpm goes through the device. The second, represented by the blue line, is the pressure reduction that occurs at static pressure (no flow). Typically, a manufacturer will make several different models of direct acting pressure reducing valves and will provide two different figures to describe the performance; one for static pressure and one for residual pressure at some common flow. Figure 8.2 combined this information into a single figure for simplicity.

One of the advantages of the direct acting pressure reducing valve is that there are manufacturers that make these devices in a right angle arrangement with the horizontal piece so that it can be connected to the riser and the threaded outlet at a right angle to the vertical so that it can be pointed in whichever direction the design professional wants for the fire department (down or slightly rotated). This device looks exactly like a hose connection for a standpipe system. The pressure reduction feature is built into the center of the valve, so it looks and acts like a regular hose connection and no additional piping is necessary when using one of these devices.

Direct acting pressure reducing valves can be factory set or they can be adjustable. Once they are adjusted, they keep that adjustment and are not intended to be readjusted during fire conditions, so both types will be treated the same in this text. If firefighters had the right tool, they could theoretically adjust the device during a fire if they wanted different performance, but the design professional should give the firefighters a compliant system to start with, so they should not need to adjust the equipment during a fire.

The pilot operated valves automatically adjust to try and maintain a set pressure on the downstream side. If a high pressure is sensed on the incoming side, they take lots of pressure away. If a low pressure is sensed on the incoming side, they adjust and take less pressure away. Pilot operated valves utilize a sensing line that connects to the inlet (upstream) side of the device, allowing the pressure to push on a piece of the valve. When there is a high incoming pressure, the piece of the valve connected to the sensing line closes the valve down so that it creates more pressure loss. When the incoming pressure is lower, the piece of the valve connected to the sensing line relaxes and allows the valve to open up, creating less pressure loss.

Another term that is used in the fire protection industry to describe a device that reduces both the static and residual pressure is the “Pressure Control Valve”. Some people use this to refer to all pressure-reducing valves whether they are direct acting or pilot operated. Others only mean to refer to pilot operated devices when they use the term “Pressure Control Valve” because it does exert some sort of control over the situation, adjusting itself to the incoming water pressure conditions. NFPA 14 has chosen not to define the term and let people use it in whatever context they want. So, if you here the term “Pressure Control Valve” being used, you may wish to clarify whether the intent is to refer to a direct acting device or a pilot operated device.

The devices that reduce only the residual pressure are called, “Pressure Restricting Devices”. They devices are typically flat disks (sometimes called “Orifice Plates”) with a hole in the middle that is smaller than the full inside diameter of the pipe. These devices cause friction loss as the water flows through them, which reduces the downstream residual pressure. When no water is flowing in the system, the static pressure from the water supply gets through the hole, so there is no reduction in static pressure. If you had the room, you could make a pressure restricting device by adding a bunch of elbows and tees to a standpipe system between the standpipe and the hose connection that would perform the same function as the orifice plate.

Master Pressure Reducing Valve Assemblies

NFPA 14 limits the number of hose connections that can be served by a single pressure reducing valve to two because of the concern that if the pressure reducing valve fails in the open position, it will make controlling the hose difficult for multiple groups of firefighters. Section 7.2.4 of NFPA 24 states that whenever more than two hose connections are served by a single pressure reducing valve, there needs to be a backup pressure reducing valve that can take over in case the first pressure reducing valve fails in the open position. This provides an extra level of reliability for multiple groups of firefighters using hoses that get their water through the same pressure reducing valve.

Once the two pressure reducing valves are installed in a row (also referred to as being installed "in series") a number of other devices need to be installed with the pressure reducing valves including a bypass, pressure gauges, control valves, pressure switches and a relief valve. Figure 8.3 shows an arrangement of equipment that would comply with NFPA 14. The pressure gauges need to be installed upstream and downstream of each pressure reducing valve so that the fact that the valve is working can be verified. The bypass around each valve is there so that each valve can be isolated in case it needs to be repaired with the water still being able to get into the rest of the standpipe system. The control valves need to be installed in both the bypass and the main line in order to be able to isolate each of the control valves from the rest of the system. The pressure switches are there so that if the valve fails in the open position, the high pressure created will be sensed at a constantly attended location. Finally, if everything fails, there is a pressure relief valve that can release some of the pressure if both pressure reducing valves fail in the open position at the same time or if one of the valves fails while the bypass is open around the other. The

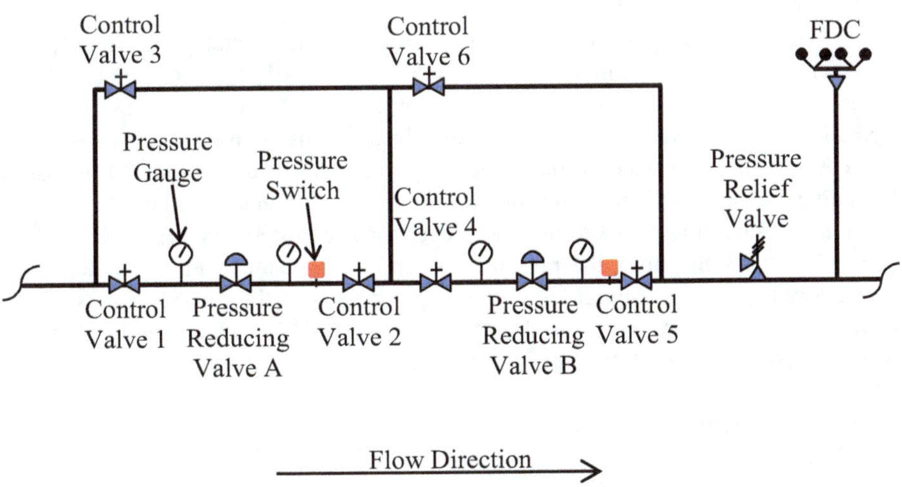

Fig. 8.3 Master pressure reducing valve assembly

entire assembly is often referred to as a "Master Pressure Reducing Valve Assembly" although NFPA 14 does not use this term.

With the master pressure reducing valve assembly, the second valve must be a pilot operated valve because you never know what pressure will be coming in from the first pressure reducing valve. If the first pressure reducing valve is operating correctly, the pressure at the second pressure reducing valve will already be reasonable, so there is no need to reduce it more. In this condition, the second pressure reducing valve can stay open and not reduce the pressure at all. But if the first pressure reducing valve fails, or if the water is coming through the bypass, then the second pressure reducing valve needs to close down and reduce the pressure. This will need to happen automatically, so the second pressure reducing valve needs to be a pilot operated valve.

The fire department connection (FDC) needs to be installed downstream of the whole master pressure reducing valve assembly so that if anything goes wrong, the fire department can still get water into the standpipe system. One of the possibilities is that the pressure reducing valves will fail in the closed position or something might get stuck in it that does not allow the water through. While this is a rare situation, the concerns are real enough that the FDC needs to be connected completely downstream of the assembly. This might take a great deal of piping if the master pressure reducing valve is somewhere high up in the building, so the location of the assembly and the FDC need to be taken into consideration when deciding if the master pressure reducing valve assembly is the best decision for any situation.

The relief valve that is installed at the downstream side of the assembly is an important safety device. Even though it should never be used, since it should only open if both of the pressure reducing valves fail in the open position at the same time, it will dump water out of the system, and therefore needs to have a drain nearby that can handle the entire flow from the system. Given that the relief valve can be a 6 in. or possible even an 8 in. pipe, a fully open relief valve at a high pressure can discharge a significant amount of water, so the drain needs to be sized to handle a great deal of flow.

The next few paragraphs will explain the use of the master pressure reducing valve assembly. Under normal conditions, Control Valves 1, 2, 4 and 5 are kept open while Control Valves 3 and 6 are kept closed. Under these conditions, Pressure Reducing Valve A will reduce the pressure in the system and Pressure Reducing Valve B will allow the water through without reducing the pressure much. If Pressure Reducing Valve fails in the open position, Pressure Reducing Valve B will close down and reduce the pressure to a safe level for firefighters needing water downstream. At the same time, the pressure switch downstream of Pressure Reducing Valve A will provide a signal letting people know about the failure. A person can then close Control Valves 1 and 2 and open Control Valve 3. This will allow the water to get to Pressure Reducing Valve B while Pressure Reducing Valve A can be taken apart to fix it.

If Pressure Reducing Valve A is working and Pressure Reducing Valve B fails, then Control Valves 4 and 5 can be closed while Control Valve 6 is opened. This allows the water to flow through Pressure Reducing Valve A and then through the bypass around Pressure Reducing Valve B. With the valve isolated, Pressure Reducing Valve B can be repaired and returned to service.

Drain Risers

There is a small chance that a pressure regulating device might fail. There is also a small chance that a pressure regulating device might be set incorrectly or the wrong device might be installed. In order to deal with these possibilities, NFPA 25 requires periodic testing of pressure regulating devices. In order to test the device, a significant flow needs to be created (about 250 gpm for Class I and Class III systems and about 100 gpm for Class II systems). Since the pressure regulating devices might be at many different locations in the building, there needs to be a plan for how the flow from the test will be handled. In the 1990 and earlier editions of NFPA 14, there was no requirement for a fixed method to handle the flow to be built into the system. After the One Meridian Plaza fire in Philadelphia in 1991, where a pressure reducing valve was partially responsible for the lack of water pressure on the fire floor, NFPA 14 was revised to contain a requirement for a drain riser.

In the 2016 edition of NFPA 14, Section 7.11 is the one that requires a drain riser when any type of pressure regulating device is installed. This might be an orifice plate that just reduces the residual pressure or a pressure reducing valve that drops both the static and residual pressures. The drain riser size will be based on the size of the device that is being tested. For 2.5 in. hose connections, the drain riser will need to be a nominal 3 in. pipe. For 1.5 in. hose connections, the drain riser can be reduced to nominal 2 in. pipe. There are some occasions, such as with the master pressure reducing valve assembly, where the device that you are testing is larger than 2.5 in. in size, in those conditions, the drain riser needs to be at least as large as the discharge outlet of the pressure regulating device. For example, if you were designing a standpipe systems with a 6 in. master pressure reducing valve assembly, you would be required to have at least a 6 in. drain riser.

The drain riser needs to terminate outside the building in a location where it is safe to flow water, or into a space within the building that can handle the discharge of the water such as a sewer connection, sump or holding tank where the water can be recycled. Each of these options has advantages and disadvantages. A connection to a sewer would have to have an air gap so that sewage does not get up into the drain riser. A sump would have to be large enough to hold the water for a short time until the sump pump can pump it away. The holding tank would need to be large enough to hold the whole flow from the system during the test, which could be quite large if there are a large number of devices that need to be tested on the same day.

The inlet connections on the drain riser need to be at every other floor in the portion of the building where there are pressure regulating devices. In the upper portions of the building, there may be no pressure regulating devices because the elevation loss has brought the water pressure down low enough that safety is no longer a concern and the pressure is below the thresholds discussed earlier. If this is the case, the drain riser does not need to extend to these upper portions of the building. Since the inlets to the drain riser are on every other floor, the person doing the testing will need about 30 ft of hose to make sure that they can reach one of the inlets. If they are testing one of the pressure regulating devices on the same level as the inlet to the drain, then only a short piece of hose will be needed (a few feet depending on the arrangement of hose connection and drain riser). But if the hose connection is at one of the floor landings that does not have an inlet for the drain, then the most the tester will have to travel to get to a drain inlet is one full flight of stairs, which is about 20 ft depending on the number of steps in the stairwell and the size of the landings. Having a few more feet of hose to account for getting to the hose connection at one end and the inlet connection on the other end is a good idea, so a test kit that includes at least 30 ft of hose would be a good idea.

The hose used in the testing as well as the drain inlets need to have the same threads as the ones used by the local fire department. All fire departments do not use the same hose threads, so make sure that you figure out which thread the local fire department uses. While the local fire department will not be performing the testing on the pressure regulating devices, it is still a good idea to use their threads for the drain inlets because their threads need to be used at the hose connection, and you don't want to have to buy your hose with different threads at each end.

See Chap. 5 of this text for more information on the installation of the drain riser including information on making sure that the drain riser has some sort of opening at the top so that air can get in during the test, allowing the water to get out of the bottom easier.

Examples Needing Pressure Regulating Devices

A total of three examples will be explored here. The first will be a Class I standpipe system using a public water main as the water supply. The second will be a Class I standpipe system using a pump and tank as the water supply. The third example will be a Class II system that needs a pressure restricting device at the hose connections. More examples will be included in Chap. 9 of this text regarding fire pumps.

First Example

In Chap. 6 of this text, a simple Class I standpipe system with two standpipes (shown in Fig. 6.2) was hydraulically calculated to need a flow demand of 750 gpm at a pressure demand of 187.7 psi at the water supply. See Fig. 6.12 for a full accounting of the hydraulic calculations for this system. In order for the standpipe system to work properly, the static pressure from this water supply would need to be higher than 187.7 psi. Although this would be very unusual, let us begin our discussion with such an example because it is the simplest, and then build to the more complicated situations involving fire pumps.

Assume that the water supply has a static pressure of 200 psi and a residual pressure of 185 psi at a flow of 825 gpm as shown in the purple line on Fig. 8.4. Also assume that the static and residual pressure reported here have already been adjusted

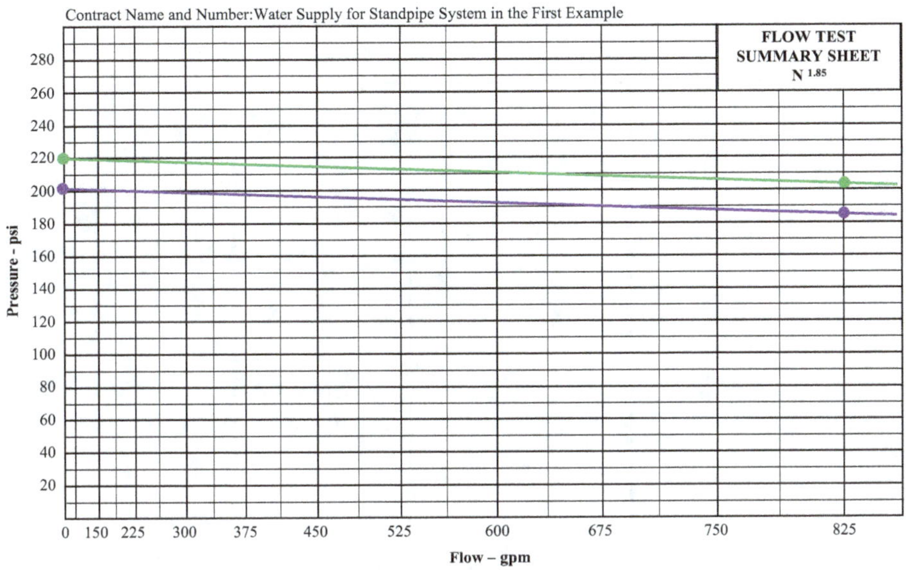

Fig. 8.4 Water supply for standpipe system in first example

down to take into account the reasonable daily and seasonal fluctuations in the water supply. This water supply will barely meet the demand of 187.7 psi at 750 gpm, which is okay since it has already been adjusted down for reasonable fluctuations. However, if we choose this water supply, the static pressure in the system will be more than 187.7 psi. During the low point in the fluctuations, it might be as low as 200 psi, but it is also possible that a fire could occur during the reasonable high pressure times as well. If we talk to the water utility and they agree that the reasonable high static pressure is 220 psi, then we need to start our pressure control evaluation from there, which is represented by the dark green line on Fig. 8.4.

In order to determine how many of the hose connections need pressure reducing valves, the analysis needs to start at the top of the standpipe system. If the static pressure is 220 psi at the water supply, then the static pressure at the top of the system will be 200.5 psi at the same time because the elevation difference between the hose connection at the top of the standpipe and the fire hydrant where the water supply was measured is only 45 ft. With an elevation change of 45 ft, the water pressure will lose 19.5 psi (45 × 0.433 = 19.5). This makes the static pressure at the top of the standpipe 200.5 psi (220 – 19.5 = 200.5). With the static pressure greater than 175 psi at the top outlet (and therefore at all of the outlets), the standpipe system will need pressure reducing valves at all of the hose connections. See Fig. 8.5 for a graphical representation of the static pressure situation in the First Example. It is worth noting that the static pressure at Node C and at Node A will be the same because without any water flowing, the pressure at the top of each independent standpipe is only dependent on the static pressure from the water supply and the difference in elevation from the water supply to the top of the standpipe. Since

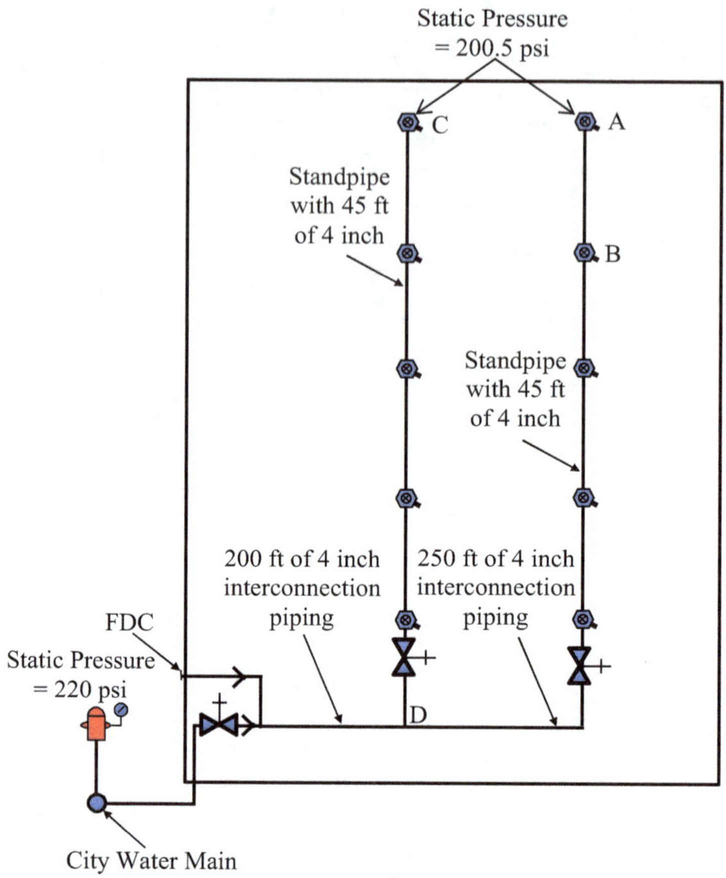

Fig. 8.5 Static pressure situation for first example using higher water supply

Nodes A and C are at the same elevation, they will have the same static pressure at the inlet to the hose outlet.

The next step is to figure out what model pressure reducing valve should be used. Assume for the moment that all of the standpipe hose connections are going to use the same pressure reducing valves and that the valve will be the one with the pressure reductions shown in Fig. 8.2. Under the static pressure condition, this valve will work just fine. For the highest valve, with an inlet pressure of 200.5 psi, the outlet pressure will be about 160 psi, which keeps the valve under the maximum pressure of 175 psi. At the lowest hose connection on the standpipe, where the static pressure is just under 220 psi at the inlet of the hose connection (because the hose connections are a few feet above the level of the fire hydrant where the water supply was measured), the outlet pressure will be just under 175 psi under static conditions, which is just barely acceptable under NFPA 14.

So far, this pressure reducing valve looks like it is working fairly well. But the next step is to evaluate it under flowing conditions. For this case, the worse-case flowing condition is at Node A, at the most remote hose connection from the water supply. We know from the hydraulic calculations that when a flow of 750 gpm is moving through the system with 500 gpm going up into the most remote standpipe the pressure loss will be 58.9 psi from the water supply to Node A as shown in Fig. 6.12 (note that the friction loss for this step should not be taken to Node O because we need the pressure at the inlet to the valve in order to select the correct pressure reducing valve, not the outlet). At the same time, there is still 19.5 psi lost to elevation. The total pressure loss between the water supply and Node A becomes 78.4 psi (58.9 + 19.5 = 78.4). Using the low water supply curve (since this is the worse-case for this analysis) with a residual pressure at the water supply of 189 psi at 750 gpm, the residual pressure at Node A will be 110.6 psi (189 − 78.4 = 110.6). The residual pressure at other nodes can be calculated as well using the information on friction loss and elevation loss from Fig. 6.12. The residual pressures at each node are shown in Fig. 8.6.

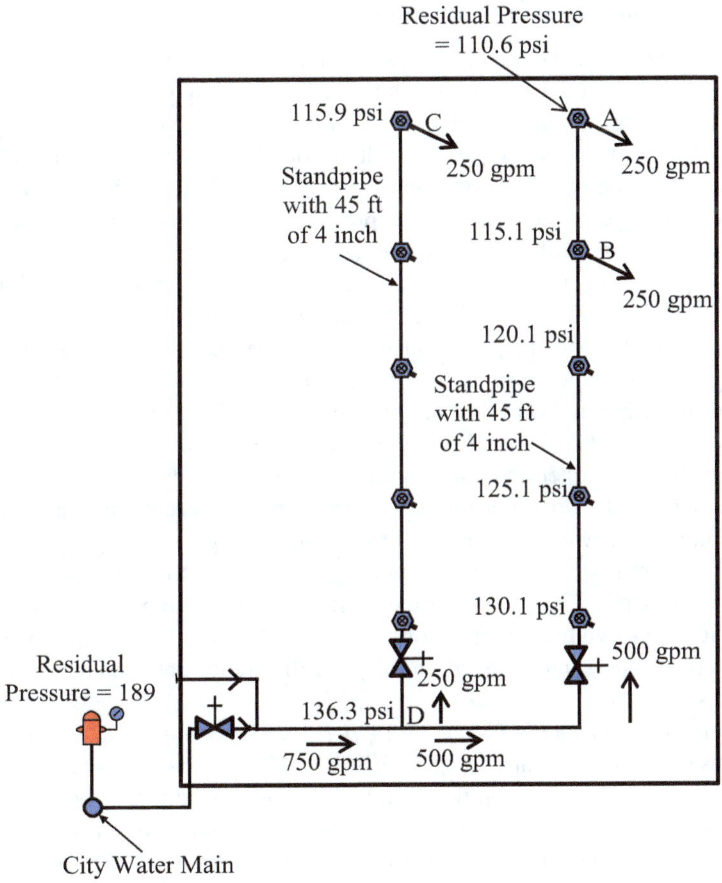

Fig. 8.6 Residual pressure at hose connections for first example using lower water supply

Looking at the information in Fig. 8.6, the most interesting residual pressures occur at the inlets to the valves at Node A and Node B (the top two hose connections at which water is flowing from the system). At Node B, the inlet pressure will be 115.1 psi. At Node A, the inlet pressure will be 110.6 psi. These pressures are low enough that they don't need a pressure reducing valve, but one has to be installed at this point because of the static pressure, so we need to evaluate the effect of such a valve. In this case, the valve represented by Fig. 6.2 would reduce the pressure at Node A from 110.6 psi to 65 psi (using the red residual pressure curve) and the pressure at Node B would be reduced from 115.1 psi to about 68 psi. Both of these pressures would be unacceptable to NFPA 14 because they fall below 100 psi.

The design professional might be able to deal with the problem by selecting a different pressure reducing valve, but it is going to be difficult to find one because in the static conditions, this valve is barely reducing the pressure in an acceptable way. A pressure reducing valve that reduces the pressure less for the residual condition is also going to reduce the pressure less for the static condition, which will mean that it does not work on the high side of the situation. So, it does not appear that a direct acting pressure reducing valve is going to work in this situation. The design professional has a few options including the following:

1. *A Pilot Operated PRV for Each Hose Connection*: Use a pilot operated pressure reducing valve on the system instead of a direct acting pressure reducing valve. During the high pressure static conditions, the valve will reduce the pressure more. During the low pressure residual conditions, the valve will reduce the pressure less, allowing the minimum pressure of 100 psi to be met. The advantage of this type of valve is that the pressure demands of NFPA 14 can be met. The disadvantage is that the manufacturers of this valve do not make one in the right angle arrangement that makes it easy to use for a hose connection, so the design professional will need to connect it to the standpipe and then add a separate hose connection. This makes for a slightly more bulky installation as shown in Fig. 8.7. The pipe to the hose connection from the riser can be 2.5 in. and the valve can be 2.5 in. because it only serves a single hose connection.
2. *A Pilot Operated PRV for Every Two Hose Connections*: A cost savings can be made if the pilot operated valves can be set to work for two hose connections. Since NFGPA 14 does not require a master pressure reducing valve assembly until more than two hose connections are served by a single pressure reducing valve, it is permissible for a single pilot operated valve to serve two hose connections as shown in Fig. 8.8. The pipe in this case needs to be a minimum or 4 in. since it serves two hose connections and the PRV needs to be 4 in. as well, so the exact cost savings may depend on the price difference between the different size devices.
3. *A Master PRV Assembly Using Pilot Operating Valves*: As long as you are going to consider 4 in. pilot operating PRV's, consider putting them on the incoming water supply rather than on the individual hose connections. The resulting system would look something like Fig. 8.9. The pilot operating valves would need to be set at 177 psi so that the static pressure at the first outlet (5 ft above the

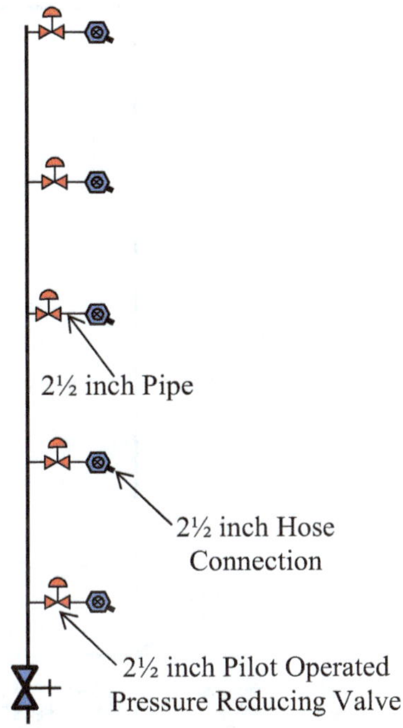

Fig. 8.7 Standpipe with pilot operated pressure reducing valves

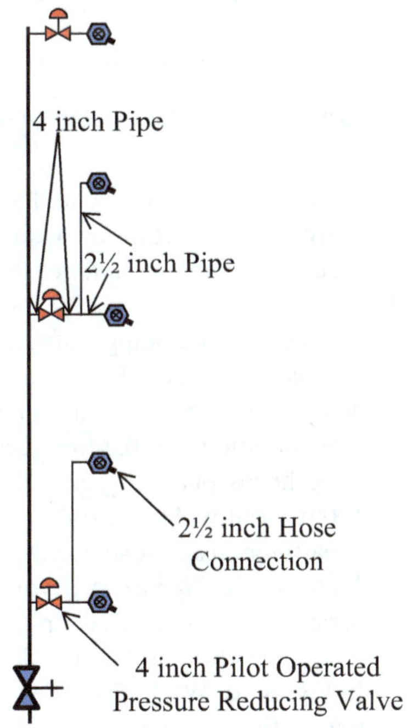

Fig. 8.8 Standpipe with pilot operated pressure reducing valves serving two hose connections

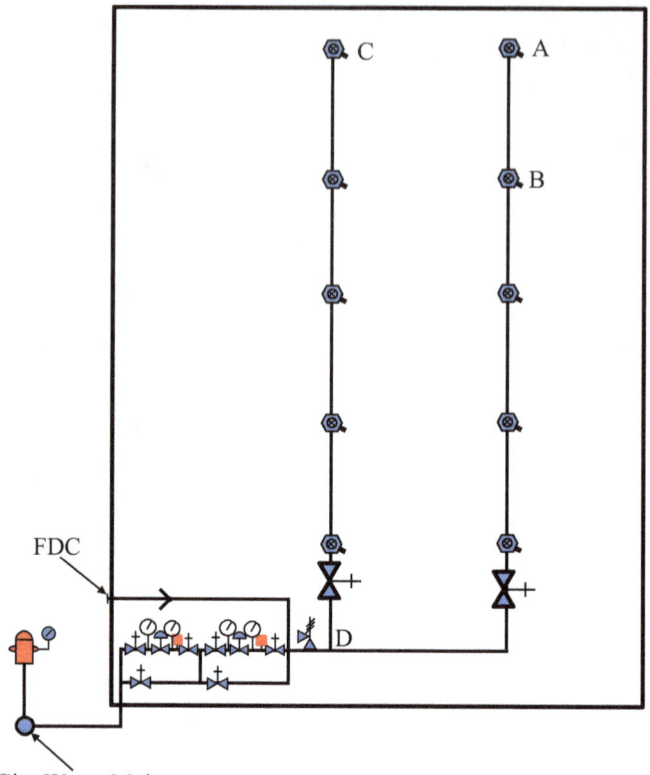

Fig. 8.9 Master PRV for standpipe system

elevation of the PRV) would be reduced below 175 psi. The friction loss from the valve and the elevation would not exceed 77 psi, so this would produce an acceptable pressure at all of the outlets of the hose connections.

4. *Ask for a Variance from NFPA 14*: Another option would be to use the direct acting pressure reducing valve as discussed above and ask the authority having jurisdiction to allow the use of these particular valves recognizing that under many conditions, the minimum pressure of 100 psi would be achieved. In the rare condition that the fire occurs in the highest point of the building and needs three hoses put into action, the pressure at each outlet might be lower than 100 psi, but will still produce usable water for many types of nozzles. If the fire department uses those nozzles, they might agree.
5. *Increase the Pipe Size*: If you go back to the original design with direct acting pressure reducing valves and change the pipe size to 6 in. for all of the risers and interconnection piping, the demand goes down to 148 psi at 750 gpm as shown in Fig. 8.10. While this does not change the water supply issue, it does reduce the friction loss between nodes, which will increase the pressure at each node, which will made the residual pressure higher at the inlet of each of the direct

Node 1	Elev 1 (ft)	K-Factor	Flow added–this step (*q*)	Nominal ID	Fittings–quantity and length	L ft	C Factor	total (P_t)	Notes
						F ft	P_f per foot (psi)	elev (P_e)	
Node 2	Elev 2 (ft)		Total flow (*Q*)	Actual ID		T ft		frict (P_f)	
O	45		250	2½	Right angle valve (31) Tee (12)	1	120	100	
						43	0.215	0	
A	45		250	2.469		44		9.5	
A	45		0	6	None	10	120	109.5	$P_e = 0.433 \times 10$
						0	0.003	4.3	$P_e = 4.3$
B	35		250	6.065		10		0.0	
B	35		250	6	Elbow	285	120	113.8	$P_e = 0.433 \times 35$
						14	0.010	15.1	$P_e = 15.1$
D	0		500	6.065		299		3.0	
D	0		250	6	Check Valve (32) Gate Valve (3) Elbow (14)	200	120	131.9	
						49	0.021	0	
T	0		750	6.065		249		5.2	
T	0		0	6	Elbow (20.1)	75	140	137.1	Equivalent Length Adjustments
					Tee (43.1)	63.2	0.014	0	C-factor = 1.33
U	0		750	6.16		138.2		1.9	$(6.16/6.065)^{4.87} = 1.08$
U								148.0	

Fig. 8.10 Hydraulic calculation for system with 6 in. pipe

acting pressure reducing valves. Figure 8.11 shows the residual pressure at each of the nodes in the system when the most remote three hose connections are calculated in accordance with NFPA 14. Once again, the most interesting residual pressures are at Node A (159.4 psi) and Node B (163.8 psi). This time, the pressure reducing valves do much better. At Node B, the valve represented by Fig. 6.2 would reduce the pressure from 163.8 psi to about 109 psi. At Node A, the pressure reducing valve would reduce the pressure from 159.4 psi to about 107 psi. Since both of these outlet pressure exceed 100 psi, this option meets the rules of NFPA 14.

Given the five options discussed above, the best one is probably going to be Option 5. The increased pipe size to 6 in. is going to increase the cost of the installation a small amount, but the other options are each going to cost something as well. Option 5 is the least objectionable because it will use less space at each of the hose connections, will not require pilot operated valves, and will not require any special permission from the fire department.

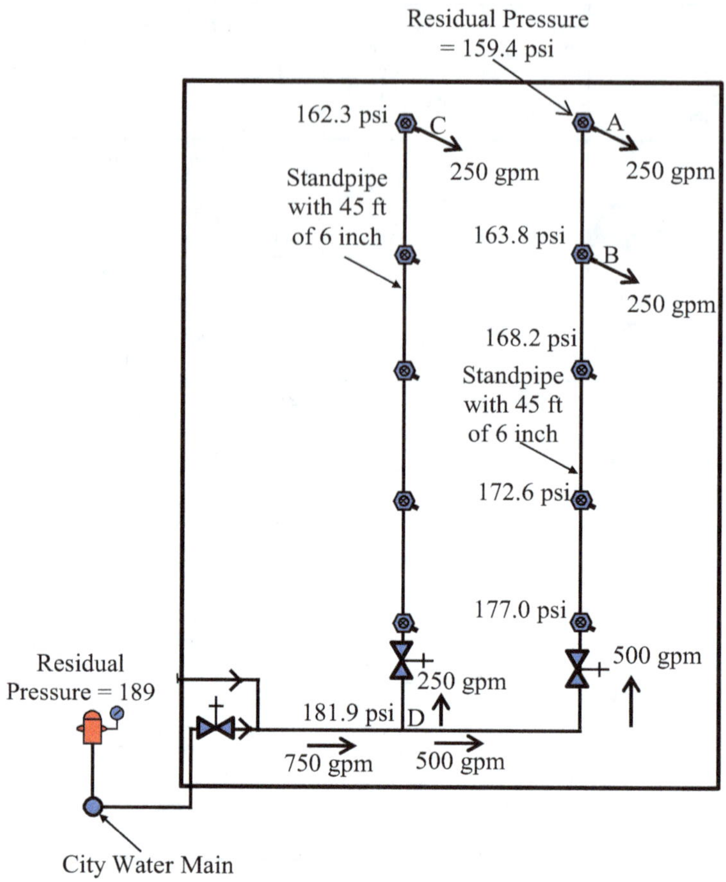

Fig. 8.11 Residual pressure at nodes in first example with 6 in. pipe

Second Example

For the second example, we will consider a Class I standpipe system with a fire pump and suction tank as the water supply as shown in Fig. 8.12. There will be three standpipes on this system with eight hose connections on each standpipe, so the total flow demand will be 1000 gpm. The tank will be sized for 30 min of water, so there will need to be 30,000 gallons above the anti-vortex plate inside the tank. If the tank has an 18 ft diameter, the amount of water in the tank will be 1903.6 gpm per ft of height. With 16 ft of water above the anti-vortex plate, the tank will have 30,458 gallons, which meets the 30 min demand of NFPA 14.

The standpipe system in our second example will have a demand of 1000 gpm at 150.4 psi, calculated to the discharge flange of the pump (Node DF) as shown in Fig. 8.13. The next step will be to take a look at the net pressure curve for the fire

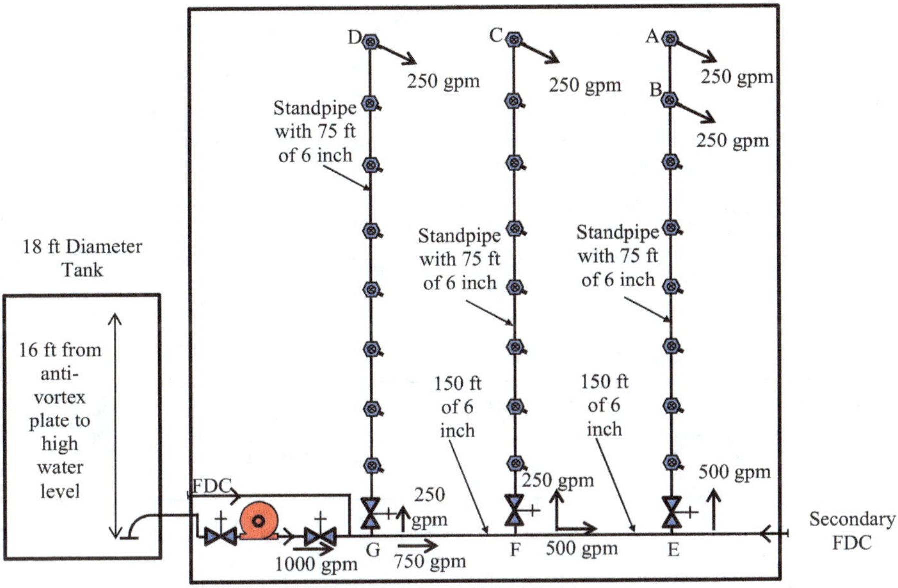

Fig. 8.12 Layout of standpipe system for second example

pump that is in the pump room, which is shown in Fig. 8.14. The pump is rated at 1000 gpm and 160 psi. The churn pressure is 190 psi and the net pressure at maximum flow is 110 psi.

When the water in the tank is near the bottom (towards the end of a fire), the elevation pressure created by the water level will be near zero. The friction loss in the suction pipe is allowed to be 3 psi at worst, so the worse-case low suction pressure will be −3 psi at a flow of 1000 gpm. The net pressure added by the pump at 1000 gpm will be 160 psi. So, the discharge pressure will be 157 psi, which exceeds the demand pressure, so this pump will be acceptable.

The next step will be to examine the maximum pressure created by the pump, which will happen in the churn (no flow) condition when the tank is full. The elevation pressure when the tank is full will be 6.9 psi (16 × 0.433 = 6.9) because the water level in the tank will be 16 ft above the anti-vortex plate in the tank. Since the net pressure of the pump at churn is 190 psi, the discharge pressure will be 196.9 psi (190 + 6.9 = 196.9). With the lowest hose connection in the system 5 ft above the level of the pump, the static pressure at this hose connection will be 194.7 psi (196 − (0.433 × 5) = 194.7), which is greater than 175 psi, so the bottom row of hose connections will need pressure reducing valves.

The question is whether all of the hose connections will need pressure reducing valves. The top row of connections is 75 ft over the level of the pump. The elevation pressure loss will be 32.5 psi (0.433 × 75 = 32.5). So, at the top of the system, the pressure is only going to be 164.4 psi (196.9 − 32.5 = 164.4), which is less than the maximum pressure allowed (175 psi), so the hose connections at the top of the system are not going to need pressure reducing valves.

Node 1	Elev 1 (ft)	K-Factor	Flow added–this step (*q*)	Nominal ID	Fittings–quantity and length	L ft	C Factor	total (P_t)	Notes
						F ft	P_f per foot (psi)	elev (P_e)	
Node 2	Elev 2 (ft)		Total flow (*Q*)	Actual ID		T ft		frict (P_f)	
O	75		250	2½	Right angle valve (31)	1	120	100	
						43	0.215	0	
A	75		250	2.469	Tee (12)	44		9.5	
A	75		0	6	None	10	120	109.5	$P_e = 0.433 \times 10$
						0	0.003	4.3	$P_e = 4.3$
B	65		250	6.065		10		0.0	
B	65		250	6	Gate Valve & Elbow	225	120	113.8	$P_e = 0.433 \times 65$
						17	0.010	28.1	$P_e = 28.1$
F	0		500	6.065		242		2.4	
F	0		250	6		150	120	144.3	
							0.021	0	
G	0		750	6.065		150		3.1	
G	0		250	6	Check Valve (32) Gate Valve (3)	50	120	147.4	
						35	0.035	0	
DF	0		1000	6.065		85		3.0	
DF								150.4	

Fig. 8.13 Hydraulic calculation for second example standpipe system

The next question that needs to be answered is how many of the hose connections will need pressure reducing valves. We know that the elevation pressure will be 4.3 psi between hose connections on each level (assuming 10 ft per story). This would make the static pressure at the second level of hose connections from the top 168.7 psi, which would not need pressure reducing valves.

The static pressure for the third row down from the top will be 173.0 psi (168.7 + 4.3 = 173). This row will not need pressure reducing valves, but the next row down will. At 177.3 psi, the fourth row down from the top and all of the hose connections below it will require pressure reducing valves. Figure 8.15 shows the hose connections that will need to have pressure reducing valves in red and the ones that do not need pressure reducing valves in blue. The maximum static pressure is shown in Fig. 8.15 for each level of hose connections in the system. Note that the pressure difference is 4.3 psi per floor except at the bottom where it is 4.5 psi between the static pressures. This is due to the rounding errors that occurred from

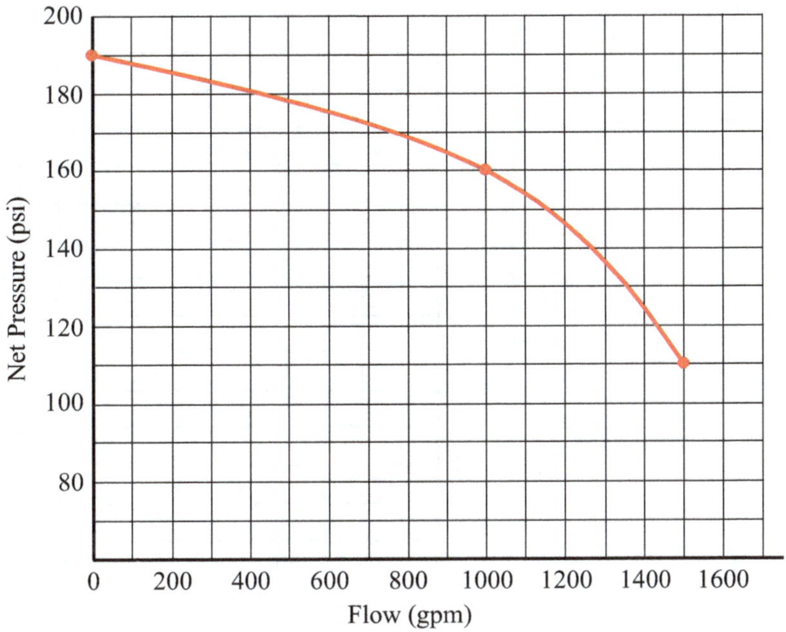

Fig. 8.14 Net Pressure curve for fire pump in second example

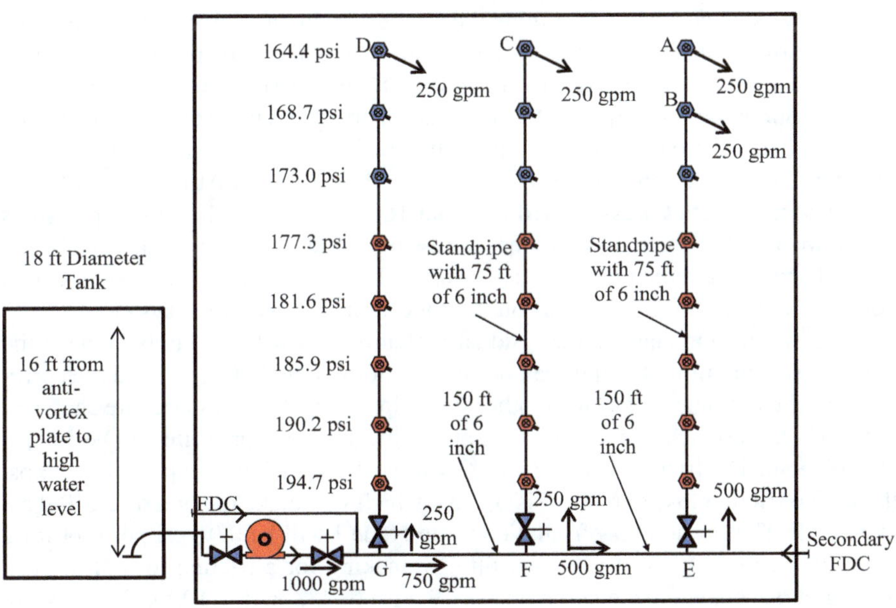

Fig. 8.15 Second example standpipe system with hose connections having pressure reducing valves in red, hose connections without pressure reducing valves in blue and the static pressure at each level of hose connections

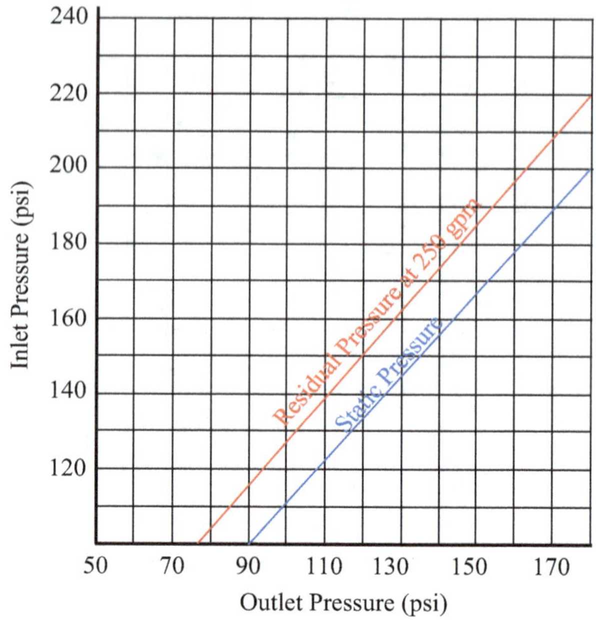

Fig. 8.16 Static and residual pressure reduction for a specific model pressure reducing valve being used in the second example

rounding 4.33 to 4.3 and counting down from the top while the bottom number was calculated straight from the pump.

The next step will be to determine the pressure reducing valve that will be used in the 15 hose connections in this system. Starting with the static pressure, the direct acting pressure reducing valve shown in Fig. 8.16 will work. The two extreme static inlet pressures are 177.3 psi at the topmost of the pressure reducing valves and 194.7 psi at the bottom. Looking at the static pressure information on Fig. 8.16 (the blue line), the outlet pressure at the bottom will be 175 psi. At the topmost hose connection, the outlet pressure will be about 160 psi. Since both of these pressures are between 100 psi and 175 psi, these pressure reducing valves are acceptable so far, and a pressure reducing valve that reduces the pressure to a lesser extent would not be acceptable since this one is on the edge at the bottom of the system.

Next, we need to analyze the residual pressure. This will need to be done at the worse-case condition of using the hose connections at Node H and I and the corresponding hose connections on the other standpipes. The inlet pressures are shown in Fig. 8.17. At the discharge flange of the pump, the discharge pressure will be 157 psi at 1000 gpm. The friction loss to Node G will be 3 psi, making the pressure 154 psi there. The friction loss to Node F is 3.1 psi, which will make the pressure 150.9 psi at Node F. The friction loss from Node F to Node I will be 0.01 psi per foot for a total distance of 212 ft of pipe and fittings, making for a friction loss of 2.1 psi. Taking into account the elevation loss of 19.5 psi ($0.433 \times 45 = 19.5$). The pressure at Node I will be $150.9 - 2.1 - 19.5 = 129.3$ psi.

Looking at the inlet residual pressure of 129.3 psi, using the red line on Fig. 8.16, the valve will reduce the pressure to 102 psi. All of the other hose connections in the

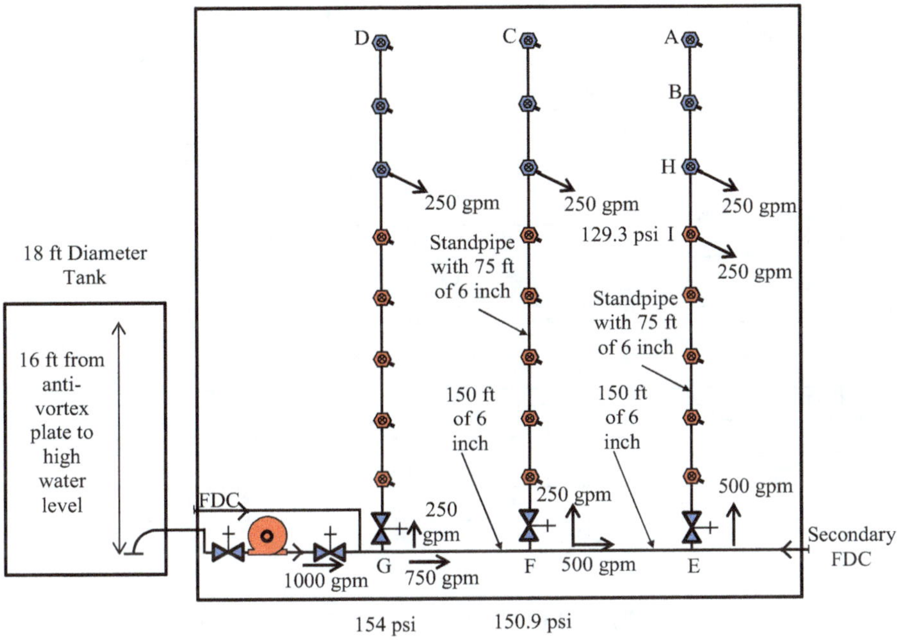

Fig. 8.17 Residual pressures at inlets for pressure reducing valves in second example

system with pressure reducing valves will act at a slightly higher pressure, so the pressure reducing valve represented by the pressure losses in Fig. 8.16 will be acceptable.

Third Example

The third example will be for a Class II system with the most remote hose station 400 ft from the water supply. Assume that the pipe from the water supply will be 2.5 in. Schedule 40 (inside diameter of 2.469 in.). If you assume that there will also be a bunch of elbows and tees to get the pipe to the remote location, you might add another 50 ft for these fittings. The friction loss of 100 gpm through 2 in. pipe is 0.039 psi/ft. Over a total of 450 ft of pipe and fittings, this would be a total friction loss of 17.5 psi of friction loss ($0.039 \times 450 = 17.5$). If you assume that the most remote hose connection is 50 ft above the water supply, you will also lose 21.7 psi ($50 \times 0.433 = 21.7$). The minimum pressure at the hose station is 65 psi, so the pressure demand at the water supply will be 104.2 psi ($65 + 17.5 + 21.7 = 104.2$).

The water supply for this system has a static pressure of 110 psi and a residual pressure of 60 psi at 500 gpm adjusted to the worse-case low pressure that is reasonable given daily and seasonal fluctuations. Speaking with the water utility, the reasonable high pressure that might be available is 130 psi static and 80 psi at

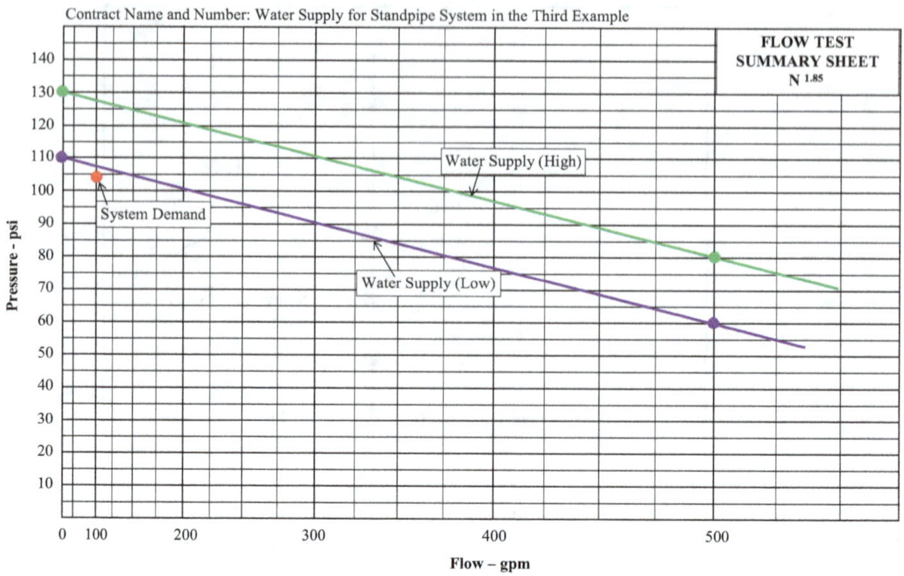

Fig. 8.18 High and low water supply given reasonable daily and seasonal fluctuations

500 gpm. The water supply is shown in Fig. 8.18 with the reasonable low pressure (purple curve) and the reasonable high pressure (green curve). This water supply will work with this system, but with a pressure greater than 100 psi, a pressure restricting device will be needed at some, if not all of the hose connections to bring the residual pressure below 100 psi with a flow of 100 gpm.

Pressure restricting devices have to be selected using a graph from the manufacturer that relates the inlet pressure to the outlet. Typically, the manufacturers provide a setting for the valve that gets the outlet to a pressure of 80 psi at a flow of 100 gpm. The device is built into the right angle hose connection so that you don't even realize that it is there unless you really know the construction of the valve. A pin holds the valve setting in place. If the fire department shows up and really needs more pressure, they can pull the pin and open up the valve completely. Figure 8.19 shows a graph that allows the user to select the correct setting for their situation.

The worse-case residual pressure at 100 gpm from the water supply will be 127 psi. With a friction loss of 17.5 and elevation loss of 21.7 to the most remote hose connection, the residual pressure there will be 87.8 psi ($127 - 17.5 - 21.7 = 87.8$), which would not require a pressure restricting device. If the closest outlet was only 50 ft from the water supply and on the same elevation, the friction loss would be 2 psi ($0.039 \times 50 = 2$) leaving a residual pressure of 125 psi ($127 - 2 = 125$). Looking at Fig. 8.19, the graph shows that the selection should be either setting seven or setting eight. Either would be acceptable, since the outlet pressure would be a little over or a little under 80 psi. Either way it would be between the acceptable values of 65 psi and 100 psi.

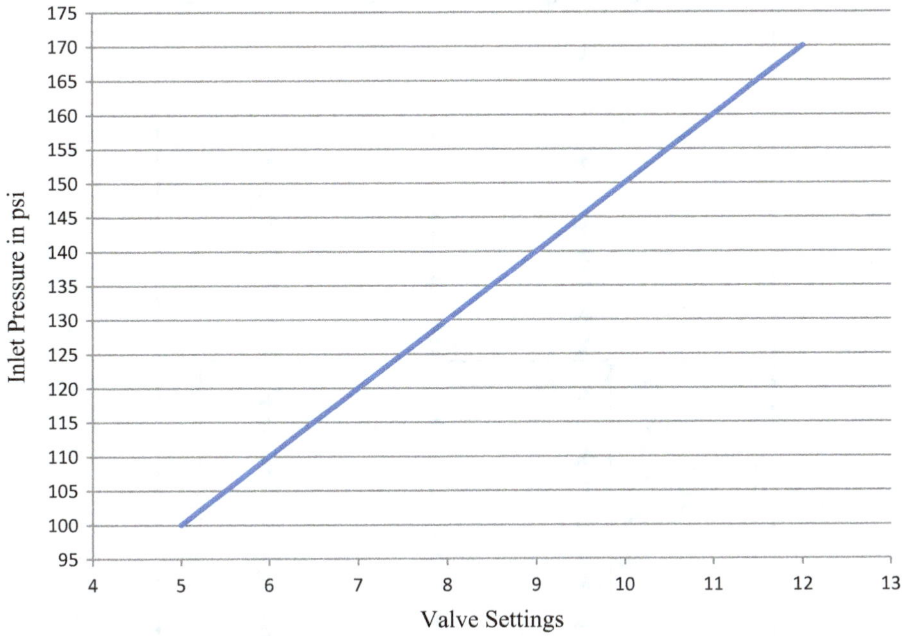

Fig. 8.19 Settings for pressure restricting device in 1.5 in. hose connection to insure 80 psi at 100 gpm at the outlet of the valve

As you get farther from the closest hose station, the residual pressure drops, so you would need to calculate the friction loss and elevation loss to other hose stations. For example, at one level up from the closest hose station, there might be an extra 10 ft of pipe and an extra 10 ft of elevation loss. This would make the residual pressure 120.3 psi, which would still need a pressure restricting device set at seven. The rest of the calculations would proceed using the same concepts.

Vertical Zones

There is another option to controlling the pressure rather than using pressure regulating devices. The excess pressure within a system is created when the standpipe system needs to overcome the pressure losses due to friction and elevation. The greater these pressure losses are, the higher the pressure demand, so a good way to manage the pressure would be to minimize the friction loss and elevation loss. In a large building, one way to do this would be to break the system into multiple zones so that each standpipe system within each zone is just like its own system. Figure 8.20 shows a high-rise building with two vertical zones. The low zone goes from the first floor to the 10th floor and the high zone goes from the 11th floor to the 20th floor.

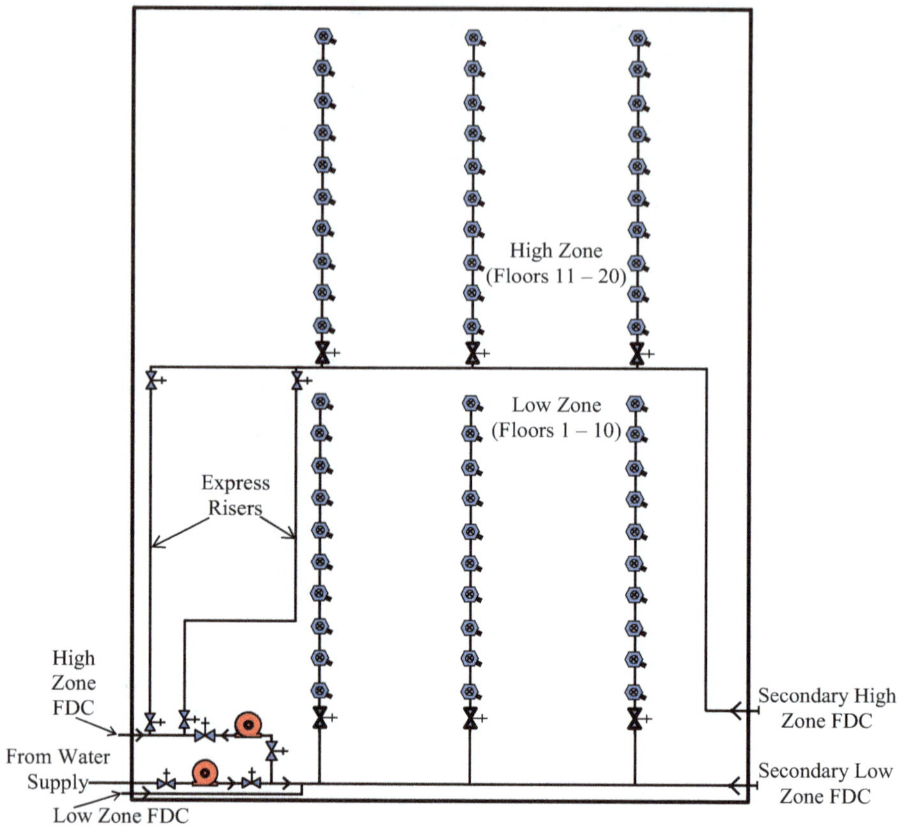

Fig. 8.20 High rise building with two vertical zones

In Fig. 8.20, the water comes from the water supply to the first pump. Once the water gets through the first pump, there are two places that it can go: into the low zone of the fire protection system or into the second pump. If the water goes into the second pump, it gets an additional boost of pressure and then goes up to the high zone. The pipes that carry water from the pump to the high zone are called "Express Risers" if they do not have hose connections on them. NFPA 14 allows the pressure in the express riser to exceed 350 psi as long as the pipe and fittings are rated to handle whatever pressure is created. In reality, a system in a 20 ft building with two vertical zones is not going to have a pressure greater than 350 psi in the express risers, but the concept would be the same for a much taller building with two or more vertical zones.

There are two express risers shown in Fig. 8.20 because NFPA 14 requires two separate pipes to get the water to the standpipes in the high zone (see section 7.9.2 of NFPA 14). The two separate pipes have to have control valves at each end so that they can be isolated from each other in case one of them fails. Individually, they need to be able to carry the flow for the upper zone, so you do not downsize these

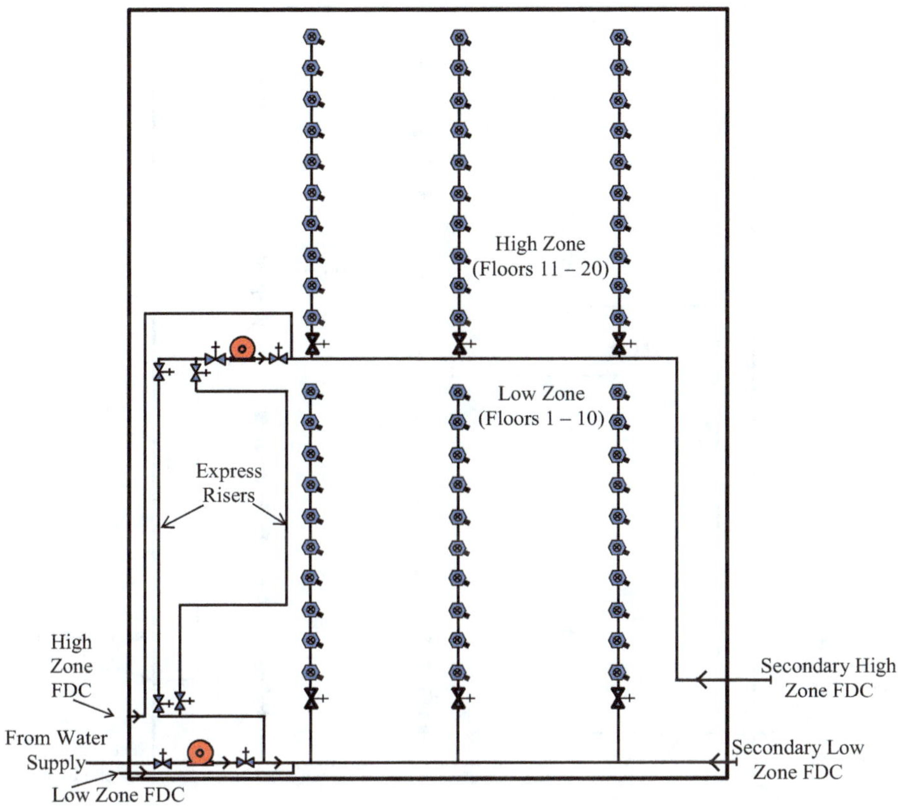

Fig. 8.21 Two zone standpipe systems with pumps in series in different pump rooms on different levels of the building

pipes to take advantage of the extra flow carrying ability of the extra pipe. In an actual installation, the two pipes would not be so close together in the building, but for the purposes of this sketch, it helped to show the two express risers close to each other.

The two pumps in Fig. 8.20 are in the same pump room. This can be advantageous for the person monitoring the pump(s) during a fire in the high zone, but can cause a high pressure in the express riser. Another option that is permitted by both NFPA 20 and NFPA 14 is to put the pumps on different floors in separate pump rooms as shown in Fig. 8.21. This design takes a great deal more thought and preparation. There would need to be some communication between the two pump rooms and the method of getting to the high zone pump room would need to be protected so that a person going to the pump room during a fire in the high zone is not exposed to that fire. For more information on putting pumps in series, see Chap. 9 of this text.

A third option would be to get rid of the express risers and use the standpipes from the low zone to feed the pump for the high zone. This option is shown in

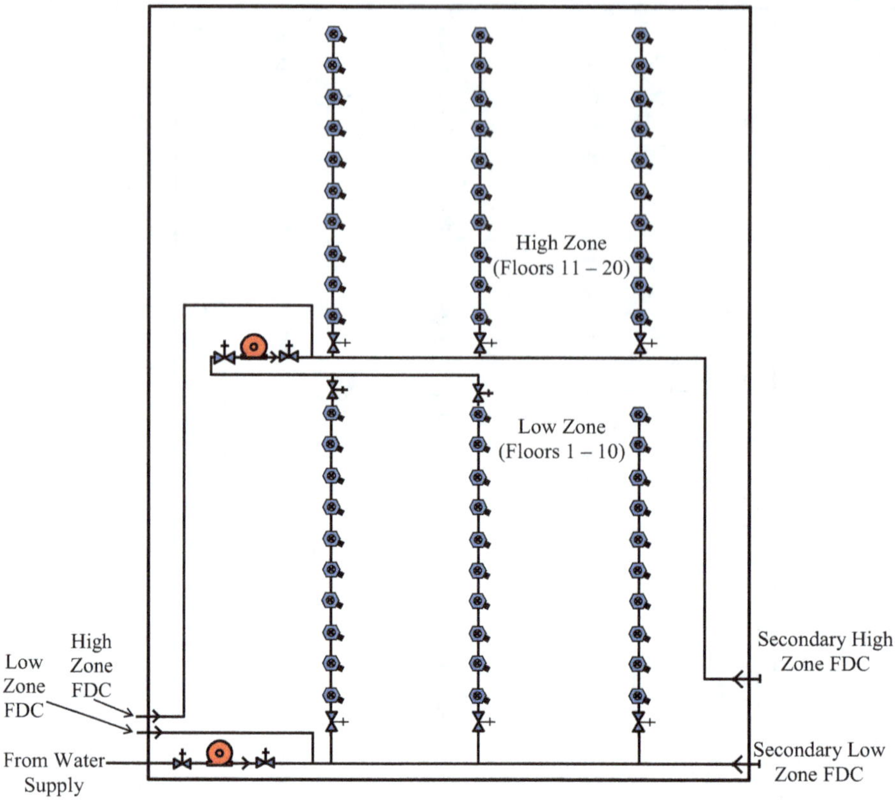

Fig. 8.22 High rise building with two vertical zones and the low zone standpipes service as the water supply pipes for the high zone

Fig. 8.22. NFPA 14 still requires two pipe to feed the high zone system, so two standpipes are connected to the suction for the high zone pump and control valves need to be installed at the top of each of these standpipes in order to be able to isolate them if there is a problem and one of them needs to be taken out of service. In this case, the two standpipes that are serving as supply pipe for the high zone may need to be oversized since they are each individually responsible for carrying the entire flow for the high zone, which even in this simple building with three standpipes per zone would require a flow demand of 1000 gpm. The configuration shown in Fig. 8.22 will only work for situations where the pressure in the standpipes carrying water to the high zone will not exceed 350 psi since these pipes are essentially acting as the express mains, but they have hose connections on them, so the maximum pressure of 350 psi applies to these pipes. This should not be a problem since there are only two zones in this building and the upper zone pump is above the standpipes carrying water in the low zone.

The chief reason for breaking up the building into vertical zones is to control the pressure. If the zones are carefully handled, the design professional can completely eliminate the use of pressure reducing valves in the system by keeping the pressure at every hose connection at 175 psi or less. For example, consider the building in Fig. 8.22. If the building was protected with a single standpipe system and not broken up into different zones, the static pressure at the bottom of the standpipes would need to be at least 187 psi just to overcome the elevation pressure of 87 psi (0.433 × 200 = 87) and the end pressure demand of 100 psi at the most remote outlet. In reality, the pressure would need to be higher from the water supply to overcome the friction loss of getting the water up to the most remote outlet. So, there is no question that many of the hose connections on this system would be at a pressure over 175 psi and would need pressure reducing valves.

But when the building is broken up into two vertical zones, each zone can be calculated separately. For the low zone, the water only needs to get from the first pump to the most remote standpipe on the 10th floor. Taking into account elevation loss (100 × 0.433 = 43.3 psi) and friction loss, which can be minimized with 6 in. pipe having a friction loss per ft of 0.010 for 500 gpm, 0.021 for 750 gpm, and 0.035 for 1000 gpm. If there are 50 ft of pipe carrying 1000 gpm, 150 ft of pipe carrying 750 gpm and 250 ft of pipe carrying 500 gpm, then the pressure demand at the pump will be 150.7 psi (100 + 43.3 + (0.035 × 50) + (0.021 × 150) + (0.010 × 250) = 150 .7). If the pump is sized carefully, you should be able to avoid any pressure reducing valves in the system.

In the upper zone of our example, the hydraulic calculations will be the same to the pump in the upper portion of the building. Depending on the pressure coming in from the pump on the lower floor, the second pump should be able to be sized so that there is no need for pressure reducing valves in the upper zone.

With one exception, NFPA 14 states that the number of pumps in the building needs to match the number of zones with each zone having its own fire pump (Section 7.9.1). The exception is for a system that uses a pump sized for the high zone and a master pressure reducing valve assembly knocking the pressure down for the low zone as shown in Fig. 8.23. Sections 7.9.1 and 7.2.4 of NFPA 14 combine to allow this master pressure reducing valve assembly option. The option of the master pressure reducing valve assembly can appear enticing because it saves the considerable cost of a fire pump and the cost of the maintenance that goes with it. But pressure reducing valves need to be maintained as well, so this does not absolve the building owner of performing inspection, testing and maintenance on significant components in the standpipe system.

Test Yourself

8.1. What is the minimum pressure required at the outlet of the hose connection on a Class I standpipe system?

(a) 50 psi
(b) 65 psi

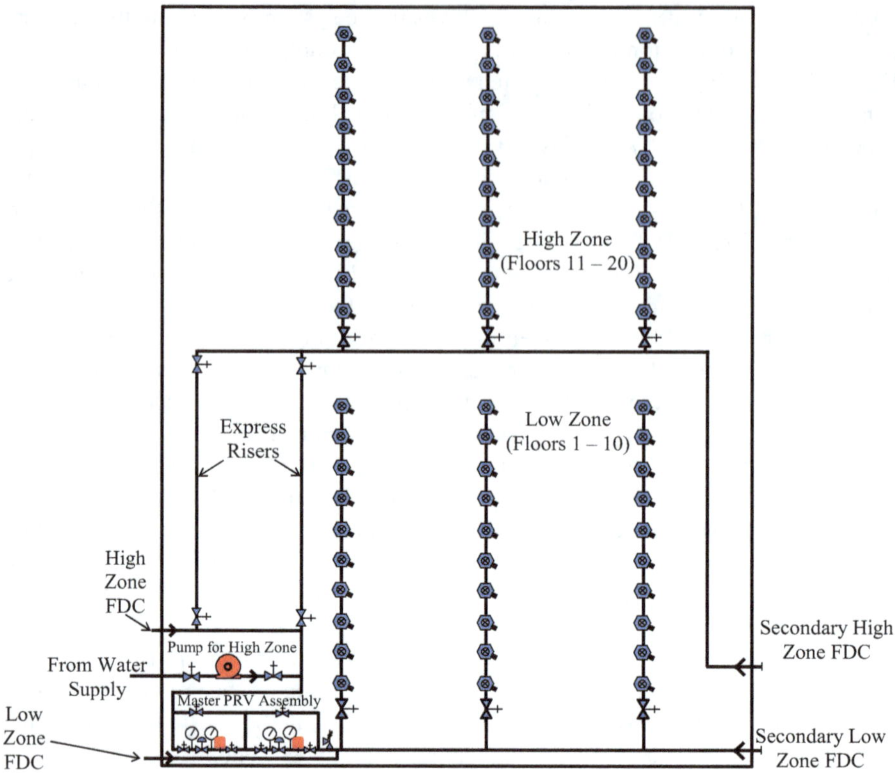

Fig. 8.23 Master pressure reducing valve assembly replacing a fire pump for the low zone in a building with multiple vertical zones

(c) 100 psi
(d) 150 psi
(e) 175 psi

8.2. What is the minimum pressure required at the outlet of the hose connection on a Class II standpipe system?

(a) 50 psi
(b) 65 psi
(c) 100 psi
(d) 150 psi
(e) 175 psi

8.3. What is the maximum static pressure allowed at the outlet of the hose connection on a Class I standpipe system?

(a) 50 psi
(b) 65 psi
(c) 100 psi

(d) 150 psi
(e) 175 psi

8.4. What is the maximum residual pressure allowed at the outlet of the hose connection on a Class II standpipe system?

(a) 50 psi
(b) 65 psi
(c) 100 psi
(d) 150 psi
(e) 175 psi

8.5. What is the maximum static pressure allowed within the pipe of a Class I standpipe system if there are hose connections on that portion of pipe?

(a) 100 psi
(b) 150 psi
(c) 175 psi
(d) 350 psi
(e) There is no maximum

8.6. What is the maximum static pressure allowed within the pipe of a Class I standpipe system if there are no hose connections on that portion of pipe?

(a) 100 psi
(b) 150 psi
(c) 175 psi
(d) 350 psi
(e) There is no maximum

8.7. What is the difference between a pressure reducing valve and a pressure restricting device?

8.8. How does the performance of a pilot operating pressure reducing valve differ from the performance of a direct acting pressure reducing valve?

8.9. For the model MD1 direct acting pressure reducing valve shown in Fig. 8.24, what will be the static outlet pressure when the static inlet pressure is 270 psi? Is this outlet pressure acceptable for a Class I system? (*The MD1 PRV discussed here is not an actual valve. The performance curves were drawn to help the reader understand how to read such curves.*)

8.10. For the model MD1 direct acting pressure reducing valve shown in Fig. 8.24, what will be the residual outlet pressure at 250 gpm when the residual inlet pressure is 220 psi? Is this outlet pressure acceptable for a Class I system?

8.11. The Class I standpipe system shown in Fig. 8.25 is in a 14 story building with 10 ft per floor. The water supply will be a tank with the anti-vortex plate at the same elevation as the pump and the high level water line 24 ft above the level of the pump. The pump will have the net pressure shown in Fig. 8.26. Answer the following questions:

Fig. 8.24 Performance of model MD1 direct acting pressure reducing valve

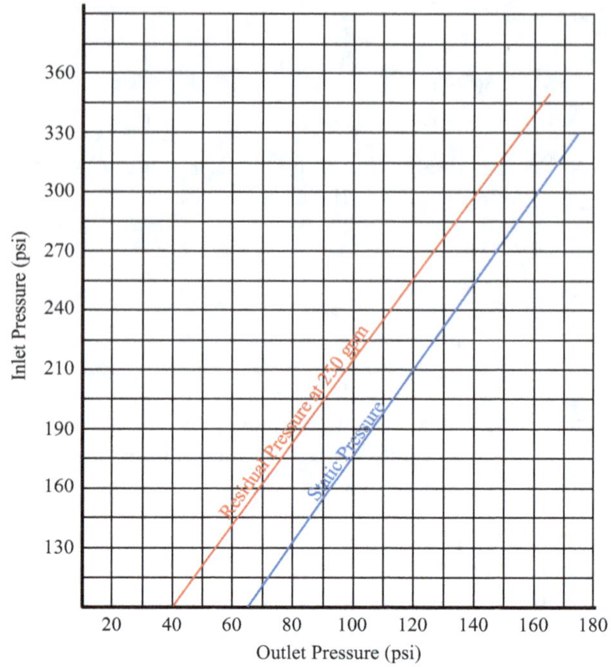

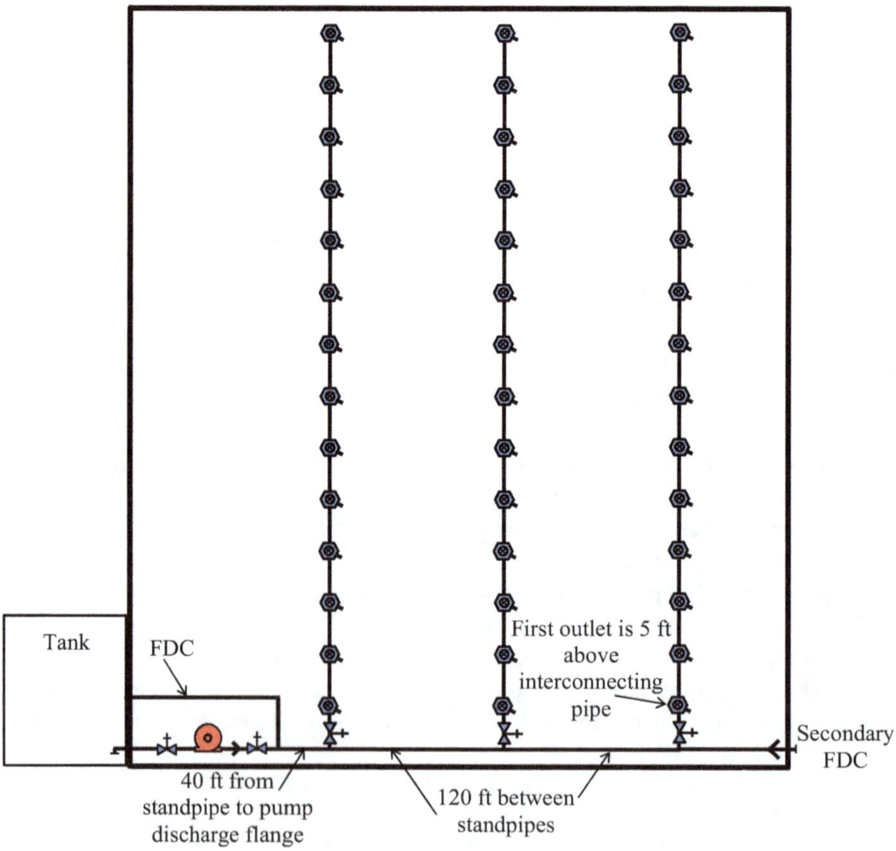

Fig. 8.25 Standpipe system for question 8.11

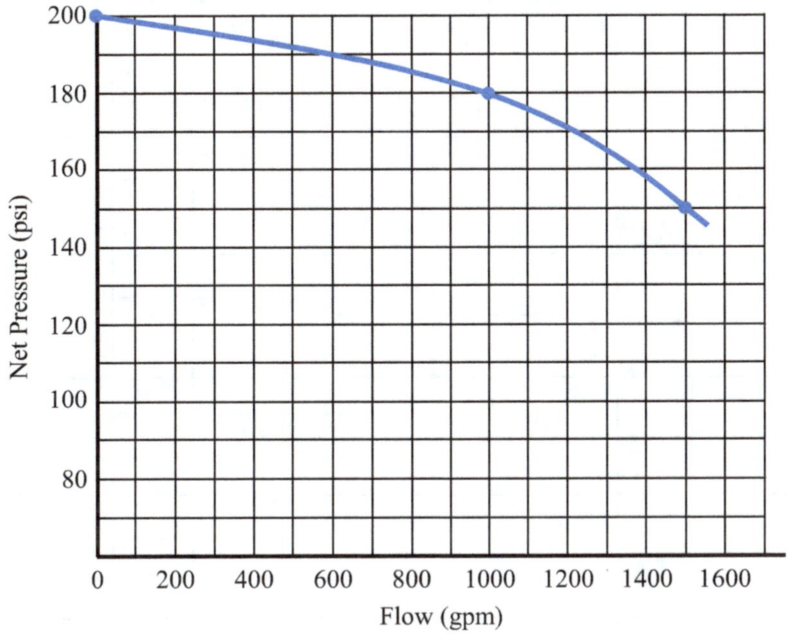

Fig. 8.26 Net pressure curve for pump in question 8.11

(a) What size pipe you would recommend for the standpipes and the interconnecting pipe?
(b) What will be the flow and pressure demand at the pump discharge flange? A blank calculation form is provided as Fig. 8.27.
(c) Which of the hose connections (if any) will require pressure reducing valves?
(d) Select a pressure reducing valve for each of the outlets in the answer to question (c) from the different model options shown in Fig. 8.28 (static pressure) and Fig. 8.29 (residual pressure at 250 gpm) and prove that they will provide acceptable outlet pressures. (*These figures simulate what such pressure reduction curves might look like from a manufacturer, but do not actually represent any specific manufacturer's product.*)

Node 1	Elev 1 (ft)	K-Factor	Flow added–this step (q)	Nominal ID	Fittings–quantity and length	L ft	C Factor	total (P_t)	Notes
						F ft		elev (P_e)	
Node 2	Elev 2 (ft)		Total flow (Q)	Actual ID		T ft	P_f per foot (psi)	frict (P_f)	

Fig. 8.27 Blank hydraulic calculation form for question 8.11

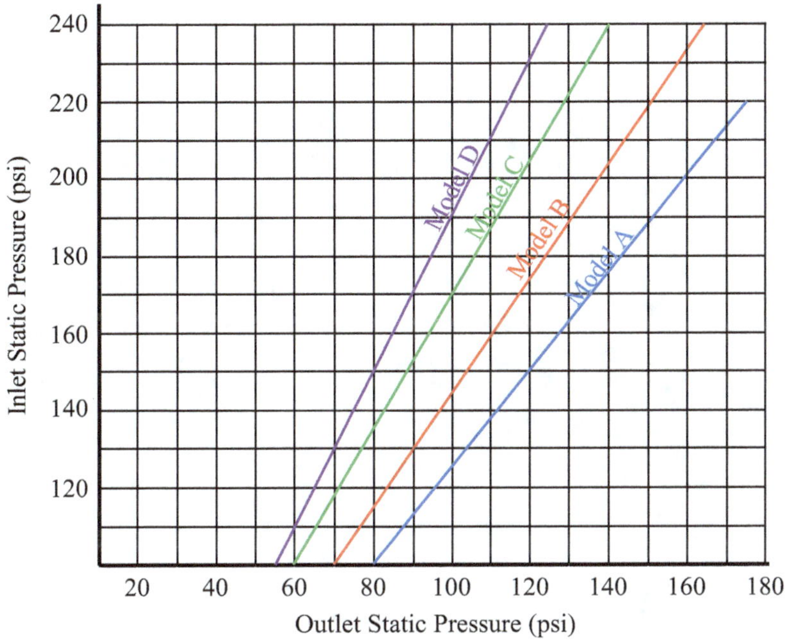

Fig. 8.28 The static pressure reductions for four models of direct acting pressure reducing valves to select from for question 8.11

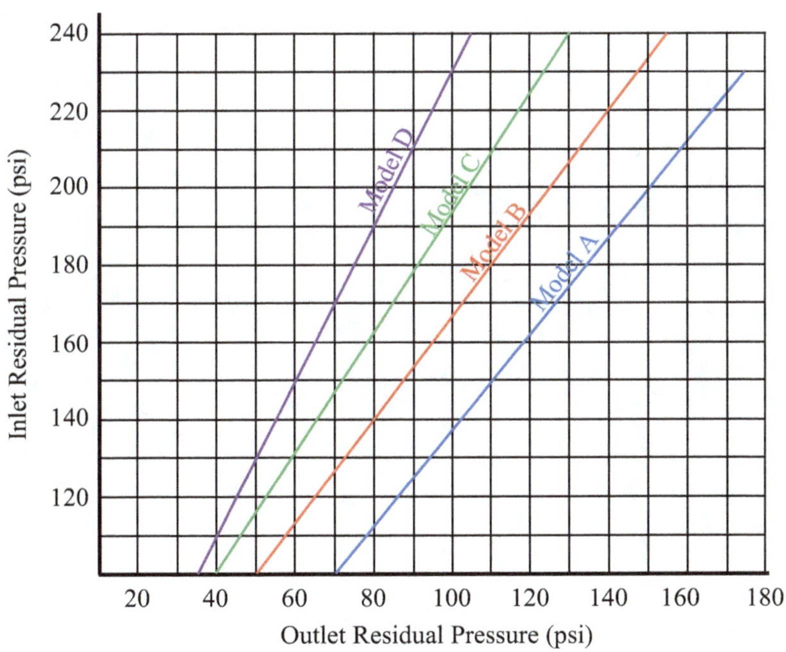

Fig. 8.29 Residual pressure reductions at 250 gpm for four models of direct acting pressure reducing valves to select from for question 8.11

Chapter 9
Pumps and Standpipe Systems

Manual standpipe systems, which are permitted for low-rise buildings, do not need fire pumps. But all automatic and semiautomatic standpipe systems need a water supply that can meet the system demand without input from the fire department. For most standpipe systems, this will mean the installation of a fire pump.

This chapter will cover the basics of fire pumps and how to meet the competing demands of providing the minimum pressure requirements of the standpipe system while making sure that you do not create too much pressure with a fire pump. For the reader that is not too familiar with fire pumps, there will be information here on how fire pumps work, how to select pumps, and how to make sure that the pump will not over-pressurize the standpipe system. For the seasoned fire protection professional, this information will be a valuable review focused on fire pumps being used in standpipe systems.

From the basics, this chapter will move into more advanced concepts. There are advanced solutions to problems in standpipe system design such as the use of variable speed drives for fire pumps will be discussed here. The recent changes to NFPA 20 regarding the concerns over putting pumps in series on different levels of the building and in different pump rooms will be discussed here.

The fire pumps that will be discussed here are all centrifugal pumps, which use centrifugal force to increase the energy of the water in the fire protection system. While there are positive displacement pumps used in fire protection, they are generally not used in standpipe systems, so they will not be discussed in this text. There is no way that all of the issues regarding fire pumps can be addressed in this chapter. There are a number of other texts that delve much deeper into the subject. The book *Pumps for Fire Protection Systems* was published by the NFPA many years ago. While it is now out of print, it is still an excellent text on the basics of fire pumps and copies of it pop up in the used book market frequently. In time, the authors of this text may be able to update this book and get it back on the market.

K.E. Isman, *Standpipe Systems for Fire Protection*,
DOI 10.1007/978-3-319-47750-3_9

Basics of Fire Pumps

Centrifugal pumps are devices that create energy by rotating a liquid. That energy is stored in the water droplets as pressure. This additional pressure helps the water move through the fire protection system and work for us when it leaves the system by overcoming the forces of the fire.

It is important to understand that fire pumps cannot create flow. There is no magical device that takes oxygen and hydrogen out of the air and makes water. If a fire protection system needs more flow than what the water supply can provide, putting a fire pump on the system will not solve the problem. In that case, an additional water supply will be needed. A fire pump might be a part of getting the water in that additional water supply to the standpipe system. But the fire pump on its own will not be able to solve the problem of not having enough flow. What a fire pump is designed to do is take the water flow that is delivered from the water supply and increase the pressure from that water supply as the flow moves towards the fire protection system (in this case, the standpipe system). There are three different pressure conditions that are important to the installation of a fire pump: suction pressure, net pressure and discharge pressure. Each of these will be defined and discussed here.

Suction Pressure: This is the pressure read on a gauge at the suction flange (or intake) of the pump. The suction pressure is a function of the strength of the water supply, the size of the suction pipe, and the difference in elevation between the pump and the water supply. The suction pressure is not affected at all by the size of the pump.

The manufacturers of fire pumps make a few assumptions when they make their pumps. One of the assumptions is that in almost every circumstance, the water will arrive at the pump suction flange at a positive gauge pressure. The one circumstance where water is allowed to arrive at the pump suction flange at a negative gauge pressure is where a pump is taking suction from a tank at the same elevation as the pump. In this case, the water is allowed to arrive at the pump suction flange at a pressure of −3 psi. This will rarely ever happen, but when you design a fire protection system, you need it to work at the beginning of the fire as well as at the end of the fire. So, when you have a tank being used as the water supply, and a fire occurs that burns for a long time and pulls almost all of the water out of the tank, the last few gallons of water leaving the tank have very little elevation pressure, and they lose energy in the suction pipe from friction loss and elevation changes, so they will technically arrive at the suction flange of the pump at a negative gauge pressure. Rather than make everyone elevate the bottom of their tanks above their pumps, NFPA 20 allows the user to put the tank at the same elevation as the pump and just have a small negative gauge pressure towards the end of the fire. Under regular testing and use conditions, the suction pressure will be positive with the tank full or nearly full. The negative pressure will only occur in the unlikely event that a fire occurs and lasts for a long time.

The suction pressure is calculated by starting with the pressure at the water supply and subtracting the friction loss of the water flowing from the water supply to the pump. Then the change in elevation between where the location where the pressure of the water supply was measured and the pump needs to be taken into account. If the pump is below the location where the water supply was measured, the elevation pressure will be added. If the pump is above the location where the water supply was measured, the elevation pressure will be subtracted. The suction pressure will need to be calculated for at least the following conditions during standpipe system design:

- At the churn condition of the pump and the highest static pressure from the water supply. This will determine the highest pressure that will ever be in the system, which is important to know so that you do not over-pressurize the system.
- At the maximum flow for the fire pump (150 % of the rated flow of the pump) during the lowest reasonable water supply pressure at that flow. This will determine the lowest suction pressure that the pump will reasonably experience. This is the pressure that NFPA 20 wants to be a positive gauge pressure. NFPA 20 requires this suction pressure to be calculated because it is possible that this pressure will occur during the acceptance test or the annual test of the pump if the test happens to be performed on a day when the water supply is at a low pressure. Since this pressure could occur, we need to make sure that the pressure does not get below the point where damage could happen to the pump.
- At the standpipe system flow demand during the lowest reasonable water supply pressure at that flow. This calculation will help you make sure that the water supply is sufficient to meet the demand of the fire protection system.

Consider a city water main that fluctuates between the high pressure represented by the blue curve in Fig. 9.1 and the low pressure represented by the red curve. This city water main feeds a 750 gpm fire pump for a Class I standpipe system with three standpipes where the fire pump is 50 ft above the level where the city water main pressures were determined as shown in Fig. 9.2. Note that the 50 ft is measured from the outlet on the fire hydrant where the pressure gauge would be placed to determine the water pressure in the main, not the elevation of the main itself. The underground pipe in this example will be 6 in. Class 50 lined ductile iron (inside diameter of 6.28) and it will be 450 ft long (already including the equivalent length of the fittings). The pipe in the pump room will be 6 in. Schedule 40 steel (inside diameter 6.065) and it will be 25 ft long (already including the equivalent length of the valve and fittings).

The suction pressure at the maximum static condition will be 115 psi from the water supply (at the high pressure condition) minus the elevation pressure loss from the 50 ft change in elevation from the water supply to the pump's suction flange. The elevation pressure loss will be 21.7 psi (50 × 0.433 = 21.7). The maximum suction pressure will then be 93.3 psi (115 − 21.7 = 93.3).

The suction pressure at the maximum flow for the pump needs to be calculated next. The pump is rated for 750 gpm, so the suction pressure needs to be calculated at a flow of 1125 gpm (750 × 1.5 = 1125). The pressure at 1125 gpm using the reasonable low water supply condition will be 46 psi. Using the Hazen-Williams formula, the friction loss in the underground pipe will be 0.028 psi per ft. Multiplying

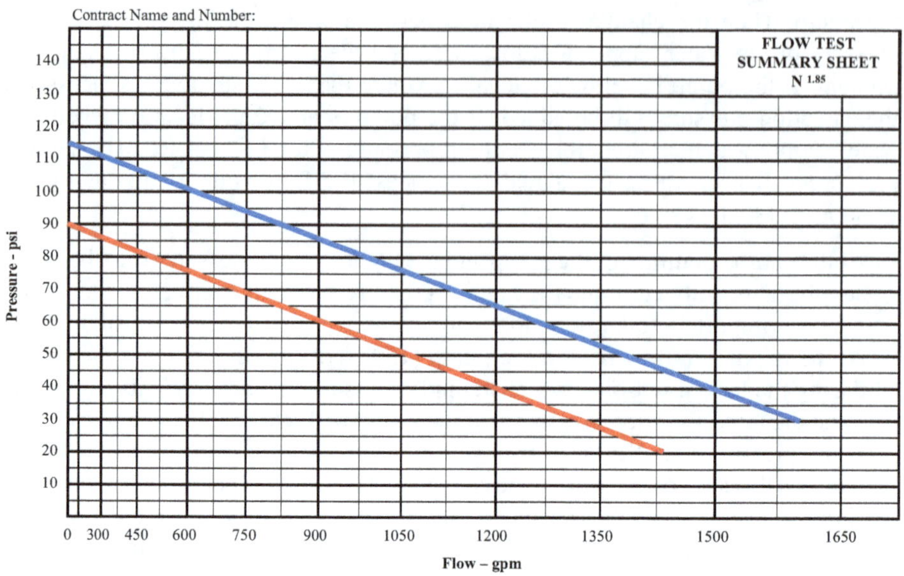

Fig. 9.1 Reasonable high and low conditions for the city water main in the suction pressure example

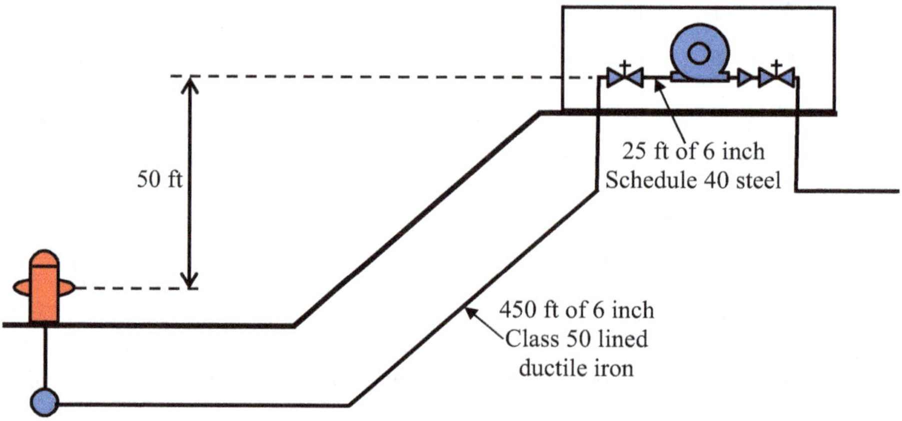

Fig. 9.2 Arrangement of city water main and pump in the suction pressure example

this by the 450 ft of pipe and fittings yields a friction loss in the underground of 12.6 psi. Also using the Hazen-Williams formula, the friction loss in the aboveground pipe in the pump room will be 0.044 × 25 = 1.1 psi. The elevation loss will occur during this flowing condition as well, so that will be another 21.7 psi loss between the water supply and the pump. This makes for a residual pressure at the pump of 10.6 psi (46 – 12.6 – 1.1 – 21.7 = 0.6). Since 10.6 psi is a positive gauge pressure, this meets the requirements of NFPA 20 and is acceptable under the requirements for standpipe system design.

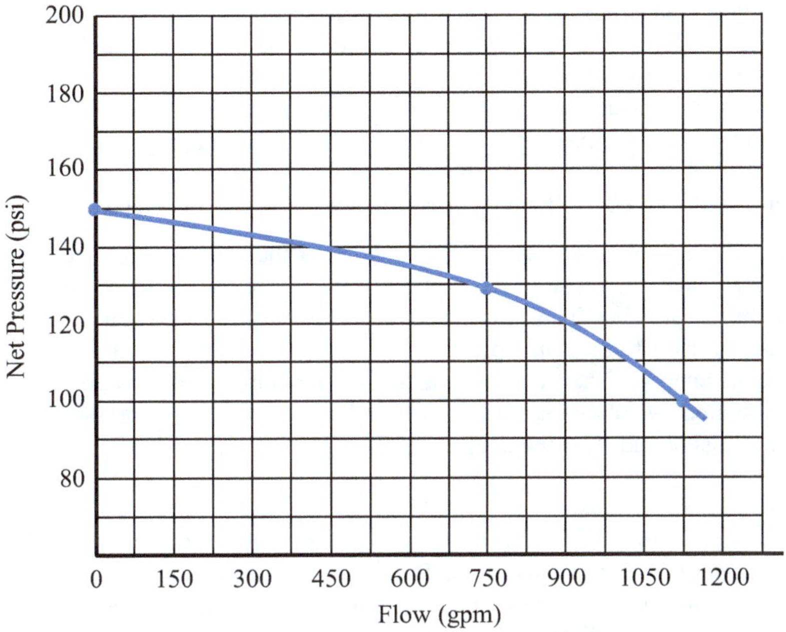

Fig. 9.3 Example of a performance curve for a pump rated at 750 gpm and 129 psi

The next suction pressure that needs to be calculated is the suction pressure at the standpipe system demand. Since the standpipe system has three standpipes, the system demand will be 1000 gpm. At a flow of 1000 gpm, the reasonable low water supply condition will provide a pressure of 54 psi. There will be 9.9 psi (0.022 × 450 = 9.9) friction loss in the underground. There will be 0.9 psi (0.035 × 25 = 0.9) in the aboveground pipe in the pump room. The elevation pressure loss will also be 21.7 psi, making the residual suction pressure 21.5 psi (54 – 9.9 – 0.9 – 21.7 = 21.5) at a flow of 1000 gpm to the pump.

Net Pressure: This is the pressure actually created by the pump. Using centrifugal force, the pump transfers energy from the driver (diesel engine or electric motor) to the water. The net pressure is an expression of that energy. The manufacturers of fire pumps produce net pressure curves for each model of fire pump that they make. Figure 9.3 shows an example of a manufacturer's performance curve for a pump rated at 750 gpm and 129 psi. Note that the pump can have more flow going through it than 750 gpm and it can have less flow going through it. At flows less than 750 gpm, the pump produces more pressure than 129 psi and at flows more than 750 gpm, the pump produces less pressure than 129 psi. This particular pump shown in Fig. 9.3 produces a net churn pressure of 150 psi, a net pressure of 129 psi at rated flow (750 gpm) and a net pressure of 100 psi at maximum flow (1125 gpm).

The performance curve that is produced by the manufacturer is produced when the pump is turning at a specific speed. If you are planning on using the fire pump with a driver that will turn at a different speed, then you need to adjust the performance of the pump. The speed that a pump turns at has a significant effect on the pressure produced by pump. The net pressure produced by the pump is related to the speed that the pump turns to the second power (squared). This means that if the pump turns twice as fast as expected, it will create four times as much pressure. If the pump turns three times as fast as expected, it will create nine times as much pressure. With such a powerful relationship between pressure and speed, make sure that you are using the pump at the speed at which the performance curve was generated.

Discharge Pressure: This is the gauge pressure at the discharge (outlet) of the fire pump. The discharge pressure can be simply calculated by adding the suction pressure to the net pressure. This is an expression of energy conservation. The energy of the water going into the pump plus the energy of the water created by the pump is equal to the energy of the water coming out of the pump. The discharge pressure can be represented by following formula where P_D is the discharge pressure, P_S is the suction pressure and P_N is the net pressure.

$$P_D = P_S + P_N$$

The discharge pressure is most often used to determine if the water supply is sufficient for the fire protection system. The easiest way to perform the hydraulic calculations is to calculate the demand of the fire protection system back to the discharge flange of the pump and then calculate the discharge pressure from the suction pressure and net pressure of the pump. If the discharge pressure at the flow demand of the fire protection system is greater than or equal to the pressure demand of the fire protection system, then the water supply will be sufficient.

The formula for the discharge pressure can be turned around to also calculate the net pressure. Rather than saying that the suction pressure plus the net pressure is equal to the discharge pressure, you can say that the discharge pressure minus the suction pressure is equal to the net pressure. The actual flow conditions can be established and the suction and discharge pressures can be read directly from the gauges at the suction and discharge flanges of the pump. These net pressures can be compared to the performance curves generated by the manufacturer to make sure that the pump is still performing the way that it is supposed to. The net pressure formula can be written as follows, using the same variables as the discharge pressure formula:

$$P_N = P_D - P_S$$

For example, consider the suction pressure example discussed earlier where the suction pressure at three different conditions (maximum static, maximum flow, and demand flow) was calculated. If the pump is the one rated at 750 gpm and 129 psi that is represented by the performance curve in Fig. 9.3, the discharge pressure at each condition can be calculated as follows:

- Maximum static: The suction pressure was calculated to be 93.3 psi. The pump churn pressure will be 150 psi as shown in Fig. 9.3. This makes the discharge pressure at the maximum static condition 243.3 psi (93.3 + 150 = 243.3).
- Maximum flow: The suction pressure was calculated to be 10.6 psi. The net pressure of the pump at a flow of 1125 gpm is 100 psi as shown in Fig. 9.3. This

makes the discharge pressure at the maximum flow of the pump (1125 gpm) 110.6 psi (100 + 10.6 = 110.6).

- Demand flow: The suction pressure was calculated to be 21.5 psi. The net pressure of the pump at a flow of 1000 gpm is about 117 psi as shown in Fig. 9.3. This makes the discharge pressure at the demand flow of the standpipe system (1000 gpm) 138.5 psi (117 + 21.5 = 138.5).

The residual pressures of 110.6 and 138.5 psi are all within the reasonable range for any type of equipment in a fire protection system. But the pressure that is a concern is the maximum static pressure of 243.3 psi, which is well above the 175 psi rating of many types of pipe, fittings and valves used in fire protection systems. There are certainly types of pipe, fittings and valves that are rated to be used at a pressure of 243.3 psi, but the design professional would need to make sure that every part of the system exposed to this pressure was rated for this pressure. As discussed in Chap. 8 of this text, if hose connections were on any of the piping where the pressure was this high, a pressure reducing valve would need to be inserted between the piping with this pressure and the outlet of the hose connection.

Selecting the Right Size Fire Pump

The process of selecting the right size fire pump for any specific standpipe system becomes an exercise in juggling. On one hand, you're trying to select a fire pump that will produce enough pressure to meet the demand of the standpipe system. On the other hand, you're trying to make sure that you don't over-pressurize the system and you don't blow away the firefighters by providing too much pressure. While balancing these two sometimes conflicting objectives, you also need to worry about the cost of the pump. As the pump size and net pressure produced goes up, so does the cost. So, the design professional is usually trying to find the smallest pump that still meets the standpipe system demand.

One of the challenges with selecting a fire pump is that the churn pressure of the pump is always higher than the net pressure at the demand of the standpipe system. Just how much higher is dependent on the manufacturer of the pump. NFPA 20 allows the manufacturer to make the pump so that the churn pressure is as high as 140 % of its rated net pressure. So, for a pump rated at 1000 gpm and 150 psi, the net pressure at churn is permitted to go as high as 210 psi. If you need the 150 psi at rated flow, you have to deal with the fact that there is a significantly higher pressure at churn. The situation gets worse if you are using a pump further down the curve. For example, you are allowed to use a pump rated at 750 gpm to meet the demand of a standpipe system that needs 1000 gpm, but when you do this, you are much further down the pump curve, so the difference between the net pressure at the system demand and the net pressure at churn is much greater. Consider a pump rated at 750 gpm that can provide a net pressure of 150 psi at 1000 gpm. Such a pump could be rated at 190 psi and would be allowed to have a churn pressure as high as 266 psi.

The examples discussed above show that it can be difficult using a pump that has a steep pump curve and barely meets the limitations of NFPA 20. While a pump might be allowed to have a net pressure at churn that is as high as 140 % of the rated net pressure, it is not required to go so high and it would be beneficial to select a pump with less of an increase. A pump that only goes to 115 % of rated pressure at churn would produce much less of a discharge pressure at the maximum static condition and would be much easier to deal with from a design standpoint. Such a pump creates less of a variation in net pressure both above and below the rated flow, making the performance curve look flatter. The flatter the pump curve, the easier it is to design the standpipe system using the pump. In order to make the pump perform with a flatter curve, the manufacturer needs to use a driver with greater horsepower, which does increase the cost of the pump, but helps to deal with the maximum pressure, which can also be costly.

Step-by-Step Process for Selecting a Pump

The first step in selecting a fire pump is to narrow down the options regarding the size (flow capacity) of the pumps that you want to choose from. NFPA 20 allows a pump to be used for a fire protection system with a demand up to 150 % of the rated flow of the pump. There are three common system demands for standpipe systems: 750, 1000, and 1250 gpm. For the 750 gpm system demand, fire pumps with ratings of 500 gpm or 750 gpm are generally used. For the 1000 gpm system demand, fire pumps with ratings of 750 gpm or 1000 gpm are generally used. For the 1250 gpm system demand, fire pumps with ratings of 1000 and 1250 gpm are generally used. Of course, larger pumps can always be used, but there is a tradeoff regarding the cost that makes most of these larger pumps not cost effective.

The rest of the process for selecting a pump will be presented here as a step-by-step process with the user following from one numbered step to the next in a linear fashion. While this makes the discussion easy from a learning standpoint, the reality is that this is much more of an iterative process. The user frequently gets to a step in this process, realizes that the particular pump that they selected will not work for some reason, and then backs up a step or two and moves back through the steps with a new pump in mind. The step-by-step process is as follows:

1. Calculate the standpipe system demand from the most remote standpipe outlet or hose station to the discharge flange of the fire pump.
2. Select a fire pump where the flow demand for the standpipe system is less than or equal to 150 % of the rated flow of the fire pump.
3. Check the net pressure of the pump at the demand flow of the standpipe system.

4. Add the net pressure from Step 3 to the suction pressure at the demand flow of the standpipe system to get the discharge pressure at the demand flow.
5. If the discharge pressure calculated in Step 4 exceeds the demand pressure calculated in Step 1, continue to the next step. If not, you have three options:
   (a) Go back to Step 2 and select a different pump.
   (b) Go back to Step 1 and alter the system to lower the demand.
   (c) Change the water supply to raise the suction pressure.
6. Check the net pressure of the pump at churn.
7. Add the suction pressure at the high static condition to the net pressure in Step 6 to get the high discharge pressure of the pump.
8. If the high discharge pressure of the pump is less than or equal to the pressure rating of the components of the standpipe system, continue to the next step. If not, you have four options:
   (a) Go back to Step 2 and select a different pump.
   (b) Select different components for the standpipe system that are rated for the high discharge pressure.
   (c) Install a master pressure reducing valve assembly in the standpipe supply downstream of the pump. The components of the standpipe system between the pump and the master pressure reducing valve assembly will still need to be rated for the pressure calculated in Step 7.
   (d) Change the water supply and lower the suction pressure. Go back to Step 4 and repeat the earlier steps to still make sure that the pump can meet the system demand.
9. Calculate the suction pressure at maximum flow and make sure that it meets the requirements of NFPA 20. If not, you have two options:
   (a) Go back to Step 2 and select a pump with a lower flow rating.
   (b) Increase the pressure available from the water supply. Go back to Step 7 and make sure that the system does not get over-pressurized.

If you've made it through all nine steps, the pump is acceptable and can be used. If you have not, then there are options sending you back to try different changes that might help make the pump acceptable. These options are not the only possible options, but are the most common. You may be able to think of other options that will also be acceptable. Not every option is available in every case. For example, several of the options would require some change being made to the increase or decrease the pressure from the water supply. If the water supply is a tank, this might be accomplished by raising or lowering the water level or bottom of the tank. But if the water supply is a city water main, getting the water utility to change the pressure may be an impossible task.

Pump Selection Example

Consider a standpipe system for a building with a flow and pressure demand of 1000 gpm at 184 psi calculated to the discharge flange where the pump will go. The following conditions exist for this standpipe system:

- The water supply is a public water utility with the reasonable high and low as shown in Fig. 9.4.
- The fire pump will be installed 25 ft above the level of the fire hydrant where the pressures in the public utility were measured as shown in Fig. 9.5.
- The suction pipe from the public water utility to the pump room will be 6 in. Class 52 lined ductile iron (inside diameter of 6.16 in.). The pipe will be 150 ft long with four standard turn elbows, a tee and a gate valve. The equivalent lengths in Table 8.3.1.3 in NFPA 14 need to be multiplied by the two conversion factors 1.33 for the c-factor conversion and 1.08 for the different inside pipe diameter (calculated as follows $(6.16/6.065)^{4.87} = 1.08$). According to Table 8.3.1.3, the elbows have an equivalent length of 14 ft each, the tee has an equivalent length of 30 ft, and the gate valve has an equivalent length of 3 ft. The equivalent lengths of the fittings, adjusted for the type of pipe is ((14 + 14 + 14 + 14 + 30 + 3) × 1.33 × 1.08 = 127.8 ft). With the total length of the underground pipe and fittings is 277.8 ft (150 + 127.8 = 277.8).
- The suction pipe inside the pump room will be 8 in. Schedule 30 steel pipe (inside diameter of 8.071 in.). The pipe will be 10 ft long with an elbow and a

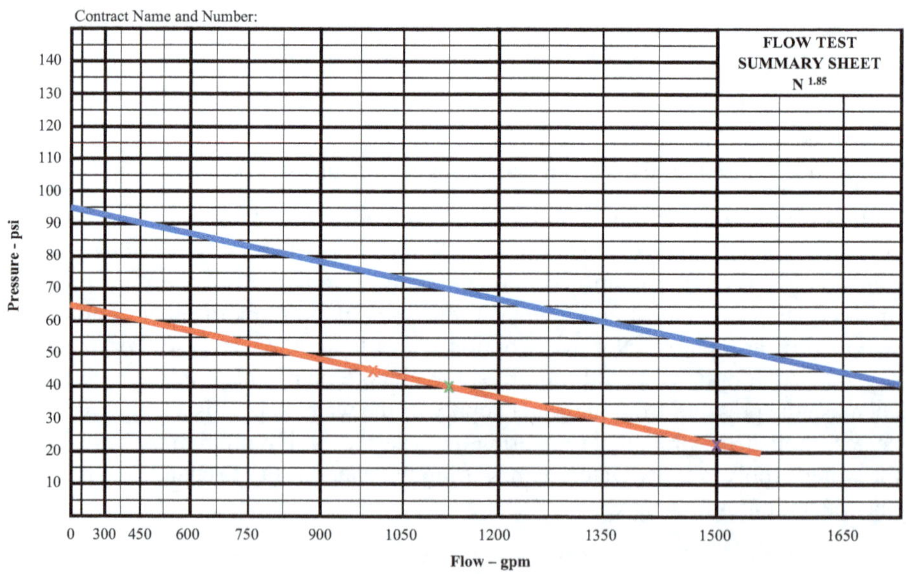

Fig. 9.4 Water supply for pump selection example with reasonable high and low fluctuations shown

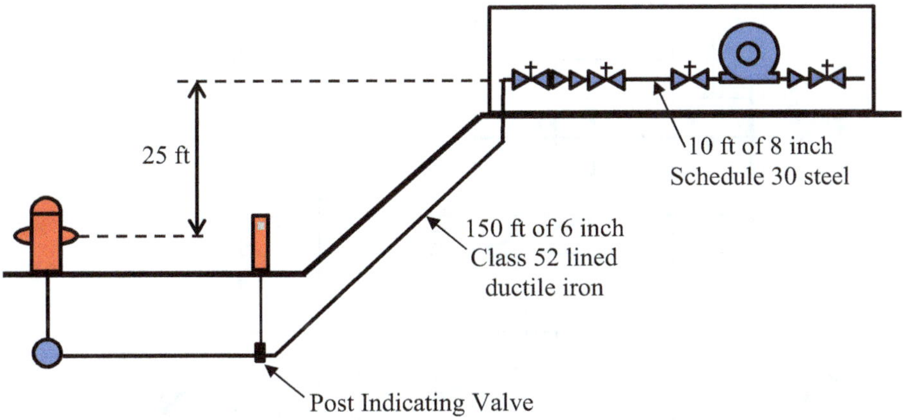

Fig. 9.5 Arrangement of public water utility, underground pipe, pump and pump room for pumps selection example

gate valve. The equivalent lengths in Table 8.3.1.3 in NFPA 14 will need to be converted for the different inside pipe diameter (calculated as follows $(8.071/7.981)^{4.87} = 1.06$). According to Table 8.3.1.3, the elbow has an equivalent length of 18 ft and the gate valve has an equivalent length of 4 ft. The adjusted equivalent length of the fittings is 23.3 ft ((18 + 4) × 1.06 = 23.3). This makes the total length of the pipe and fittings in the pump room 33.3 ft (10 + 23.3 = 33.3).

- There will be an 8 in. backflow preventer installed in the pump room with a friction loss of 7 psi at all flows.

There are two fire pumps under consideration for this standpipe system. The first is a pump rated at 750 gpm and 175 psi with the manufacturer's performance curve as shown in Fig. 9.6. The other is a pump rated at 1000 gpm and 175 psi with the manufacturer's performance curve as shown in Fig. 9.7. Before selecting one of these pumps for our standpipe system, the suction pressure at a number of conditions needs to be calculated.

High Suction Pressure Under Static Conditions: The high static pressure from the water supply is 95 psi (using the blue water supply curve on Fig. 9.4). The elevation pressure loss due to the 25 ft change in elevation will be 10.8 psi (0.433 × 25 = 10.8). The suction pressure under static conditions will then be 84.2 psi (95 − 10.8 = 84.2).

Suction Pressure at Standpipe System Demand: When 1000 gpm is flowing to the pump, the low residual pressure from the water supply is 45 psi, represented by the red "X" on the water supply graph on Fig. 9.4. The elevation loss between the water supply and the pump will be 10.8 psi. The suction pressure will be 21.1 psi (45 − 10.8 − (0.021 psi/ft × 277.8 ft) − (0.009 psi/ft × 33.3 ft) − 7 = 21.1).

Suction Pressure at Maximum Flow for the 750 gpm Pump: The maximum flow for the pump rated at 750 gpm is 1125 gpm (750 × 1.5 = 1125). The low residual pressure from the water supply is 40 psi, represented by the green "X" on the water

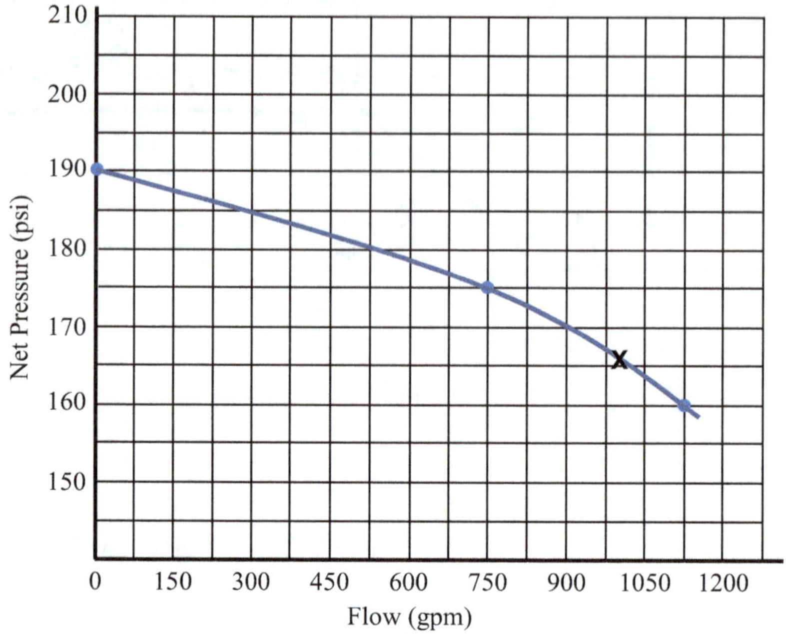

Fig. 9.6 Fire pump rated at 750 gpm and 175 psi being considered for standpipe system in example

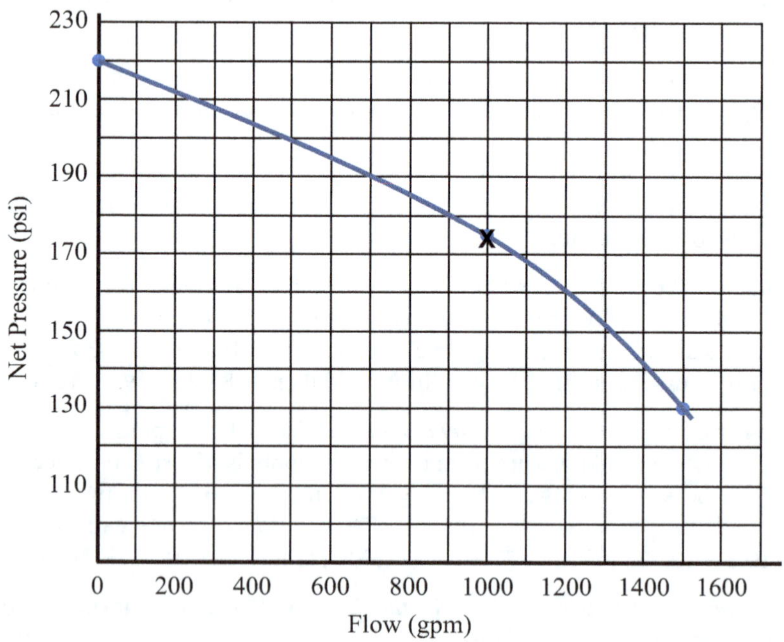

Fig. 9.7 Fire pump rated at 1000 gpm and 175 psi being considered for standpipe system in example

supply graph on Fig. 9.4. The elevation loss between the water supply and the pump will be 10.8 psi. The suction pressure will be 14.3 psi (40 – 10.8 – (0.027 psi/ft × 277.8 ft) – (0.011 psi/ft × 33.3 ft) – 7 = 14.4).

Suction Pressure at Maximum Flow for the 1000 gpm Pump: The maximum flow for the pump rated at 1000 gpm is 1500 gpm (1000 × 1.5 = 1500). The low residual pressure from the water supply is 23 psi, represented by the purple "X" on the water supply graph on Fig. 9.4. The elevation loss between the water supply and the pump will be 10.8 psi. The suction pressure will be −8.2 psi (23 – 10.8 – (0.046 psi/ft × 277.8 ft) – (0.019 psi/ft × 33.3 ft) – 7 = −8.2).

Using these suction pressures and the information in Figs. 9.6 and 9.7, the nine step procedure can now be followed for each of the pumps to see which would be the best option:

1. *Calculate the standpipe system demand*: The demand has been calculated as 1000 gpm at 184 psi.
2. *Select a fire pump*: The 750 gpm pump shown in Fig. 9.6 will be examined first.
3. *Net pressure at demand flow*: According to Fig. 9.6, the net pressure of the pump at 1000 gpm is 167 psi (represented by the black "X" on the figure).
4. *Discharge pressure at demand flow*: 167 + 21.1 = 188.1 psi.
5. *Does discharge pressure (Step 4) exceed demand pressure (Step 1)*: Yes, continue to the next step.
6. *Net pressure at churn*: 190 psi according to Fig. 9.6.
7. *High discharge pressure*: 190 + 84.2 = 274.2 psi. While this is more than 175 psi, and will require pressure reducing valves at the hose connections. Since the high static pressure is less than 350 psi, it should be possible to meet all of the high and low pressure requirements of NFPA 14.
8. *Pressure rating of the components below high discharge pressure*: Yes. Components rated for 300 psi will be used.
9. *Suction pressure at maximum flow*: The suction pressure at a flow of 1125 gpm will be 14.4 psi as calculated above. This is a positive pressure, which is acceptable.

The pump rated at 750 gpm will work for the standpipe system. All of the requirements of NFPA 14 and NFPA 20 can be met. The process now needs to be repeated for the pump rated at 1000 gpm as shown in Fig. 9.7. The nine steps are as follows:

1. *Calculate the standpipe system demand*: The demand has been calculated as 1000 gpm at 184 psi.
2. *Select a fire pump*: The 1000 gpm pump shown in Fig. 9.7 will be examined second.
3. *Net pressure at demand flow*: According to Fig. 9.7, the net pressure of the pump at 1000 gpm is 175 psi (represented by the black "X" on the figure).
4. *Discharge pressure at demand flow*: 175 + 21.1 = 196.1 psi.
5. *Does discharge pressure (Step 4) exceed demand pressure (Step 1)*: Yes, continue to the next step.

6. *Net pressure at churn*: 220 psi according to Fig. 9.7.
7. *High discharge pressure*: 220 + 84.2 = 304.2 psi. While this is more than 175 psi, and will require pressure reducing valves at the hose connections. Since the high static pressure is less than 350 psi, it should be possible to meet all of the high and low pressure requirements of NFPA 14.
8. *Pressure rating of the components below high discharge pressure*: Yes. Components rated for 320 psi will be used.
9. *Suction pressure at maximum flow*: The suction pressure at a flow of 1500 gpm will be −8.2 psi as calculated above. This is a negative pressure, which is generally not acceptable. NFPA 20 does contain an allowance for a pump to be used in situations where the water supply cannot provide 150 % of rated flow at a positive gauge pressure (see sections 4.15.3.1 and 4.6.2.3.1) as long as the demand flow of the system can be provided. This pump falls into that category, and could be used with some additional provisions such as a sign in the pump room warning people what the maximum flow of the pump is.

So, either one of these pumps would work. They both produce more than the demand pressure at the demand flow. But there are a number of items that point to the 750 gpm pump being the better choice for this situation. First, the maximum pressure produced by the 750 gpm pump is lower than 300 psi while the maximum pressure produced by the 1000 gpm pump is greater than 300 psi. There are many methods of joining steel pipe (which is probably what the standpipe system pipe will be) that are rated for 300 psi without additional cost. Once the pressure rating exceeds 300 psi, extra heavy pattern fittings are required, which is the case with the 1000 gpm pump.

The suction pressure with the pump rated at 750 gpm is an acceptable pressure while the suction pressure with the pump rated at 1000 gpm is a concern. While the concern can be handled, it would mean making some special provisions for the system while wasting some water during the system testing to get the system up to maximum flow. All in all, the better choice in this case is the pump rated at 750 gpm.

Variable Speed Pumps

The examples in Chap. 8 and so far in this chapter have shown that there are a number of challenges to selecting a fire pump to work with a standpipe system. The competing challenges of getting a pump to produce a sufficient amount of pressure while at the same time not over-pressurizing the system can be extremely difficult to reconcile. A combination of a fire pump and pressure reducing valves, can make the whole system work, but this option can be expensive. Another option is to use a fire pump with a variable speed drive.

The variable speed drive on a fire pump uses the relationship between speed and net pressure to produce a constant pressure on the discharge side of the pump. If the discharge pressure is lower than the set pressure, the pump speeds up. If the discharge pressure is higher than the set pressure, the pump slows down. All of this happens automatically, faster than a human could adjust the settings. The relationship between the speed of the pump and the net pressure is that the net pressure produced by the pump is related to the speed squared. The easiest way to show this relationship is to use the following formula:

$$P_{N2} = P_{N1}\left(\frac{N_2}{N_1}\right)^2$$

Where P_{N2} = The net pressure at the second speed condition
P_{N1} = The net pressure at the first speed condition
N_2 = The second speed
N_1 = The first speed

The following is an example of how to use the formula. Consider a fire pump that produces 110 psi at 1760 rpm. How much pressure would it produce at a speed of 2000 rpm? The answer is:

$$P_{N2} = P_{N1}\left(\frac{N_2}{N_1}\right)^2 = 110\left(\frac{2000}{1760}\right)^2 = 142$$

So, when a pump that is turning at a speed of 1760 rpm produces a pressure of 110 psi, that pump will produce a pressure of 142 psi at a speed of 2000 rpm. The increase in speed from 1760 to 2000 is only a 14 % increase in speed, but it creates an increase in pressure of 29 %. This is a powerful relationship that can be used to help meet all kinds of fire protection system demands while not over-pressurizing systems.

Consider a fire pump rated at 750 gpm and 155 psi that is being used to supply a standpipe system with a demand of 165 psi at 1000 gpm. The manufacturer's performance curve for the pump is shown in Fig. 9.8, which was generated at a speed of 1760 rpm, but the pump will be connected to a variable speed driver to maintain the discharge pressure at 165 psi. For this example, assume that the maximum suction pressure under static conditions will be 60 psi and the suction pressure at the system demand flow (1000 gpm) will be 20 psi.

Without the variable speed drive, the discharge pressure of the pump would be 245 psi (60 + 185 = 245), which would normally require pressure reducing valves at the hose connections. But with a variable speed drive, the discharge pressure can be maintained at 165 psi in this churn condition. To calculate how much the pump will need to slow down to make this happen, you can turn the formula that relates the speed to the net pressure around by solving for the second speed condition as follows:

$$N_2 = N_1\sqrt{\frac{P_{N2}}{P_{N1}}}$$

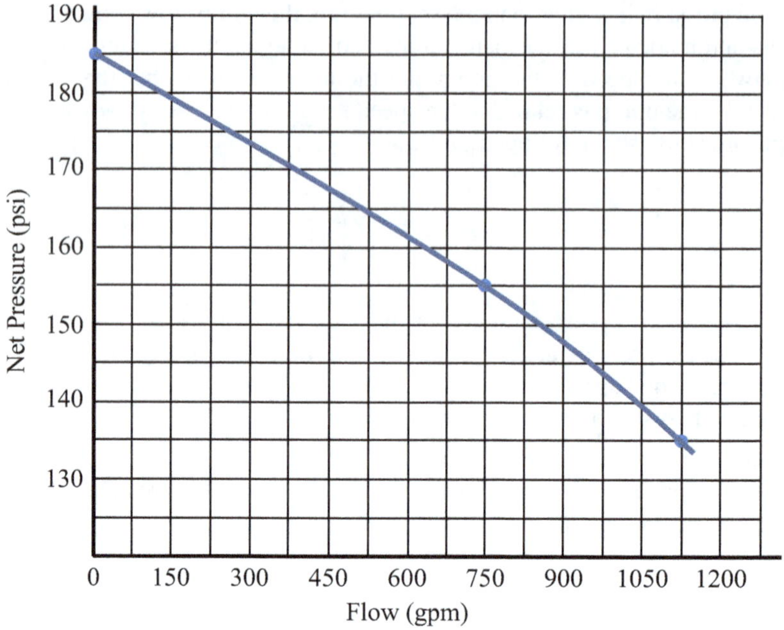

Fig. 9.8 Manufacturer's performance curve for 750 gpm rated pump being used in variable speed drive example

In this situation, the pump would slow down to 1326 rpm. This is calculated as shown below with the target net pressure of 105 psi (165 – the suction pressure of 60 = 105). The formula is shown below:

$$N_2 = N_1\sqrt{\frac{P_{N2}}{P_{N1}}} = 1760\sqrt{\frac{105}{185}} = 1326$$

At the system demand of 1000 gpm, the discharge pressure needs to be 165 psi. With a suction pressure of 20 psi, the pump will need to produce 145 psi of net pressure. According to Fig. 9.8, the pump will only produce 143 psi if it turns at 1760 rpm. But the pump can speed up to 1810 rpm to cover the difference. You can calculate the speed that the pump will need to achieve as follows:

$$N_2 = N_1\sqrt{\frac{P_{N2}}{P_{N1}}} = 1760\sqrt{\frac{145}{143}} = 1810$$

Diesel engines can be designed to be variable speed pumps simply by taking the governor off of the engine and replacing it with a sensing line and control mechanism. Diesel engines are originally designed to run at variable speeds. In order to use them with fire pumps that only are designed to run at one speed, the manufacturer puts a governor on the engine to keep it running at a constant speed (or as close to a constant speed as possible given the load on the engine from the flowing water). By taking the governor off the engine, it can run at a wide variety of speeds. A sensing line connected to the discharge side of the pump on one end and a digital pressure sensor at the other end can be used to speed up or slow down the pump. Firefighters have been doing this for 60 or 70 years with the pumps on fire trucks, but it is only since the late 1990's that NFPA 20 has allowed fixed fire pumps to be listed for use with variable speed diesel engines. There is at least one manufacturer of diesel engines that has achieved a listing for a variable speed drive for a fire pump.

It is a little more difficult to design an electric motor to operate at variable speeds. An electric motor will only operate at a single speed based on the frequency of the Alternating Current (AC) electricity. In the United States, the frequency is 60 Hz, which will keep an electric motor running at a constant speed. So, in order to get the motor to run at different speeds, the controller changes the AC current from 60 Hz to DC current with no frequency. The electricity is then converted from DC current back to AC current at a frequency different from 60 Hz to either speed up or slow down the motor. All of this happens in less than a second to control the pressure on the discharge side of the pump. There is at least one manufacturer of electric motors and controllers that has achieved a listing for fire pumps for their variable speed driver, which is also called a variable frequency driver because of the way that it uses different electric frequencies to control the speed.

Using a Pump with Multiple Water Supplies at Different Suction Pressures

So far, the examples in this text have only shown one water supply for the standpipe system. But there are a number of conditions under which there will be more than one water supply for a standpipe system. The possibility exists that these water supplies will have different static and residual pressures, making it much more difficult to select a pump that will provide the system pressure demand regardless of which of the water supplies it is drawing from while not over-pressurizing the system. The following is a partial list of reasons why more than one water supply would be installed on a standpipe system:

1. The first water supply cannot provide the flow necessary for the system.
2. The building owner chooses to put in a secondary water supply for their own reasons.
3. The insurance company wants a second water supply to improve the reliability of the system and offers a discount to encourage the building owner to add a secondary water supply.
4. The code or standard being enforced requires a secondary water supply to improve reliability. Many codes and standards do this for high-rise building or important buildings. Section 403.3.2 of the International Building Code requires

that the fire pump be connected to two separate water utility mains if the building is greater than 420 ft in height. While this section applies to pumps used for sprinkler systems, it would apply to standpipe systems as well if the sprinkler and standpipe systems are combined. Both NFPA 14 and NFPA 20 have rules for two water supplies in very tall high rise buildings that will be discussed more in Chap. 10 of this text.

5. The code requires a secondary water supply due to concerns that seismic activity will take out the public water main. Section 403.3.3 of the International Building Code requires a secondary water supply for the sprinkler system of a high rise building in Seismic Design Category C, D, E or F. While this is only for the sprinkler system, and the secondary supply only needs to be sized for the sprinkler system, if the pump is providing water to both the sprinkler and standpipe systems, it will affect the performance of the standpipe system as well.

Consider the following example where there are two water supplies for a fire pump providing water to a Class I standpipe system with a demand of 1000 gpm at 165 psi. The first supply is the city water main (Water Supply 1) with the low and high pressure situations shown in Fig. 9.4. The elevation of the pump will be the same as the fire hydrant where the water main pressures were measured. The suction pipe and fittings will be equivalent to 200 ft of 6 in. schedule 40 steel pipe. The second supply will be a tank inside the building (Water Supply 2) with the bottom of the tank 10 ft above the pump and the high water level in the tank 26 ft above the pump. The suction pipe and fittings will be equivalent to 40 ft of 6 in. schedule 40 steel pipe. If the manufacturer's performance curve from the pump is the one used previously in Fig. 9.6. The various suction and discharge pressures related to the pump will be as follows:

Water Supply 1

- High static pressure: 95 + 190 = 285
- Discharge pressure at system demand: 45 − (0.035 × 200) + 167 = 205 psi

Water Supply 2

- High static pressure: (0.433 × 26) + 190 = 201.3
- Discharge pressure at system demand: (0.433 × 10) − (0.035 × 40) + 167 = 169.9 psi

Both water supplies can meet the demand of the standpipe system, and both water supplies are going to create situations where pressure control valves are going to be needed for the hose connections. The challenge is going to be that the city water main creates a high pressure of 285 psi while the tank only creates a high pressure of 201.3 psi. So, sizing the pressure reducing valves is going to be difficult. A valve that reduces the pressure sufficiently at 285 psi might reduce it too much when the pump is taking water from the tank.

The pressure reducing valves that you have to choose from for this system are shown in Fig. 9.9 (residual pressure) and Fig. 9.10 (static pressure). Starting with the static pressure (Fig. 9.10), you will not be able to use Model A or Model B

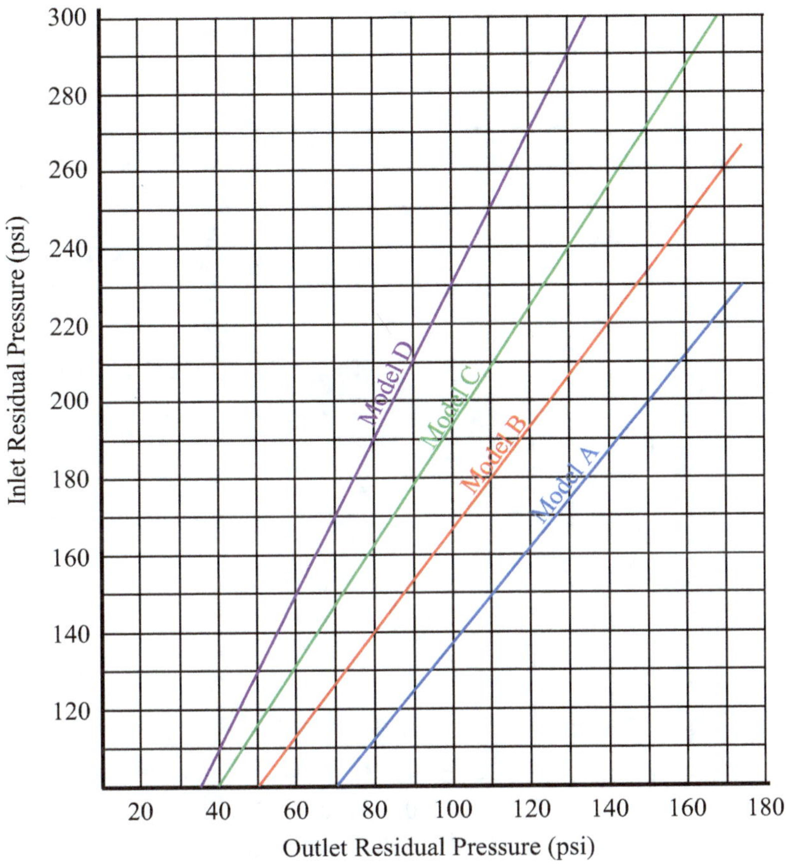

Fig. 9.9 Residual inlet and outlet pressure for direct acting pressure reducing valves under consideration for two water supply example

because they cannot handle an inlet pressure as high as 285 psi. Model C or Model D might work for the static pressure, at least at the bottom of the system where the pressure is high, but the problem comes with the residual pressure (Fig. 9.9). Even at the bottom of the system, both Model C and Model D cannot be used. At an input pressure of 205 psi (when the water is coming from the city main), the outlet pressure from Model C is about 107 psi, which is just barely acceptable. But when the input pressure is 169.9 psi (when the water is coming from the tank), the Model C valve will reduce the pressure down to about 85 psi, which is too low. The Model D valve won't work at all because even at the high inlet pressure of 205 psi, the pressure will be reduced to about 87 psi, which is too low.

So, there isn't a pressure reducing valve that can meet all of the requirements of NFPA 14 when the water is coming from the two different water supplies in this example, and this example is a fairly common set of two water supplies. How do you then design the fire protection system so that it can work? The next few bullet points explain the various options and why they may, or may not, work:

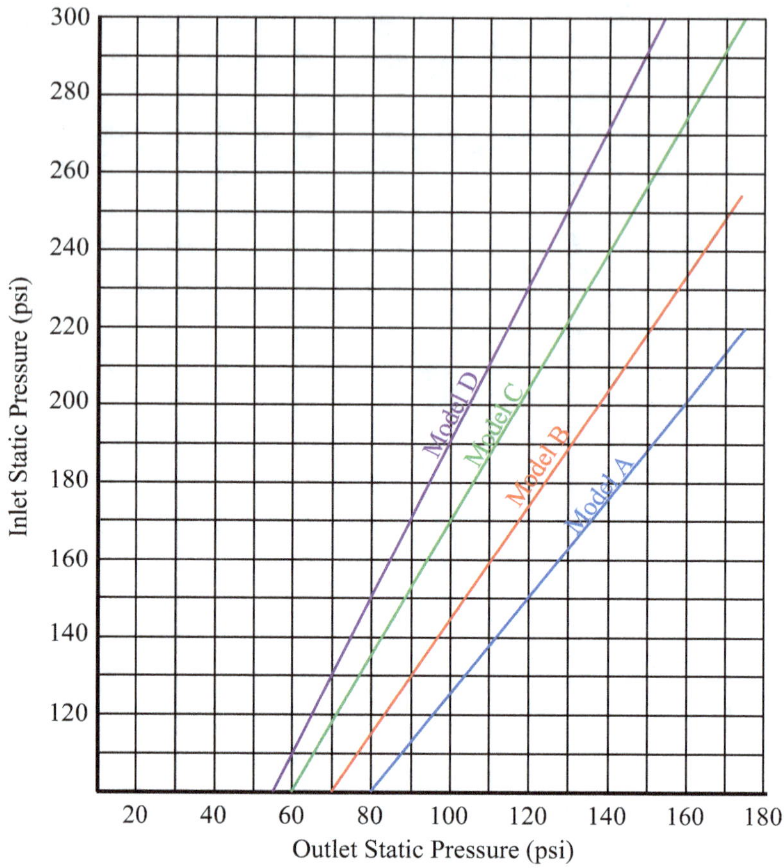

Fig. 9.10 Static inlet and outlet pressure for direct acting pressure reducing valves under consideration for two water supply example

- *Pressure reducing valve in suction pipe*: One of the early suggestions for this problem was to put a pressure reducing valve in the suction pipe from the city water supply. This would drop down the pressure that gets to the pump to be similar enough to the second water supply (the tank) that the discharge pressures would be similar and a pressure reducing valve could be found to work with both situations. While this sounds like a pretty good option, the NFPA 20 committee outlawed this installation in the mid-1990's. The committee was concerned that the pressure reducing valve might drop the pressure into a negative gauge pressure before the water got to the pump. Another concern is that the pressure reducing valve would fail in the closed position and starve the pump for water while it was running, which would destroy a fire pump. So, this option is not one that you will be able to select.
- *Break tank from city main*: One option that is permitted by NFPA 20, NFPA 22, and NFPA 14 is to put a break tank between the city water main and the fire

pump. A break tank is a tank that will not hold the entire duration demand, but will be refilled as fast as the water leaves the tank. NFPA 22 has a series of installation requirements for break tanks in order to insure the reliability of the refill mechanism. Since the break tank creates the same situation as the secondary water supply (the tank), the pump and the pressure reducing valves can be sized to work with both water supplies.

- *Pilot operated valve at hose connections*: This option was discussed in Chap. 8. While there don't appear to be any right angle pilot operated pressure reducing valves that can be installed right at the standpipe as a part of the hose connection, there are 2.5 in. pilot operated pressure reducing valves that can be installed at the standpipe and the hose connection can be installed downstream of that valve. The pilot operated valve will adjust to different inlet pressure conditions depending on which water supply is delivering water to the pump at the time.
- *Master PRV after discharge control valve*: A master pressure reducing valve, as discussed in Chap. 8, can be installed downstream of the control valve for the pump. It needs to be after the discharge valve because the NFPA 20 committee does not allow pressure reducing valves in the discharge pipe of the pump, which extends from the pump discharge flange to the pump discharge control valve. Downstream of the pump discharge control valve, the pipe is under the jurisdiction of NFPA 14, which allows pressure reducing valves, but the master pressure reducing valve assembly needs to be used because there are more than two hose connections downstream of the valve. Both of the pressure reducing valves in the assembly will need to be pilot operated valves to deal with the variable pressures coming from the different water supplies.
- *Variable speed driver*: As discussed earlier in this chapter, the variable speed pump can take advantage of the relationship between the speed of the pump and the net pressure produced by the pump to speed up or slow down to control the pressure. While the pump installation is more expensive with this solution, the rest of the system can remain unchanged, which might make this the least expensive of the options.

Multiple Vertical Zones

In Chap. 8 of this text, the concept of dividing the building into different vertical zones was introduced as a way to control the pressure. In most circumstances, each vertical zone needs its own fire pump. There are at least three different ways that the fire pumps can be installed:

1. Using a different fire pump and tank for each zone. None of the zones are interconnected and the standpipe systems in each zone are like isolated standpipe systems in their own buildings.
2. Pumps in series with the pumps in the same pump room.
3. Pumps in series with the pumps in different pump rooms.

The concept of two or more pumps being in series is where one pump feeds water directly into another pump. The water goes through one pump and picks up the pressure increase from the net pressure of that pump, then goes into the next pump and picks up the pressure increase from the net pressure of that pump as well. By putting several pumps in series, the water pressure can be significantly increased.

When the pumps are installed in series in the same pump room, it will be easiest for the person that is sent to the pump room to monitor the pumps. While the pumps are running, a person needs to make sure the pumps are not overheating, that the relief valves that are supposed to operate have operated, that the control valves are in the correct position, and that the pump is running at an acceptable speed. When pumps are in the same room, a single person can handle this job.

Another advantage to having fire pumps in the same pump room is that the second pump in the series will not be damaged if the first pump fails to start. In the rare condition where the first pump does not start, the water will still get to the second pump at a positive gauge pressure. This would not necessarily be the case with pumps in different pump rooms on different floors. If the pump starts and does not get water, the pump will be severely damaged, so if the second pump starts before the first, damage can occur if they are separated and too far apart.

After years of debate and heated discussions, the NFPA Committee on Fire Pumps finally came to a consensus on how to install fire pumps in series with the pumps in different pump rooms on different floors of the building. The rules deal with improving communication between the pump room and making sure that situations in each pump room are known in the other pump room. The rules are in section 4.20.2.2, 4.20.2.8, 4.20.2.9, and 4.20.2.10 of NFPA 20 and are summarized as follows with Fig. 9.11 illustrating the situation:

- Any pump needs to be manually started or stopped from any of the pump rooms.
- The suction and discharge pressures from each pump need to be displayed in each of the pump rooms.
- The alarms and signals from all of the pumps need to be annunciated in each of the pump rooms. The alarms and signals that need to be annunciated include:
  - Pump running
  - Loss of phase
  - Phase reversal
  - Controller pulling power from alternate power source
  - Alternate power circuit breaker open
  - Pow suction pressure
  - Control switch in “Off” or “Manual” position
  - Trouble on controller or engine
  - A two-way communication system needs to be included between the pump rooms so that people in each pump room can talk to each other. The wiring between the pump rooms needs to meet the survivability requirements of NFPA 72.

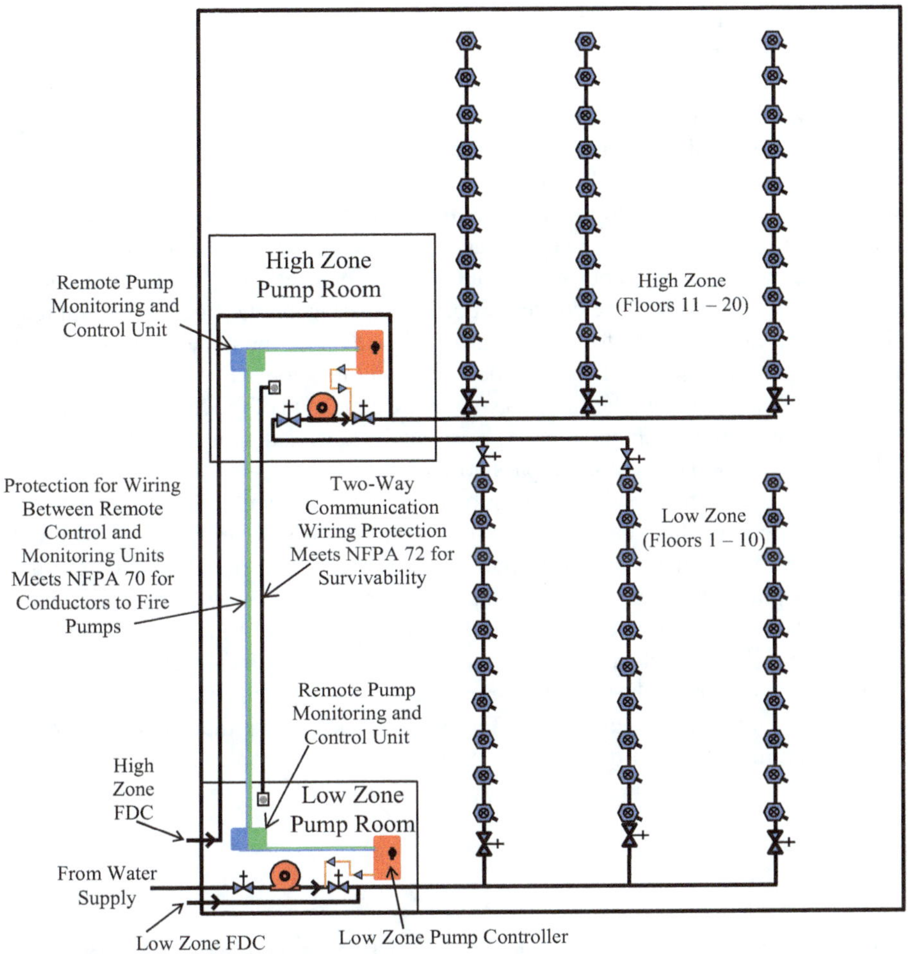

Fig. 9.11 Pumps in series in different pump rooms

- The wiring that runs between pump rooms to meet the requirements discussed above needs to be protected from fire just like the power (feeder) cables that bring electricity to electric motor driven pumps. This requires that the wires meet one of the following:
  - Encased in 2 in. of concrete.
  - Encased in a fire rated assembly dedicated to the circuit wiring that has a 2-h fire resistance rating. Note that by using the words "dedicated to the circuit" the committee is not allowing the walls of a shared pipe chase or electrical chase to count for this option.
  - Installed in a listed electrical circuit protective system with a minimum 2-h fire rating.

Many of the items above regarding the monitoring of pumps at a remote location such as another pump room somewhere else in the building can be accomplished right now. Fire pump controller manufacturers are working on the ability to connect controllers to the Internet so that they conditions they monitor can be viewed anywhere in the world. Changes were made to the 2016 edition of NFPA 20 to add Annex C Fire Pump Controller Connectivity. This new annex outlines how the manufacturers should approach making the information stored by the controller available on the Internet for those people that should have access to that information and how to keep those people out who should not have access. Central to the issue on controller connectivity is the concern about a hacker using the Internet to damage a fire pump remotely. Therefore, right at the beginning of the annex (section C.1.2) the NFPA committee clarifies that their intent is to keep all controller functions completely isolated from the Internet so that the condition of the fire pump (starting, stopping or putting in manual mode) cannot be changed remotely regardless of whether someone has the correct password or not.

While the information above helps to annunciate what conditions have occurred at remote locations like another pump room, the very first requirement of NFPA 20 for putting fire pumps in different pump rooms stated that the ability to start and stop every pump in series needed to be possible from each pump room. This remote start and stop ability has yet to be worked out by pump and controller manufacturers. As the technology develops, it is likely that it will be worked out while this text is still on the market as the manufacturers address the dual concerns of being able to legitimately control the pumps remotely while making sure that malicious forces don't do damage to the fire protection systems that keep people and property safe.

Test Yourself

9.1. If a water supply delivers water to the suction flange of a pump at 30 psi at 750 gpm and the pump produces 127 psi at 750 gpm, what will be the discharge pressure of the pump at 750 gpm?

9.2. While testing a fire pump, you start the pump and open a hose connection to create flow downstream of the pump. With the water flowing, the pressure gauge at the discharge flange reads 162 psi and the pressure gauge at the suction flange reads 44 psi. What is the pressure being produced by the pump at this flow?

9.3. Figure 9.12 shows the reasonable high and low pressure conditions for a water supply taking into account daily and seasonal fluctuations in water usage. If a standpipe system with a demand of 1000 gpm is going to be connected to this water supply, answer the following questions regarding this water supply:

(a) What is the maximum static pressure at the water supply?

(b) What is the reasonable expected low residual pressure available from this water supply at the standpipe system demand flow?

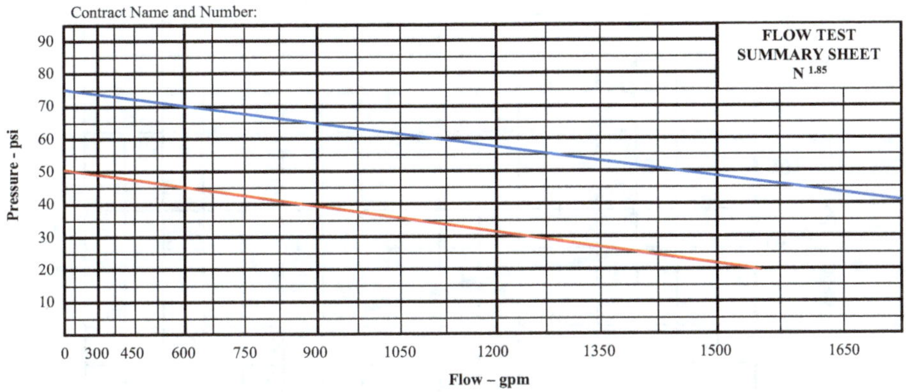

Fig. 9.12 Water supply for test yourself questions

(c) What is the reasonable expected low residual pressure available from this water supply at the maximum flow for a 750 gpm pump that might be connected to this water supply?
(d) What is the reasonable expected low residual pressure available from this water supply at the maximum flow for a 1000 gpm pump that might be connected to this water supply?

9.4. The manufacturer's performance curve of a 750 gpm fire pump is shown in Fig. 9.13. The manufacturer's performance curve of a 1000 gpm fire pump is shown in Fig. 9.14. The standpipe system for which these pumps are under consideration to supply has a flow demand of 1000 gpm. What is the churn pressure, rated net pressure, and net pressure at system demand from each of these pumps?

9.5. The fire pump, water supply and standpipe system arrangement are shown in Fig. 9.15. The Class I standpipe system has a demand of 1000 gpm at 180 psi. The underground suction pipe is 50 ft of 6 in. Class 52 lined ductile iron with four standard turn elbows, a tee, and a gate valve. The pipe in the pump room is 10 ft of 8 in. Schedule 30 steel with an elbow and a gate valve. Answer the following questions:

(a) If you use the 750 gpm rated pump, what will be the maximum discharge pressure?
(b) If you use the 750 gpm rated pump, what will be the reasonable low discharge pressure at the standpipe system demand?
(c) If you use the 750 gpm rated pump, what will be the reasonable low suction pressure at maximum flow for the pump?
(d) If you use the 1000 gpm rated pump, what will be the maximum discharge pressure?
(e) If you use the 1000 gpm rated pump, what will be the reasonable low discharge pressure at the standpipe system demand?

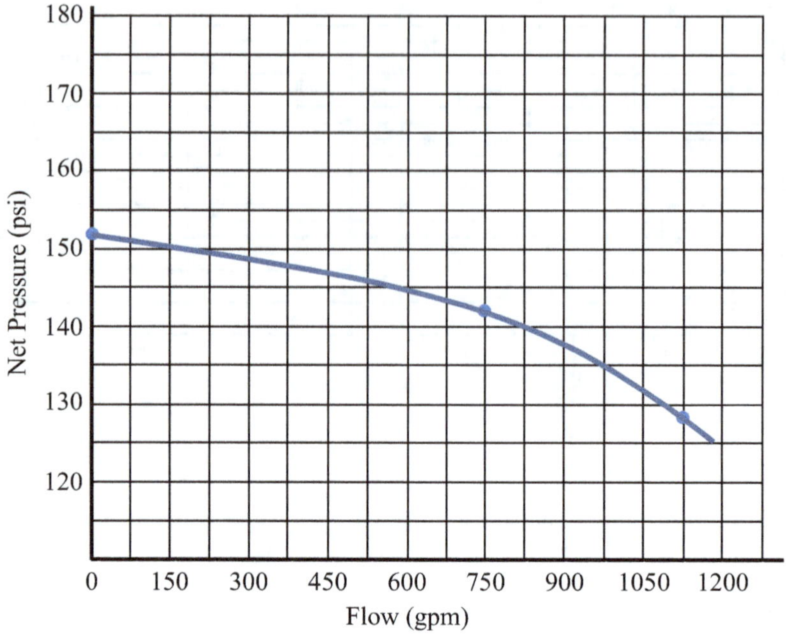

Fig. 9.13 750 gpm fire pump for test yourself questions

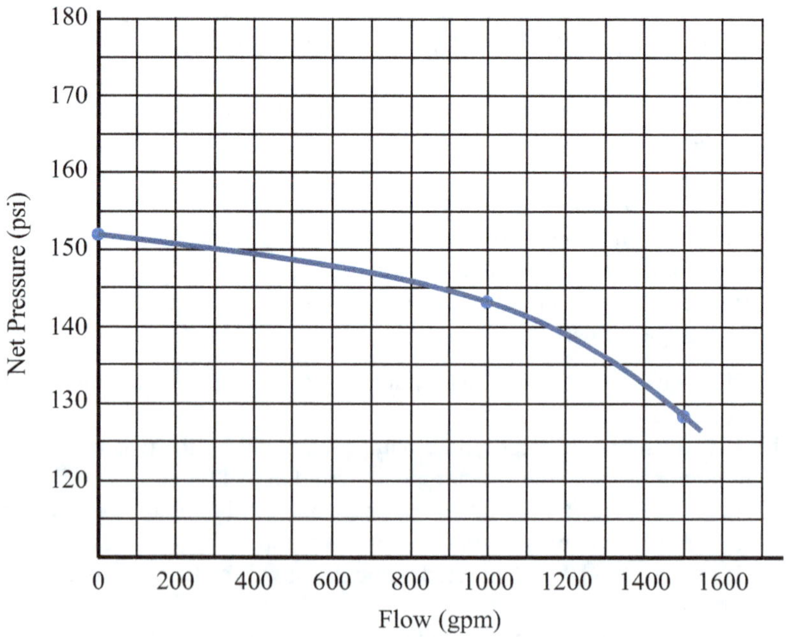

Fig. 9.14 1000 gpm fire pump for test yourself questions

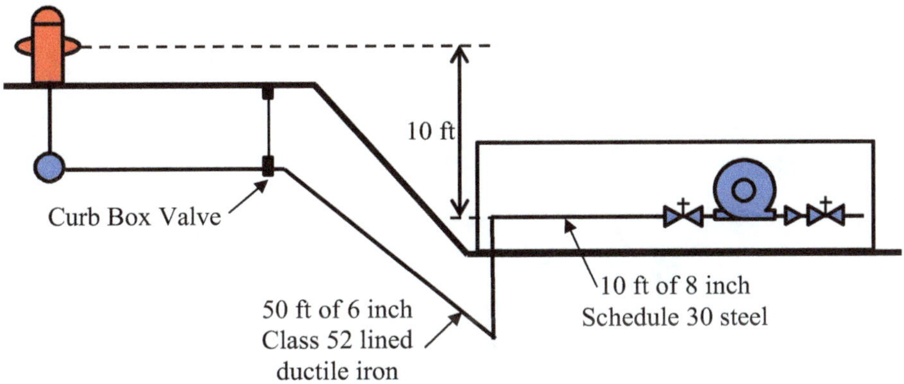

Fig. 9.15 Water supply, pump and suction pipe arrangement for question 9.5

(f) If you use the 1000 gpm rated pump, what will be the reasonable low suction pressure at maximum flow for the pump?
(g) If you had only these two pumps to pick from, recommend one and support your answer with reason(s).
(h) Which, if any, of the hose connections on the standpipe system will need pressure reducing valves?

Chapter 10
High Rise Buildings

There are two categories of high rise buildings that will be discussed in this chapter. There are your regular high rise buildings, and then there are your super high rise buildings. From a fire protection standpoint, the regular high rise buildings are the ones where the fire department is not going to be able to fight the fire from the ground. These buildings are too tall to use fire department ground ladders and even too tall to use the ladders built into the ladder trucks. Rescue and firefighting operations need to be conducted within the building. For most fire departments, this starts to be an issue around seven or eight story buildings.

Super high rise buildings are even taller. These are the buildings where the fire department cannot even pump water from the street to the top of the building because the combination of elevation pressure, friction loss and pressure demand exceeds the pressure capacity of the pump on their trucks and/or the pressure rating of their hose. For these buildings the water supply for fighting the fire needs to be completely within the building with sufficient redundancy to provide firefighters with a safe working environment while they do their jobs.

Every fire department has a different opinion as to how tall a building needs to be in order to fall into this category of "super high rise" building. The situation depends on the pumps on the fire trucks, the pressure rating of the hose they use between the truck and the fire department connection, and the water pressure available from the city water mains. There are some fire departments that consider any building 200 ft tall to be beyond their ability to provide water from street level. There are some fire departments that will allow buildings up to 350 ft before they want the special provisions for super high rise buildings to kick in. The International Building Code has special provisions that begin when a building is 420 ft high, although this does not prevent a shorter building from still being considered a super high rise as we are thinking about it.

This chapter will be split into two sections: the requirements for regular high rise buildings and the requirements for super high rise buildings. All of the requirements for regular high rise buildings will also apply to super high rise buildings and will

K.E. Isman, *Standpipe Systems for Fire Protection*,
DOI 10.1007/978-3-319-47750-3_10

not be repeated in that section. Only the special rules for super high rise buildings will be discussed in that portion of the chapter. NFPA 14, NFPA 20 and the International Building Code all have a variety of rules that apply to both high rise buildings and super high rise buildings. This text will explain them all separately and then try to put them together for a comprehensive view of all of the requirements that need to be met at the same time when all of these codes and standards are applicable.

Regular High Rise Buildings

Each of the three documents that we are working with in this Chapter (NFPA 14, NFPA 20 and the International Building Code) all have the same definition for a high rise building, "A building where the floor of an occupiable story is greater than 75 ft above the lowest level of fire department vehicle access." There are a couple of items worth noting in this definition. First is the idea that the measurement is made to the floor of the highest story of the building, not the roof. So this is not really a measurement of building height. The second item worth noting is that the 75 ft is not measured from the base of the building, it is measured from the lowest surface where a fire truck can pull up near the building. If the building is on a level street, there is no difference in this elevation. But if the building is constructed into a hillside and the fire department cannot get close, then the measurement starts at the street where the fire truck can get access, not the base of the building as shown in Fig. 10.1. If the fire trucks can get to the building from multiple sides, it's the side that is the lowest where the measurement starts.

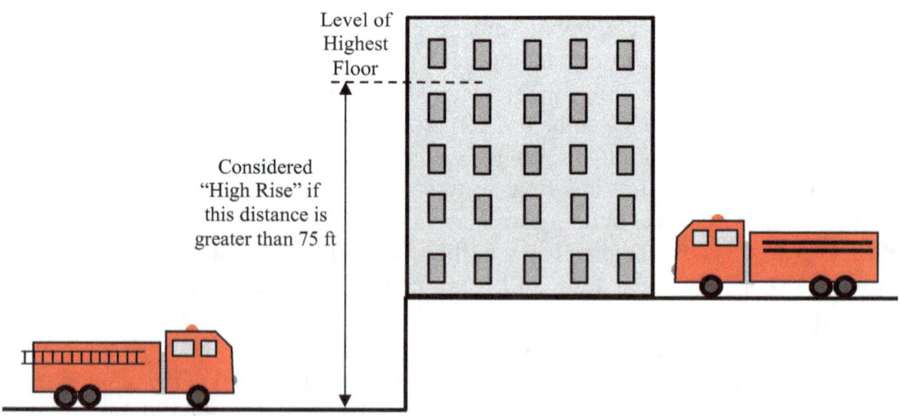

Fig. 10.1 Measurement of distance to determine whether a building is "high rise"

Rules from NFPA 14

NFPA 14 starts out by eliminating some of the types of standpipe systems that can be designed in high rise buildings. Section 5.4.1.2.1 requires that all Class I standpipe systems in high rise buildings be either automatic or semiautomatic. This means that manual systems are not allowed in high rise buildings.

Section 6.1.2.2 requires a different level of protection for the pipes of a standpipe system in a high rise building. This concept will be explored in greater detail in Chap. 11 of this text, which will be devoted to explaining how to protect the pipes against a wide variety of threats including earthquakes and fire.

Section 7.12.2 requires that each zone of a high rise building have two remotely located fire department connections. This allows the fire department to supplement the automatic water supply from different fire trucks on different sides of the building. In many cities, this will allow the fire department to take water from different water mains, conserving the water pressure as they fight the fire. Section 7.12.2.1 does allow the local fire department to waive the requirement for a second fire department connection if they want to. If you plan on using this option, make sure to get the fire department's agreement in writing before finalizing your system design.

Rules from NFPA 20

The more recent editions of NFPA 20 (since 2010) have a whole chapter on high rise buildings; Chap. 5. Most of this chapter is focused on super high rise buildings, but a few of the provisions apply to that seven or eight story that just meets the definition of a high rise. Interestingly, even though there is a special chapter for high rise buildings in NFPA 20, not all of the rules that apply to high rise buildings are in that chapter. There are a few other rules scattered around other places in NFPA 20.

Section 4.13.1.1.1 requires that the fire pump room or building be separated from the building for which it is providing fire protection by 2 h fire rated construction or a distance of 50 ft. The 2 h fire rated construction is intended to apply to the walls of a pump room in the same building as the fire protection system. During a fire, NFPA 20 requires that a person be in the pump room monitoring the operation of the pump. If you were the person being sent to the pump room while the building was on fire, wouldn't you want there to be some substantial separation between you and the rest of the building that is burning while you are in the pump room? The 2 h fire rated construction is that substantial separation.

The 50 ft separation is for the pump house. As long as the pump house is 50 ft away from the building that has the fire protection systems fed by the pump, it does not need to have any specific fire resistance rating as far as NFPA 20 is concerned (although building codes or local zoning rules may have some applicable requirement for the structure). The distance of 50 ft was selected because of the reduction of radiant heat at this distance and the results of a study by the insurance industry that showed that in the rare circumstance of a building collapse from a fire, the debris doesn't tend to spread out more than 40 ft. Therefore, a person in a pump room 50 ft away should be pretty safe. Admittedly, this study was done with mostly non-high rise buildings since the collapse of a high rise building due to fire is an extremely rare event.

This section (4.13.1.1.1) has a direct effect on another provision of fire protection, the fire resistance rating of the walls for the corridors leading to the pump room. Put yourself in the shoes of the person being sent to the pump room. Ideally, the pump room would have a door directly to the outside so that in the event of a fire, you could get out of the building using the regular egress for the building and then go into the pump room from an exterior door.

But not all pump rooms are on the ground floor with one or more of the walls being an outside wall. Many pump rooms are in basements of buildings or up in a high zone of a building where you have to travel up or down a stairwell to get to the level of the pump room. Within the stairwell itself, you are protected, but once you have arrived at the level of the pump room and need to leave the stairwell, NFPA 20 wants you protected until you get to the pump room, which has the 2 h fire rated walls. So, NFPA 20 has some relatively new sections that require a protected path from the stairwell to the pump room (4.13.2.1.1 and 4.13.2.1.2). The protected path needs to be constructed of fire resistive materials with a rating equivalent to the pump room enclosure rating. So, in a high rise building, this protected path would also need a 2 h fire resistance rating to the walls.

Section 5.3 references NFPA 22 for water tanks that are used as a part of the water supply for a fire pump in a high rise building. In addition, this section recognizes that water tanks in a high rise building can serve the domestic usage as well as the fire protection usage. However, they never want the domestic usage to infringe on the amount of water necessary for fire protection, so they require the domestic connection to be above the level of the water necessary for fire protection as shown in Fig. 10.2.

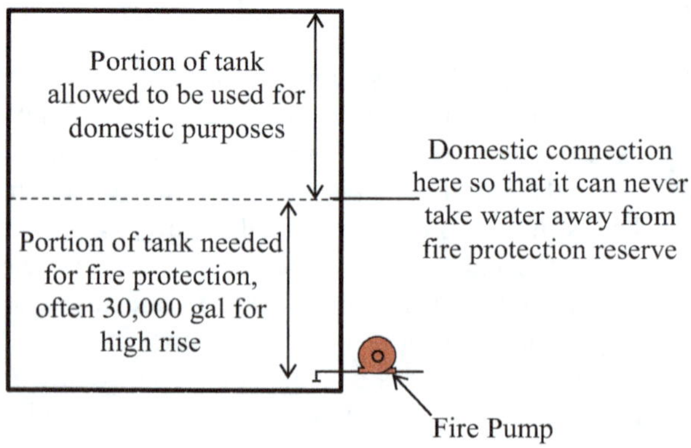

Fig. 10.2 Combined domestic and fire protection tank in a high rise building

Section 5.4 requires the design professional to think about how they are going to test the pump for the situation where the pump is taking suction from a tank. NFPA 20 mandates that the water be recirculated into the tank. This is an issue of water conservation as well as a practical issue of making sure that other methods are not tried that have failed in the past. The committee in this case has put their foot down and said that this is the only acceptable way to test a fire pump in a high rise building that has a tank.

The water that discharges from the pump during the test needs to be piped back to the tank. Within this piping system, there needs to be a method to determine the flow. One way to meet NFPA 20 is to install a flow meter in this piping. If you do that, then section 4.21.2.10 requires a method for checking the calibration of the meter to also be installed. The method of checking the calibration can be a test header that discharges the water back into the tank, but the discharge needs to be through a calibrated nozzle and there needs to be a place to put a pitot gauge. Figure 10.3 shows a plan view of an acceptable arrangement while Fig. 10.4 shows an elevations view.

Looking at Fig. 10.3, you may notice that the discharge from the test header back into the tank is not right above the suction pipe and the pump. The discharge has been taken back almost (but not quite all the way) around the tank to the opposite side. This is to prevent the frothed up water that occurs as the test water is returned

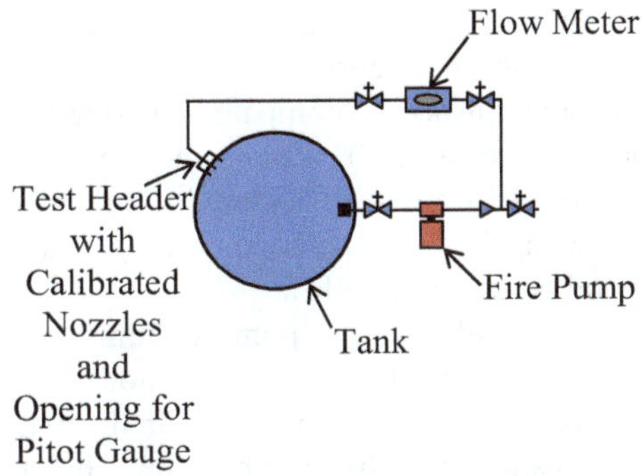

Fig. 10.3 Plan view of flow meter and testing arrangement for high rise building with a pump and tank

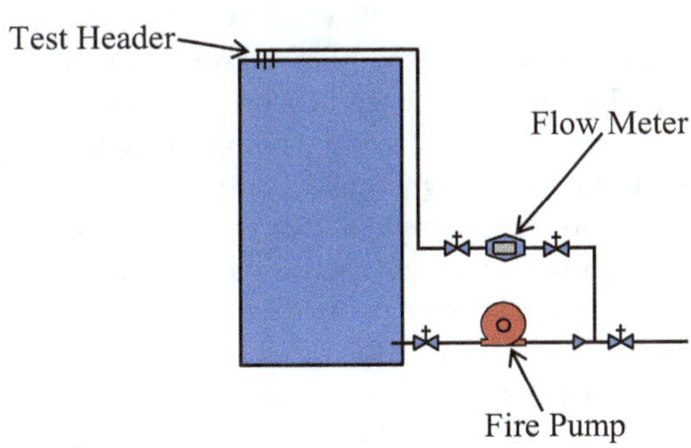

Fig. 10.4 Elevation view of flow meter and testing arrangement for high rise building with a pump and tank

to the tank from entraining air and having that air get sucked back into the suction of the pump. This placement is required by NFPA 20 in section 4.21.2.9, which says, "When discharging back into a tank, the discharge nozzle(s) or pipe shall be located at a point as far from the pump suction as is necessary to prevent the pump from drafting air introduced by the discharge of test water into the tank."

The challenge with section 4.21.2.9 is that there is no direct way to make a calculation to determine exactly how far is far enough to be sure that air won't get into the pump. The variables are the height of the tank, the diameter of the tank, the pressure of the water just before it returns to the tank, the angle of the nozzle as it directs water back into the tank, and the diameter of the nozzle or pipe that discharges back into the tank. But there is no formula that this author is aware of that relates these variable that would allow the design professional to calculate exactly how far the discharge back into the tank needs to be from the pump suction.

Many years ago, the discharge from a test loop was returned to the tank far enough under the water line that air would not get entrained, so this situation was something that did not need to be considered. But more recently, there has been concern for calibrating the flow meters in place, so NFPA 20 now requires there be a way to stick a pitot gauge in the flow stream before the water goes back into the tank, which means that there needs to be an air break to atmosphere. In order to make this work, the discharge needs to be above the water line in the tank. When the water flows across the top of the tank, air bubbles get pulled in. If those air bubbles get down into the suction for the pump, they could do damage to the pump.

Without a direct way to calculate the distance needed for the discharge to be away from the suction for the pump, it is appropriate to look at other sources. There are some insurance companies who recommend that the discharge be at least 15 ft away or at least 90° around the tank away from the suction for the pump. While these values are not calculated from a formula, they are derived from experience in watching how far the water travels when it discharges back into the tank and are good guidance to follow.

Ultimately, the fire pump committee is most interested in making sure that the discharge from the meter does not take the shortest route back to the tank right above the suction for the pump. In the long run, the proof can be provided during the acceptance test when the flow into the tank can be observed to show that it is not getting back to the suction for the pump.

One of the reasons that Fig. 10.3 does not show the discharge from the test going back into the tank exactly 180° away from the suction for the pump is that this might be a bad position depending on the diameter of the tank. If you put the discharge exactly 180° away from the pump suction, the discharge is aimed right at the pump suction, which might make it easier for the air to be pushed toward the pump. If the water flow at an angle across the tank, it will have to take a more circuitous path towards the pump and the air should float back to the top of the tank before getting sucked back into the pump suction.

Looking at Fig. 10.4, you can see that the test header is up near the top of the tank. In order to stick a pitot gauge in the flow stream, you will need to get up to that elevation. Thinking about this in advance can be helpful. Designing a platform or ladder to be able to access the location will help in performing the acceptance test and will make periodic testing and maintenance more affordable for the owner.

Figure 10.4 also shows three discharge nozzles from the test header, which would be appropriate for a fire pump rated at 750 gpm. These nozzles could be underwriters playpipes or other nozzles with known c-factors for converting pitot gauge readings into flow. Just past the playpipe or nozzle, there needs to be an opportunity to get a pitot gauge into the flow stream. The nozzles are shown sloping down towards the water in the tank, but cannot discharge below the water level so that there is a break to atmosphere.

The last rule that applies to all high rise buildings in NFPA 20 is in section 5.5. This section requires all high rise buildings with electric motor driven fire pumps to have a backup power supply or a back-up fire pump (such as a diesel pump) in case of a power outage. This rule used to just apply to super high rise buildings, but a few cycles ago, someone pointed out to the fire pump committee that many building codes required back-up power for fire pumps in all high rise buildings, not just super high rise, so NFPA 20 was changed to agree with these other codes.

Section 9.6.2.1 of NFPA 20 requires the generator that is used to provide backup power to be a Type 10 generator. This means that the generator has to recognize that the loss of power has occurred and start itself so that it produces sufficient electricity to carry the electrical load in 10 s. This is a typical requirement for emergency systems like firefighting systems. While non-emergency systems can wait a minute or two to determine if the power has really gone out or just fluctuated briefly before it comes back on, the emergency systems need to minimize how long they are without power.

Section 9.6.2.3 of NFPA 20 requires that the fuel tank for the generator be sized for 8 h of fire pump operation at rated flow. This requirement seems excessive given that standpipes have 30 min durations and most sprinkler systems have 30–60 min durations. The longest sprinkler system duration that this author can think of is a 3 h demand, so an 8 h duration would be plenty to keep the fire protection system operational through an entire fire, and to be able to put the system back in service following the fire, even if the fuel can't be refilled in the tank right away.

The 8 h requirement for the fuel tank of a generator was an attempt by the NFPA fire pump committee to try and equalize the requirements for diesel engines and electric motors. The basic rules that are used to size the fuel tank for diesel engines require that fuel tank to be sized for 8 h worth of fuel because the fuel tank is not kept full all of the time. Weekly or monthly testing of the diesel fire pump uses fuel and the owner of the building should not have to refill the tank immediately after testing, so the agreement has been made that the fuel tank for a diesel engine can go down to 2/3 full before the tank needs to be filled again. This level of fuel in the tank should allow a little over 5 h of running time for the engine, which would be able to handle the duration of almost all fire protection systems and still allow the system to be returned to service after a fire and before the tank is filled again. So, there is some logic behind the 8 h of run time for the diesel engine fuel tank, and this logic was copied for the electric generator tank to keep the types of drivers somewhat on an equal level of performance.

The 8 h fuel requirement is not shared by the International Building Code. That document has a different duration requirement for the generator, which will be discussed in the next section of this text.

Rules from the IBC

The International Building Code (IBC) has a number of sections that need to be checked when designing a standpipe system for a high rise building. Section 403 applies to most high rise buildings, Section 905 applies to standpipe systems, and Section 913 applies to fire pumps. In this chapter of this text, we will focus on Section 403. Section 403 of the IBC applies to all high rise buildings except the following:

- Air traffic control towers, which need to be designed in accordance with Section 412.3.
- Open parking garages, which need to be designed in accordance with Section 406.5.
- The portion of a building containing an A-5 occupancy, which is defined in Section 303.6 as a portion of a building for participation or viewing of outdoor activities. Examples of A-5 occupancies include amusement parks, bleachers, grandstands and stadiums. Note that only the portion of the structure being used for the A-5 occupancy is exempt from following the high rise rules. Other portions of the building containing locker rooms, offices, storage areas, and meeting rooms are not exempt.
- A special industrial occupancy as discussed in Section 503.1.1. This exemption applies to buildings that have to be very tall in order to accommodate some piece of machinery or industrial equipment like a crane or a foundry. While these buildings are tall, there is very little occupant load at all in the building and the occupant load at the upper levels is almost non-existent since the only time people go up to these levels is to repair or maintain the machinery.
- High hazard buildings (H occupancies), which have their own requirements scattered around the IBC.

Most of the special rules in the IBC for standpipe systems and their water supplies apply to super high rise buildings, which will be discussed later in this chapter. For regular high rise buildings, a few requirements apply, which have already been discussed in this text, but will be expanded on a little here.

Requirement for Standpipes: It should be obvious that standpipe systems are required in high rise buildings. The concepts discussed in Chap. 3 of this text apply to a larger set of buildings than just high rise, but all high rise buildings fall into the category of buildings that need standpipe systems. Still section 403.4.3 spells out the requirement for a standpipe system in a high rise building.

Secondary Water Supply in Seismic Zones: Technically speaking, the IBC does not require a secondary water supply for the standpipe system. But it does require a secondary water supply for the sprinkler system in a building that is within an area of the world that is considered seismic design category C, D, E or F, so buildings that use the water supply for both the sprinkler and standpipe systems need to pay attention to this rule.

Seismic design categories C, D, E and F include portions of about 40 of the states in the United States. Seismic activity, which used to be considered just a "California problem", has become much more of a concern almost everywhere. Section 1613.3.5 of the IBC explains how to determine the seismic design category using three variables:

1. The mapped spectral response acceleration parameter at a 1-s period, which is called S_1 in most structural engineering references.
2. The spectral response acceleration parameters SDS and SD1 calculated using formulas in Section 1613.3.4 in the IBC.
3. The Risk Category of the building. This depends on how the building is being used. Table 1604.5 of the IBC has a list of which buildings fall into each of the four categories with Risk Category I being the least risky and Risk Category IV being the most. Interestingly, a tank used for storing fire protection water and a pump house automatically fall into Risk Category IV, even if the building that they are providing fire protection for is a lower Risk Category.

If you are in an area of the United States and are unsure of what the Seismic Design Category is for a particular structure, the information for placing it into a category like the spectral response acceleration parameters described above can be obtained from the United States Geological Survey at their website www.usgs.gov using the longitude and latitude of the property where the building is going to be constructed. Since the structural engineer on a project will need this information for the design loads of the building, you can sometimes find this information in the structural specifications for the building or you can call the structural engineer and see if they will share the data with you.

Secondary Power Supply: Section 403.4.8 requires a secondary or emergency power supply, which needs to be designed to handle the load of all fire pumps in accordance with Section 403.4.8.4(6). If there are redundant fire pumps in the building, the power supply is only required to be sized to handle one set of the pumps. It would be too much to require a generator to be able to handle both sets of redundant fire pumps simultaneously. Section 403.4.8 references Section 2702 for the secondary power to the fire pump, so this section needs to be explored as well.

The IBC has two different types of secondary power supplies. The first is called "emergency power", which needs to provide electricity within 10 s of the primary power being lost. The second is called "standby power", which needs to provide electricity within 60 s of power being lost. Section 403.4.8 calls for "emergency power" for fire pumps in high rise buildings, which is the more stringent of the two levels of requirements. For more information on the difference between emergency power and standby power, see section 2702.1.3.

The electrical load of the fire pumps (as well as the other items being run by the emergency power source) needs to be carried for 2 h. This is an interesting requirement considering that the water supply for the standpipe system will only be 30 min and the water supply for the sprinkler system is typically 30 or 60 min for the occupancies that are usually found in high riser buildings. Nevertheless, the duration of the emergency power needs to be 120 min. With the emergency power typically being a generator, this has to do with the amount of fuel that will need to be stored to continue to run the generator for 120 min.

Note that the requirement in the IBC for the amount of fuel required to run the generator for a certain duration contradicts the same requirement from NFPA 20. While the IBC requires 2 h of fuel, NFPA 20 requires 8 h. Resolving this conflict will be discussed in the next section of this text.

Merging Rules for Regular High Rise Buildings

Most of the rules that have been discussed here from NFPA 14, NFPA 20 and the International Building Code (IBC) have been complementary. It will be fairly easy to comply with each of these rules, with one exception. As pointed out during the previous sections in this chapter, there is one requirement in the IBC that contradicts a rule in NFPA 20, the requirement for the duration that the generator needs to run. NFPA 20 requires that the generator run for 8 h while the IBC requires that the generator run for only 2 h. When forced to meet both the IBC and NFAP 20, which most jurisdictions in the United States require, which duration do you provide?

One way to solve the problem would be to meet the more stringent of the two requirements. If you were to provide 8 h worth of fuel, then you would also be providing at least 2 h worth of fuel, thereby meeting both requirements. But it turns out that you are not required to meet the more stringent of the two requirements.

Section 102.4.1 of the IBC specifically states, "Where conflicts occur between provisions of this code and referenced codes and standards, the provisions of this code shall apply." This is a very important statement within the IBC. It establishes the IBC as the primary legal document and the rules of this primary document override the rules of any referenced standard, even when the referenced standard is more stringent than the code. So, since the code requires 2 h of fuel and the standard requires eight, the user is only required to provide 2 h of fuel in order to comply with the law.

Certainly the user should consider providing a longer duration of fuel. A true fuel analysis should take into account the duration of most power outages in the area of the building. There are portions of the world where the power supply is very reliable, yet events like hurricanes and blizzards take out the power for days at a time. In such a situation, a 2 h power supply is not likely to be extremely helpful. Since fire pumps don't start as soon as the power goes out, it might be good to size the fuel for the standby loads (such as the pump controller and the alarm system that powers the waterflow switches) for the longest recent power outage and then add the fuel needed to run the fire pump for 30 min (the standpipe water supply duration). Such a performance-based method of sizing the pump could also take into account more fuel to put the system back in service following a fire if power has not been restored yet. Since such a performance-based approach is likely to exceed the blanket 2 h requirement in the IBC, it would not appear to need special permission from the authority having jurisdiction.

Super High Rise Buildings

There are two ways to look at the definition of "super high rise" buildings as discussed in the introduction portion of this chapter. The NFPA standards refer to buildings that are so tall that they have floors for human occupancy, "beyond the pumping capability of the fire department apparatus." In some jurisdictions, this has been declared to be buildings over 200 ft in height. In other jurisdictions, the limit is even higher. It all depends on the fire department, the vehicles that they have chosen to buy, and the equipment that they use to support standpipe operations like hose that goes from the fire truck to the fire department connection. NFPA 20 also uses the term "Very Tall Buildings" in section 5.6, which is intended to have the same meaning as we are discussing here for super high rise buildings.

The International Building Code (IBC) does not use the term "super high rise", but does have special requirements for buildings that are more than 420 ft high. It is not clear exactly where the threshold level of 420 ft came from. If you consider that the maximum rated pressure of many components in a fire protection system is 175 psi, the elevation of the water that could be achieved by that pressure would be 404 ft (175/0.433 = 404). Perhaps the 420 ft limit to the building height is recognition that firefighters could pump water to the floor of a 420 ft building by pumping into the fire department connection at 175 psi (if the ceiling was 16 ft over the floor). This could be at least one explanation for how the 420 ft height was determined to be the proper threshold for these rules.

Another explanation regarding the 420 ft threshold is that there have been some significant egress studies that have shown that once a building exceeds 40 stories in height, people cannot be expected to egress using the stairs only. Even a person in good shape gets tired going down 40 flights of steps. So, back in 2009, the IBC adopted a variety of new egress provisions for buildings over 420 ft in height, since this was considered the possible height of a 40 story building. Since these special egress requirements kicked in at this height, it made sense to connect other (non-egress) provisions at this height so that the code was easier for everyone to follow.

Regardless of how the IBC got to its 420 ft threshold, the reality exists that there are two different sets of rules that need to be followed in the NFPA and IBC documents, which creates three categories of high rise buildings as follows:

1. Buildings that meet the definition of high rise, but can still be supported by fire department vehicle access. These buildings only need to meet the regular high rise rules discussed in the previous section of this chapter.

2. Buildings that are less than or equal to 420 ft in height, but cannot be supported (at least in the upper portions of the building) by fire department operations at the street. These might be buildings between 200 and 420 ft in height. These buildings would need to meet all of the rules for high rise buildings discussed earlier, plus the rules in NFPA 14 and NFPA 20 for super high rise buildings that will be discussed next.
3. Buildings that are more than 420 ft in height. These buildings will need to meet all of the rules for regular high rise buildings, plus the rules for super high rise buildings in the NFPA standards, plus the specific rules in the IBC that will be discussed later in this portion of the text.

NFPA 14 Rules for Super High Rise Buildings

NFPA 14 does not have many special rules for very tall buildings, but there are a few and they are important. Section 9.1.4 requires an auxiliary water supply in the form of stored water in the building if the fire department pumpers cannot supply the required system demand through a fire department connection. This auxiliary water supply is required to be "high-level water storage" with "additional pumping equipment". The term "high-level water storage" is not defined in the standard, but many authorities interpret it as meaning that the water storage cannot be in the lower floors of the building. Instead, the water storage needs to be in the upper portion of the building that is above the level of the ability of the fire department to provide water through the FDC. This increases the reliability of the system because you do not have to rely on the water getting up through the building during a fire, it will already be there.

Section 9.1.4 also offers the user an option to the high-level water storage. The section allows "other means acceptable to the AHJ". The term AHJ is used here to mean the authority having jurisdiction, which refers to any person in a position of authority to say "yes" or "no" to an idea. This could be a building official, fire official, insurance authority, or other person that has some say in how the building is being constructed. The standard does not provide any information or criteria by which the AHJ would make a decision on a proposed idea. In general, such ideas would need to provide some equivalent type of protection with roughly the same level of reliability as the high-level water storage and pumping equipment. It is hard to imagine what ideas might work in such a case, but the standpipe committee wanted to allow some creativity.

An annex note (A.7.9.3) indicates that one possibility is a fire department connection in series with a low zone or mid-zone pump in order to get water to the high zone as an auxiliary water supply. This would violate the rule in NFPA 14 and NFPA 20 about not installing a fire department connection on the suction side of a fire pump (see section 6.4.3.1 in NFPA 14). The rule has been in place for a long time in order to prevent the system from being over-pressurized. If the water goes into the fire department connection at a high pressure coming out of the fire truck, and then into a low zone or mid-zone pump, the additional increase in pressure from the second pump might produce a pressure that is too high for the piping. It is possible that such an arrangement could work to get water to the high zone, but the low zone or mid-zone fed by the pump might need to be isolated to prevent damage. This would have to be done by closing normally open valves, which should only be done during an emergency. So, such an arrangement might work, but it would need to be planned out carefully and might require manual closing and opening of some valves.

NFPA 14 contains some redundancy, so it is no big surprise that section 7.9.3 says the same thing as section 9.1.4. The language is exactly the same including the reference to "high-level water storage" and the allowance for other means acceptable to the authority having jurisdiction. It is not clear why the committee thought that the language needed to be in the standard twice, but it is there.

Section 7.12.1.1 allows the fire department connection to be eliminated if the high zone of the building is so high that the fire department equipment will not be able to pump water to that zone anyway. In this case, the water to fight the fire will be contained within the building and the fire department should not have to pump into a fire department connection anyway.

NFPA 20 Rules for Super High Rise Buildings

Section 5.6 of NFPA 20 is the one that focuses on very tall buildings. This section contains requirements for tanks, refill mechanisms for tanks, and redundancy in the water supply. Each of these topics will be discussed in further detail.

Tanks

NFPA 20 does not require the use of tanks in order to meet the demands of the fire protection system, but it does contain a series of rules about how the tank needs to be designed if it is going to be used. A tank within the building solves many of the problems regarding the sizing of fire pumps and the control of the pressure within the portion of the standpipe system served by the pump. The suction pressure at the pump is much less variable when the water comes from a tank because the only variable is the height of the water above the suction flange of the pump. You do not have to worry about the huge fluctuations that can occur when water is provided to a pump from a city water main if you are using a tank to provide the water. The city water main can be used to fill the tank, but this can happen at any pressure during a time when there is no fire.

Even though NFPA 20 does not require tanks, NFPA 14 does (see the previous portion of this chapter) under most conditions. The only way that you can meet NFPA 14 in a super high rise building without a tank as a part of the high zone water supply is to put together an option that is acceptable to the AHJ. As previously discussed in the NFPA 14 portion of this chapter, it is not clear what this entails, and this may be a very difficult option to select. Most people end up using tanks in order to meet the requirement of NFPA 14 when designing standpipe systems for super high rise buildings.

Section 5.6.1.1 of NFPA 20 says that if you are going to use a tank, you can't just use one. If the tank needs to be taken out of service for repair or maintenance, then you don't want to lose all of your fire protection. So, Section 5.6.1.1 says that you need to split up the water storage into multiple tanks that are valved separately. This way, if you have to take one of the tanks out of service for repair or maintenance, you still have the water in the other tank to work with. The tanks don't need to each hold the demand of the fire protection system. Instead, each tank is allowed to hold up to 50 % of the demand so that with one tank out of service, you have not lost more than 50 % of the water you need for fire protection. For example, you could use two tanks with each tank having at least 50 % of the water supply demand. Or, you could use three tanks with each tank having 33.3 % of the water supply demand. A third option would be to use three tanks with one tank having 50 % of the water demand, one tank having 40 % of the water demand and the third tank having 10 % of the water demand. The number of combinations is infinite and each of them is acceptable as long as when one tank is out of service, you still have at least 50 % of the water demand available from the other tank(s).

It is possible to purchase a single tank that has separate compartments that are valved separately. Such a tank can have one portion drained and being repaired or maintained and the other portion of the tank can remain in service. Since this single tank really acts like two different tanks, NFPA 20 allows this alternative without question as long as the different portions of the tank act just like separate tanks.

Refill Mechanisms

Although the tanks are sized to meet the duration demand of the standpipe system (30 min of flow), each tank (or compartment of a tank) is also required to have an automatic refill that puts water back into the tank at the rate of the standpipe system flow demand (Section 5.6.1.4). This makes the water supply an infinite duration, assuming that everything works correctly.

In addition to the automatic refill, each tank (or compartment of a tank) also needs a manual refill that can meet the flow rate of the standpipe system. During a fire, if the manual fill is not working, someone can go open the manual fill and keep the water flowing beyond the 30 min duration. This is a little more difficult to control because there is no automatic shut off for the manual refill valve, and you don't know how fast the water is going out, so a person needs to monitor this valve continuously while having some knowledge of the water level in the tank and how fast it is changing in order to use this manual refill mechanism safely.

There are two options for connecting the automatic and manual fill valves provided in Section 5.6.1.5. The first option is to connect them to a standpipe riser that is supplied with a back-up pump. The second option is to connect them to a reliable domestic riser that can supply the flow demand of the standpipe system. The minimum size of the connection for the automatic and manual fill mechanisms is difficult to calculate, but not impossible. Just how small the connection could be depends on the water pressure available, the distance from the riser to the refill mechanism, and the number of fittings in the piping for the connection to the refill mechanism. One way to select the size is to use pipe that is the same size as the suction piping going to the pump. If the water supply can provide the system flow demand to the pump (which it is required to do by NFPA 20), then it should also be able to provide the same flow to a tank refill mechanism.

Another way to size the refill mechanism is to do a hydraulic calculation. The hardest part about this calculation is figuring out the pressure at the refill mechanism that will achieve your required flow. There are a number of ways to do this. The following is one method that works pretty well, using formulas already familiar to the people in the fire protection industry.

Since this is a tank in a high rise building, the pump needs to return the water to the tank in a manner where the flow can be measured (see the discussion on NFPA 20 requirements for regular high rise buildings). If we assume that the method that will be used is a test header with three underwriters playpipes already built in that have a 2.5 in. diameter at the outlet (as shown in Fig. 10.3), you can calculate backwards through the formula that converts pitot gauge readings into flows in order to figure out what velocity pressure is necessary to create a flow of 334 gpm out of each nozzle, which would be a total flow of 1000 gpm as follows:

$$P_V = \left(\frac{Q}{29.83cd^2}\right)^2 = \left(\frac{334}{29.83(0.9)(2.5)^2}\right)^2 = 4\,psi$$

Note that the nozzle coefficient of 0.9 that is used here is a bit low for an underwriters playpipe, but there are a number of conservative assumptions that will be made in this calculation and rather than try and be too precise, the goal here is to just show that a certain size pipe and arrangement will work for the refill situation.

This formula shows that a very low velocity pressure can create the flow we need. This velocity pressure outside the piping system converts almost completely to total pressure as you follow the water back to the piping system. Even if you get incredibly conservative and say that there is some energy lost as the water goes from total pressure in the piping to velocity pressure as it exits the nozzle and you assume that 20 % of the energy is lost, then a velocity pressure of 4 psi would convert to a total pressure of 5 psi in the piping (4/0.8 = 5), so we will assume that any delivery system that can get 5 psi to the piping system just prior to the nozzles will be able to deliver 1000 gpm to the nozzles.

If we tie the refill mechanism into the test header and deliver 5 psi to the nozzles, then we will have met our goal. The flow of 1000 gpm in 4 in. schedule 40 steel pipe creates a friction loss of 0.259 psi per ft. If the refill mechanism is being fed from a standpipe 150 ft away with four elbows, three tees, and a butterfly valve between the connection to the standpipe and the connection to the test header, the equivalent length of the fittings will be 112 ft (40 + 60 + 12 = 112), so the total length of pipe and fittings will be 262 ft. The total friction loss between the standpipe and the test header will be 68 psi (0.259 × 262). This means that as long as the standpipe riser can provide 1000 gpm at 73 psi (68 + 5 = 73), the 4 in. refill mechanism should be fine for returning 1000 gpm to the tank when it is connected to the test header arrangement. Since the standpipe riser is designed to have a minimum of 100 psi, this should work out fine.

Redundancy

Section 5.6.2 requires that every zone that is above the pumping ability of the fire department have either a backup fire pump or some auxiliary means of providing the full fire protection system demand that is acceptable to the AHJ. Once again, the committee has not explained what criteria the "auxiliary means" would need to have. They just say that they want the ability to provide the whole system demand and they want the AHJ to approve whatever gets designed and installed in advance. This is where engineers have the ability to get creative.

Figure 10.5 puts together the concepts of the tank requirements, refill mechanism requirements, and redundancy requirements in an attempt to summarize all of the rules from NFPA 20. The equipment in this figure is one possible way to meet all of the requirements of NFPA 20 for super high rise buildings, while also meeting the explicit requirements of NFPA 14. It is not the only way to meet all of the rules of both of these standards, especially when you consider the clauses in NFPA 14 and NFPA 20 that give the AHJ great authority to accept almost anything.

Figure 10.5 shows a 50 story building that has been broken up into five vertical zones. Each zone covers ten floors with three standpipes (flow demand of 1000 gpm) and has been carefully designed so as not to require any pressure reducing valves. Each of these five zones acts like its own standpipe system, with pumps taking suction from tanks within the pump room and distributing that flow to the three risers on the ten floors. In order to show all five zones, the pump rooms are too small to show detail, so Fig. 10.6 is a plan view showing how a typical pump room in this building would be arranged. The typical pump rooms in this building will be in Zones 2, 3, 4 and 5. The pump room in Zone 1 would be similar with the fill lines coming from the city water main rather than the standpipe risers from the lower floors. The fill line labeled "Tank Refill Line 1", represented by the blue line in Fig. 10.5, serves the automatic fill for Tank 1 in the pump room and the manual fill for the Tank 2 in the pump room. The fill line labeled "Tank Refill Line 2", represented by the red line in Fig. 10.5, serves the automatic fill for Tank 2 in the pump room and the manual fill for Tank 1. These two refill lines help meet two requirements from NFPA 14 and NFPA 20:

- NFPA 14 requires two connections between the standpipe system in one zone and the system in the zone below it. For situations where pumps are installed in series, this is typically met with pipe the size of suction piping. But since the use

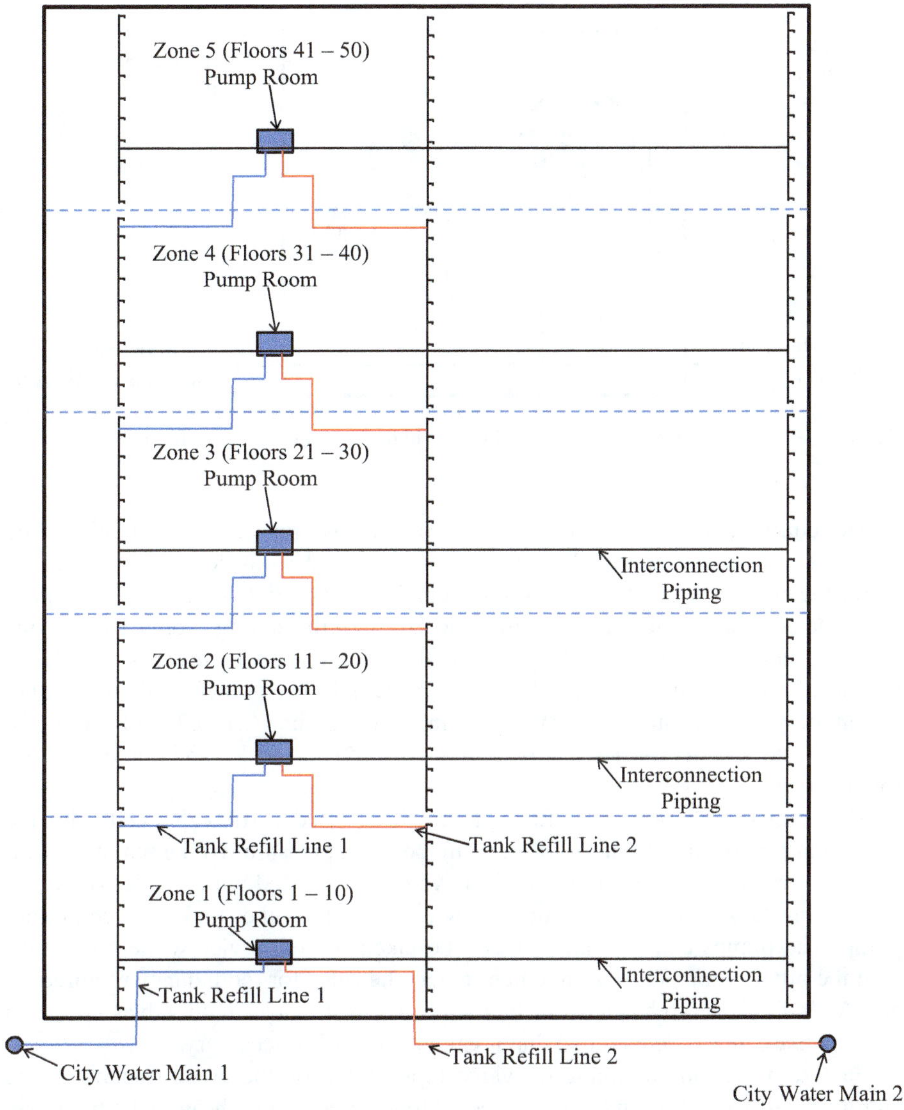

Fig. 10.5 One arrangement of equipment that meets the super high rise requirements of NFPA 14 and NFPA 20

of tanks in this example eliminates the concept of pumps in series, the pipes can be smaller and will still provide the flow demand of 1000 gpm.

- NFPA 20 requires each tank to have an automatic and a manual refill line and that the lines come from separate standpipe risers. The use of these two refill lines from different risers meets that requirement.

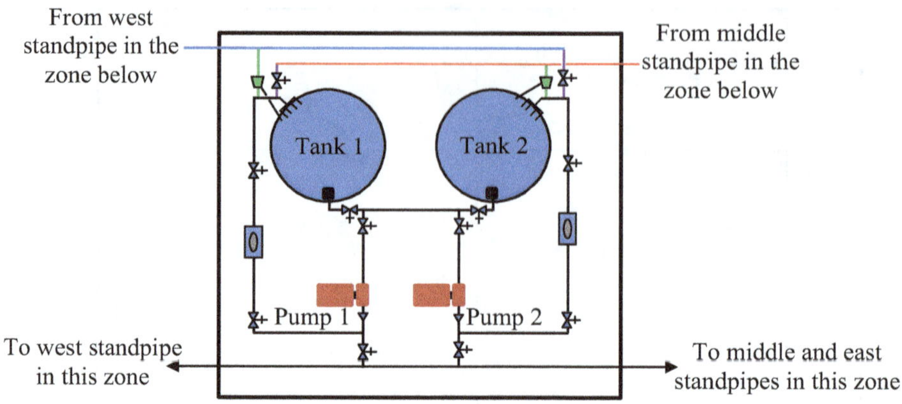

Fig. 10.6 Plan view of typical pump room in super high rise building shown in Fig. 10.5

The equipment in Fig. 10.6 has been color coded for better explanation. The tanks are shown in blue. Each tank needs to be sized for at least 15,000 gallons above the anti-vortex plate so that the system demand of 30,000 gallons (1000 gpm times the 30 min duration) is met. Since the height of the ceiling is only 10 ft on this floor, the tanks can't be too tall. A tank with a diameter of 19 ft can store 2120.9 gallons of water for every foot of height. If the tank is 7.6 ft tall, it will hold 15,058 gallons of water if the anti-vortex plate is 6 in. off the ground (2120.9 × 7.1 = 15,058). This still leaves some space at the ceiling for the test header and some working room.

The pumps are shown in red. Each pump will be rated at 1000 gpm and will have a very flat pump curve to minimize how much extra pressure will be put in the system at churn. The pumps and the tanks have been provided with a series of valves so that one tank can be taken out of service while the other tank can feed either pump. The pumps can also individually be taken out of service while both tanks feed the other pump. This arrangement meets the rules for redundancy required by NFPA 20. Each pump has a meter that leads to a test header with valves upstream and downstream of the meter so that it can be isolated if necessary.

Each tank has an automatic fill valve represented by the green trapezoid. The automatic fill valve for Tank 1 is connected to a water supply through the blue line, which comes from the standpipe on the west side of the building in the zone below the one in which the pump room is serving. The automatic fill valve for Tank 2 is connected to the water supply through the red line, which comes from the middle standpipe in the zone below the one in which the pump room is serving. The black line that goes from the automatic fill valve to the tank is a sensing line that uses the pressure in the tank to keep the fill valve closed. When the water level in the tank drops, the pressure near the bottom of the tank goes down, lowering the pressure. The automatic fill valve senses this drop in pressure through the sensing line and automatically opens, sending water back into the tank. Note that each of the automatic fill valves is receiving water from different standpipes at the lower level, so if one standpipe is out of service on the level below, it will not affect both connections and the system above will still get all the water it needs.

Each tank also has a manual fill valve, represented by the purple line in Fig. 10.6. There is a simple control valve on this line that can be opened manually if the automatic fill valves fail to operate. The manual fill line for Tank 1 is connected to the red line, which is fed from the standpipe in the middle of the building in the zone below the one which the pump room is serving. The manual fill line for Tank 2 is connected to the blue line, which is fed from the standpipe on the west side of the building in the zone below the one where the pump room is serving. Each of the manual fill lines takes its water supply from a different riser than the automatic line, so if the standpipe connected to the automatic valve is shut down, the other one for the tank will be available.

IBC Rules for Super High Rise Buildings

The International Building Code has three requirements for the sprinkler systems in buildings that are more than 420 ft tall. If the sprinkler system shares the water supply with the standpipe system, these rules will also apply to the standpipe system. The rules can be found in Sections 4033.3.1, 403.3.1.1 and 403.3.2.

Sections 403.3.1 and 403.3.1.1 have to do with the risers that serve as the supply for the sprinkler system. The language in Section 403.3.1 is rather confusing, so it has been repeated here so that it can be carefully explored:

> **403.3.1 Number of sprinkler risers and system design.**
> Each sprinkler system zone in buildings that are more than 420 ft in building height shall be supplied by no fewer than two risers. Each riser shall supply sprinklers on alternate floors. If more than two risers are provided for a zone, sprinklers on adjacent floors shall not be supplied from the same riser.

The first part of the section says that it applies to "Each sprinkler system zone". The terminology is not clear here. There is no definition of a "sprinkler system zone". It is most likely that this is referring to the vertical zones that are used to control the pressure in standpipe systems. An alternate interpretation would be the alarm zones within a building, which are required to be on each floor. The first interpretation is better because the language makes less sense when you try to consider each floor as a separate "zone" and then try and apply the rules of this section.

So, if you believe that the concept of a "sprinkler system zone" is the same as the vertical zone used for the standpipe system, then the next part of the section says that the zone has to have at least two risers and that each riser be used to supply sprinklers on alternate floors. The idea here is that if one of the risers is taken out of service, the fire might burn from one floor to the next, but when it gets to the next floor, the sprinklers on that floor might be able to control the fire since they are fed from a different riser.

The last sentence in the section appears to be redundant. It says that sprinklers on adjacent floors cannot be supplied from the same riser, which they can't be if the

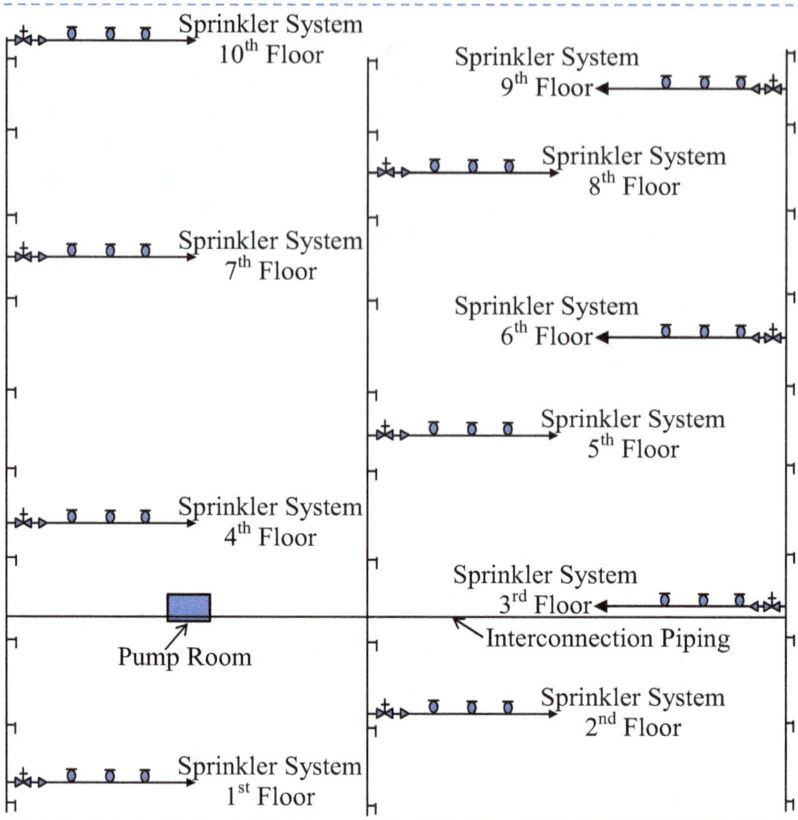

Fig. 10.7 Use of standpipe risers to supply sprinkler systems on alternate floors of a super high rise building

earlier part of the section (using the multiple risers to supply sprinklers on alternate floors) is already complied with. This is an example of the challenges that arise from trying to write and revise a code as complex as the IBC. Figure 10.7 is an example of one vertical zone of a building that meets one reasonable interpretation of this section. This figure assumes that the sprinkler system is either light hazard or ordinary hazard and that the building does not exceed 52,000 ft^2 per floor. Each of the other zones in this building would follow a similar pattern as long as the sprinkler system on the first floor of the zone above the one shown is not also fed from the riser on the left.

Section 403.3.1.1 requires the risers to be placed in the exit stairwells and that they be remotely located in the building in accordance with the egress rules. This is another example of redundancy in the Code. The stairways already have to be placed remotely in accordance with the egress rules and the standpipe systems need to be installed in the stairwells in accordance with both the IBC and the NFPA 14. There is no requirement for the sprinkler system to use the standpipe risers, but it does make sense.

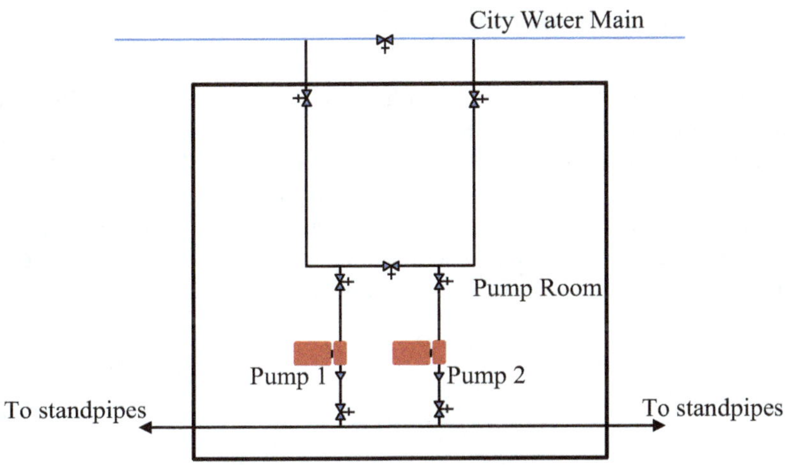

Fig. 10.8 Connection to a single main so that it can be isolated

Section 403.3.2 has to do with connecting the system to the water mains. Specifically, it says that for buildings more than 420 ft in height, the fire pumps need to be connected to at least two separate mains with each connection sized to handle the demand of the fire protection system. One of the problems with this section is that it does not recognize what to do when fire pumps are fed from tanks as shown in Fig. 10.5. The fill connections from separate mains should fulfill the requirements of this section, even though they are not directly connected to the pumps because they fill the tanks that feed the pumps.

There is an exception to this requirement. The IBC allows a connection to a single water main if the main can be isolated so that the water supply can continue into one of the connections. Figure 10.8 shows an example of one such arrangement that should meet this rule. Note that the two connections continue to the common piping for the redundant pumps so that any single supply pipe can provide water to any single fire pump.

Another one of the challenges with this section of the Code is that it really only makes sense to apply the rule to the first set of pumps. The pumps being used in the upper zones of the building should not have to be connected all the way down to the street main, but technically, there is not exception for pumps serving upper zones of the building a fed by tanks with refill lines as shown in Fig. 10.5. The logical interpretation would be that the two different fill lines for the tanks meet this rule.

The arrangement in Fig. 10.8 assumes that the water main is a part of the loop that can be fed from either side. If there is a break on the left side of the control valve on the main, that control valve can be closed and the control valve on that side of the supply pipe in the pump room can be closed in order to isolate the break and still provide water into both of the pumps. The same type of isolation can be performed for a break in the main on the right side of the control valve.

Putting the Super High Rise Building Rules Together

Figure 10.5 does a pretty good job of putting all of the rules of the NFPA standards and the IBC together for super high rise buildings. The high level water storage required by NFPA 14 is there. The redundancy required by NFPA 20 is there. The tanks are separated and valved to be able to supply either pump. Both automatic and manual refill valves are provided for both tanks and both refill valves for each tank are fed from different risers. Of course, you only need to put the rules together in the circumstance that the building is over 420 ft in height.

Test Yourself

10.1. Which of the following buildings is the lowest to be considered a high rise by the model codes?

(a) When there is a floor 55 ft above the lowest level of fire department vehicle access.
(b) When there is a floor 75 ft above the lowest level of fire department vehicle access.
(c) When there is a floor 200 ft above the lowest level of fire department vehicle access.
(d) When there is a floor 350 ft above the lowest level of fire department vehicle access.
(e) When there is a floor 420 ft above the lowest level of fire department vehicle access.

10.2. At what building height does the International Building Code require two connections to the city water main for the fire pumps?

(a) When there is a floor 55 ft above the lowest level of fire department vehicle access.
(b) When there is a floor 75 ft above the lowest level of fire department vehicle access.
(c) When there is a floor 200 ft above the lowest level of fire department vehicle access.
(d) When there is a floor 350 ft above the lowest level of fire department vehicle access.
(e) When there is a floor 420 ft above the lowest level of fire department vehicle access.

10.3. A building in which seismic design categories needs a secondary on-site water supply?

(a) A, B, C, D
(b) B, C, D, E

(c) C, D, E, F
(d) B, C, D, E, F
(e) A, B, C, D, E, F

10.4. Name two types of buildings that the IBC exempts from the high rise rules, even if they meet the definition of high rise.

10.5. In a building too high for the fire department to supply water to the upper floors, two or three tanks are going to be installed to meet the duration demand. The standpipe flow demand is 1000 gpm. Which of the following is an acceptable design for the tanks?

(a) Two tanks: 18,000 gallons and 12,000 gallons
(b) Two tanks: 21,000 gallons and 10,000 gallons
(c) Three tanks: 16,000 gallons, 7000 gallons and 7000 gallons
(d) Three tanks: 18,000 gallons, 12,000 gallons and 5000 gallons
(e) Three tanks: 18,000 gallons, 12,000 gallons and 2000 gallons

Chapter 11
Hanging, Bracing and Protection of System Piping

There are three related concepts being discussed in this chapter: the hanging or support of pipe, the bracing of pipe to resist seismic forces, and the protection of system piping from potentially destructive forces. The common theme in all three of these concepts is "protection". The hanging or support of pipe protects the pipe against gravitational forces. The bracing rules actually encompass everything necessary to protect the pipe against damage during earthquakes. The rest of the protection concerns deal with a variety of potentially destructive forces like mechanical damage, fire and freezing.

Hanging and Support of Standpipe Systems

NFPA 14 only has one section that deals with the rules for supporting the standpipe system piping against gravitational forces. Section 6.5 just says, "Support of system piping shall be in accordance with NFPA 13." That one statement could be the end of the discussion, which would make this a very short chapter, but it would be doing a tremendous disservice to the reader. The assumption that the reader actually knows the rules of NFPA 13 would be inappropriate, and it will be a valuable experience to examine these rules with standpipes in mind rather than sprinklers.

While considering the rules for supporting standpipe system piping, it is important to remember that the rules for NFPA 13 were developed with a typical fire sprinkler system in mind. The typical fire sprinkler system has piping running in two horizontal directions with relatively little vertical pipe. The pipes running in two different horizontal directions are branch lines (the pipes to which sprinklers are directly attached), which typically run parallel to the main structural members, and mains (large pipe connecting branch lines), which typically run perpendicular to the main structural members of the building, making them also running perpendicular to the branch lines. The hangers on the branch lines frequently also carry some of the gravitational load of the mains.

K.E. Isman, *Standpipe Systems for Fire Protection*,
DOI 10.1007/978-3-319-47750-3_11

In a standpipe system, the piping is very different from sprinkler systems. Most of the pipe in a typical standpipe system is vertical with horizontal pipe only being used to interconnect vertical standpipes. The horizontal pipe does not necessarily run perpendicular to the main structural members of the building, so it may be more difficult to attach the hangers to the building structure. In addition, the standpipe does not have pipe running perpendicular to the mains that help support it, so the mains have to support themselves, which will mean that the hangers will be closer together than in sprinkler systems, which might mean the use of trapeze hangers more often than with sprinkler systems.

The last difference between sprinklers and standpipes that needs to be pointed out here is the use of the term "branch line". NFPA 13 uses this term to refer to the smaller pipes that are directly connected to sprinklers whereas NFPA 14 uses this term to refer to the pipes that connect an individual hose connection to a standpipe. In a sprinkler system, branch lines are typically 2 in. in size or smaller. In a standpipe system, branch lines have to be 2.5 in. in size or larger. The entire section of NFPA 13 that applies to branch lines should not be used in a standpipe system just because the standpipe committee has chosen to use the same term and apply it to pipes with a different definition.

The hanging and bracing rules in NFPA 13 are found in Chap. 9, which divides the rules into two categories: general rules and installation rules. This text will follow a similar structure. The general rules are in section 9.1 and establish some performance-based criteria. In addition, the general rules discuss the materials that are allowed to be used for hanging and support. The installation rules are in section 9.2, which covers the maximum allowable spacing between hangers or supports and the maximum allowable unsupported lengths of pipe.

General Support Rules

There are four basic concepts for supporting the standpipe system discussed in NFPA 13. The standard goes on for about 30 pages (with additional supporting materials in the annex) on prescriptive methods for applying these four concepts, but a Professional Engineer (PE) is allowed to do whatever they want, as long as they adhere to these four basic ideas:

1. The hanger needs to be able to support five times the weight of the water filled pipe plus an additional load of 250 lbs.
2. The points of support (places on the building where the hangers are attached) need to be adequate to handle the load described in concept 1.
3. The maximum allowable spacing between hangers has already been calculated given the beam strength of the pipe and is included in a table (one table in traditional foot-pound units in one table in Metric units).
4. Hanger components need to be ferrous.

Each of these items will be discussed in more detail, starting with the load. It is important to note that the hanger load calculation is intended to be an ultimate strength calculation, meaning that the strength of the hanger is considered to failure without any additional safety factors applied. Multiplying the weight of the pipe by five is supposed to be one safety factor and then adding the additional load of 250 lb is an additional safety margin. It would be inappropriate to compare the load calculated in accordance with NFPA 13 to a load that has been calculated with a safety margin.

For example, consider a hanger where the manufacturer has determined that the components will fail at 1200 lb. The manufacturer then applies a safety factor of four and advertises that this hanger can support a weight of 300 lb. If you calculate the weight of your water filled pipe, multiply by five and add 250, coming up with a total load of 1150 lb, you may think that you cannot use a hanger being advertised to hold 300 lbs. But in fact, you can use this hanger. If you go back to find the ultimate strength of this hanger, it is 1200 lbs. Your load of five times the weight of the water filled pipe plus 250 lb only comes to 1150 lb, so this hanger is acceptable.

The same can be said for the building structure. The point at which the connection is made to the building structure needs to be able to handle the load of five times the weight of the water filled pipe plus 250 lb. Again, this is an ultimate strength calculation and needs to be compared to how strong the structure is before a safety factor or margin has been applied. The fire protection design professional should work carefully with the structural engineer to make sure that loads are accounted for appropriately.

Typically, the hanger is physically connected to the structural member of the building. Section 9.2.1.3.1 of NFPA 13 requires the system to be "substantially supported from the building structure". The use of the word "substantially" is interesting. The committee may have meant this to be interpreted as "greatly" or "most importantly", which are some of the meanings of "substantially". But another meaning of the word "substantially" is "for the most part", which would imply that every hanger is not required to be connected directly to the building structure as long as most of them are. This could allow hangers connected to non-load bearing parts of the building as long as they can handle the loads discussed here. Ultimately, the PE will need to evaluate these conditions and make the decision as to which part of the building they will connect their hangers. The performance-based clause at the beginning of the chapter allows them to use almost anything that can handle the load.

The additional 250 lb safety margin is worth discussing. This value comes from the idea that while the pipe is being installed near the ceiling, the installer (typically referred to as the "fitter" in the industry) is in a potentially dangerous situation. The fitter should be working from a safe platform following all OSHA rules for work in high spaces, but if the fitter begins to fall, they should be able to grab the pipe and hang on without the pipe hangers breaking. This is where the extra 250 lb comes from, the maximum expected weight of a fitter. It is only expected that one fitter at a time would be falling and grabbing for the pipe, so this additional load of 250 lb is only supposed to be added to the building structure at one location at a time, not cumulatively added at every point of connection from the standpipe system to the building.

The PE has been given great latitude by NFPA 13 to pretty much make their hangers as they want and ignore most of the prescriptive rules of the standard. But there are two prescriptive rules that the PE is not allowed to violate. The first is the hanger spacing table. The NFPA Committee on Hanging and Bracing has spent considerable time calculating the beam strength of steel, copper, CPVC and ductile iron pipe and applying appropriate safety factors, which is how the values in the table have been determined. The committee does not want another engineer substituting their calculations with the calculations of the committee, so the table needs to be followed no matter what. If the PE is using a different type of pipe (such as brass or copper-nickel), then the PE will need to do some calculations and stay consistent with the safety factors that went into the table.

The second item where the committee allows no deviation from the prescriptive rules is the requirement for the hanger to be ferrous (made with iron, which includes steel). There are two reasons for the ferrous rule. The first is that other materials, while they may have the necessary strength, may be combustible. When exposed to fire, they may not support the pipe long enough for the fire protection system to be used. For a standpipe system, it may take the fire department a long time to get to the scene and prepare to fight the fire considering that the fire might be in the upper portion of a building and that rescue work needs to be accomplished first. After the firefighters start using the standpipe system, it may be some time before the fire is under control. So, the hangers in a standpipe system need to hold up to significant fire exposure and still hold their strength. Even if the material is not combustible, other noncombustible metals tend to reach their yield strength at lower temperatures than steel, which might lead to unacceptable performance.

In recent years, two exceptions to the requirement that hangers be ferrous have been added to NFPA 13. The first is that nonferrous hangers can be used where the hanger has been fire tested and listed for use in fire protection systems (see Section 9.1.1.6.2). The second is a recognition that the structural members of a building can support the sprinkler pipe on their own without a separate hanger. For example, in a building of wood or concrete construction, a hole could be placed in the wood or concrete and the pipe could be run through the hole, resting on the bottom of the hole. The wood or concrete (which is not ferrous) would be the support for the pipe. This is permitted by NFPA 13 (see Section 9.1.1.6.3) however, the structural engineer will need to approve the location and size of the hole.

Use of Listed Materials

There are three parts to every hanger as shown in Fig. 11.1. There is the part that attaches to the pipe (the blue part at the bottom of Fig. 11.1), the part that attaches to the building (the black part at the top of Fig. 11.1), and the part that connects the first two parts to each other (the blue rod in Fig. 11.1). In general, NFPA 13 requires the first two parts to be listed (Section 9.1.1.5.1), while the third part does not need

Fig. 11.1 Hanger example

Fig. 11.2 Sprinkler hanger with bottom beam clamp and adjustable swivel ring

to be listed as long as it is properly sized. There are some exceptions to this rule that could use some explanation. Figure 11.2 shows a sprinkler pipe hanger with a bottom beam clamp at the top, an adjustable swivel ring at the bottom and rod in between that is threaded at each end so that it can screw into the beam clamp and the adjustable swivel ring.

The first exception is that mild steel rods are not required to be listed (Section 9.1.1.5.2). These rods need to be sized in accordance with one of the tables in Section 9.1.2, but the rods themselves are not required to be listed. The rods can be attached to the building structure and bent to set up a cradle as shown on the left side of Fig. 11.3 (which would be called a "U-hook) or they can be wrapped around the

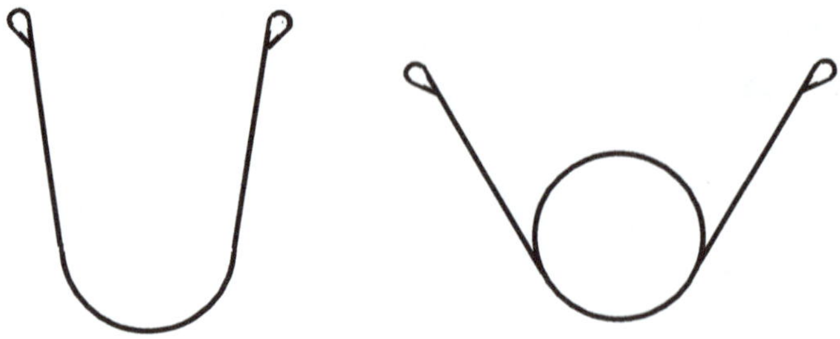

Fig. 11.3 U-hook and wraparound U-hook

Fig. 11.4 Sprinkler pipe supported by U-hook

pipe as shown on the right side of Fig. 11.3 to form a "Wraparound U-hook". The U-hook is used when gravity is the only concern. The wraparound U-hook is used to support the gravity load when some upward movement of the pipe is also a concern such as controlling the reaction force when a sprinkler discharges or restraining the movement of piping during an earthquake. Note that you can bend the rods yourself to make such hangers if the rod is not threaded. If the rod is threaded, it is not allowed to be bent (Section 9.1.2.6) because the threaded section is weaker and once bent, does not retain its strength for holding up the pipe. Figure 11.4 shows a sprinkler pipe supported by a U-hook in the middle of the picture. Figure 11.5 shows a sprinkler pipe supported by a wraparound U-hook in the middle of the picture.

Fig. 11.5 Sprinkler pipe supported by wraparound U-hook

The second exception to the listing rule (Section 9.1.1.5.3) is an exception for fasteners that attach the hanger to the building structure. Fasteners that go into concrete need to meet Section 9.1.3. Fasteners that go into steel need to meet Section 9.1.4. Fasteners that go into wood need to meet Section 9.1.5. Each of these sections contains rules for using the fasteners and selecting the correct size. Even though Section 9.1.1.5.3 says that fasteners are not required to be listed when installed in accordance with their proper section, once you get into the section you find that there are types of fasteners that do need to be listed. For example, Section 9.1.3.9.1 requires powder-driven studs to be listed, even though they are in section 9.1.3. So, if you are planning on using powder-driven studs, you need to make sure that they are listed for fire protection system use and that you use them in accordance with all of the restrictions of their listing. The exception to the listing requirement is meant to handle more common fasteners such as lag bolts and through bolts.

There are a wide variety of fasteners that can be used with concrete. The fasteners can be placed when the concrete is poured (which takes a great deal of coordination with other contractors on the site) of they can be installed after the concrete has set (which are called "post-installed anchors). The basic idea for these fasteners is to get an object firmly embedded in the concrete with a female threaded connection. A rod can then be threaded up into the connection for form two parts of the hanger. The bottom of the hanger can be threaded on to complete the assembly.

One of the challenges with using concrete for a building material is that all types of concrete are not equal in terms of the stress they can withstand. The actual amount of load that the concrete can handle depends on exactly how the concrete is mixed and exactly how much water is added, and every batch is slightly different. So, NFPA 13 requires that before certain fasteners are used in concrete, like powder-driven studs, the user needs to take a sample of the concrete and test it to verify that it can hold a specific minimum load (Section 9.1.3.9.3). Also, if you are planning on using powder-driven studs as a fastener into concrete and the building is in an active earthquake zone, you may need to make a new plan. Section 9.3.7.7 of NFPA 13 limits regularly listed powder-driven studs to areas where the horizontal force created by the earthquake is 0.5 times the weight of the water filled pipe. If you want to use powder-driven studs in an area where a greater size earthquake is expected, you will need to find a powder-driven stud specifically listed for use in earthquake areas.

More examples of different types of hangers that can be used are shown in Figure A.9.1.1 of NFPA 13. There are a variety of clamps and hooks that attach to some piece of the structural member. There are flanges that attach to the ceiling and there are attachments that connect to the side of the structural member. Most of these hanger pieces are designed to drop vertically to support a piece of horizontal pipe.

There are very few support pieces that hold up vertical pieces of pipe. The riser clamp, shown in the top row of Figure A.9.1.1 as the second object from the right can work very well to hold vertical pipe. The pipe is run vertically through a hole in the floor. The clamp is connected tightly to the pipe just above the hole. The ends of the clamp extend beyond the hole, transferring the load to the floor. Figure 11.6 shows a standpipe riser with a riser clamp at the point where the red riser penetrates the gray floor landing.

Fig. 11.6 Standpipe with riser clamp

Trapeze Hangers

All of the hangers discussed so far for supporting horizontal pipe have been designed to have a connection at the structural member and then drop straight down to the pipe. But what do you do when you need to hang the pipe in a location that is not directly below a structural member? The answer is to use a trapeze hanger to bridge between structural members as shown in Fig. 11.7. In order to use a trapeze hanger, each of the pieces of the hanger needs to be strong enough to support the standpipe piping individually.

In Fig. 11.7, the standpipe interconnection piping that runs between standpipes needed to be run in a location that was not directly below a structural member. In order to support this pipe, a trapeze member, represented by the red line, is supported from two structural members. A hanger can then be dropped straight down from the trapeze to support the standpipe piping.

One important aspect of using trapeze hangers is sizing the trapeze member to be able to support the load of the piping it supports. Since most of the people that use NFPA 13 are not structural engineers, the committee has simplified the process for everyone. NFPA 13 uses an engineering concept called a "section modulus" to help people determine whether the trapeze can support the piping. There are two tables in NFPA 13: one for the pipe that is being supported and the other for the item that is going to be used for the trapeze member. When selecting an item to be used as a trapeze member, just make sure that the section modulus for the trapeze member is greater than the section modulus for the pipe being supported.

Table 9.1.1.7.1(a) is the table that provides the section modulus for the pipe being supported. The top half of the table applies to Schedule 10 steel. The bottom half of the table applies to Schedule 40 steel. If you are using some other type of pipe for your standpipe system, you will need to do some calculations. The notes at the bottom of the table provide some information on how the calculations in the table have been performed so that an engineer could perform similar calculations for other types of pipe.

Table 9.1.1.7.1(b) provides the section modulus for three products that are common to find for trapeze members: Schedule 10 steel pipe, Schedule 40 steel pipe, and angle iron. Other items are permitted to be used for trapeze members, but

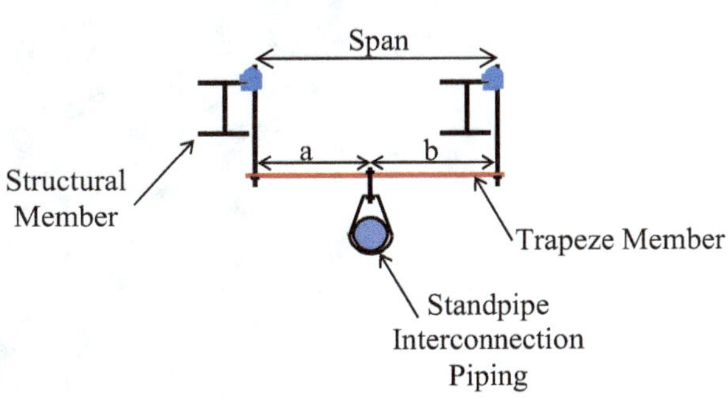

Fig. 11.7 Trapeze hanger between structural members

someone will need to calculate the section modulus to determine whether the pipe can be supported. Some people get confused because Schedule 10 and Schedule 40 pipe appear in both tables. Since the pipe is common on the job site, it is common to find trapeze members made out of the pipe. Just remember to use Table (a) for the water filled pipe that needs support and Table (b) for the trapeze member spanning between structural members.

For example, consider a situation where the 6 inch Schedule 40 standpipe interconnection piping needs to go somewhere between structural members that are five feet apart. According to Table 9.1.1.7.1(a), the section modulus for 6 in. Schedule 40 pipe between a 5 ft span is 0.72. Then you need to determine what you will use for a trapeze member. You can use any material with a section modulus of 0.72 or better. According to Table 9.1.1.7.1(b), you could use 3 in. Schedule 10 pipe, 2.5 in. Schedule 40 pipe or 2.5 × 2.5 × 0.5 in. angle iron. Of course, you could use anything larger than these as well.

The method for sizing trapeze members discussed above involves a worse-case assumption that the pipe being supported is approximately in the middle of the span (in other words, the distance "a" in Fig. 11.7 is approximately equal to the distance "b"). If the pipe being supported is not near the middle of the span, you are allowed to take advantage of this and size the trapeze member a little smaller by calculating a shorter effective span (L) using the formula (where "a" and "b" are the distances shown in Fig. 11.7):

$$L = \frac{4ab}{a+b}$$

Figure 11.8 shows an example of a trapeze member for a fire protection main very close to one of the structural members. In this case, the fire protection main

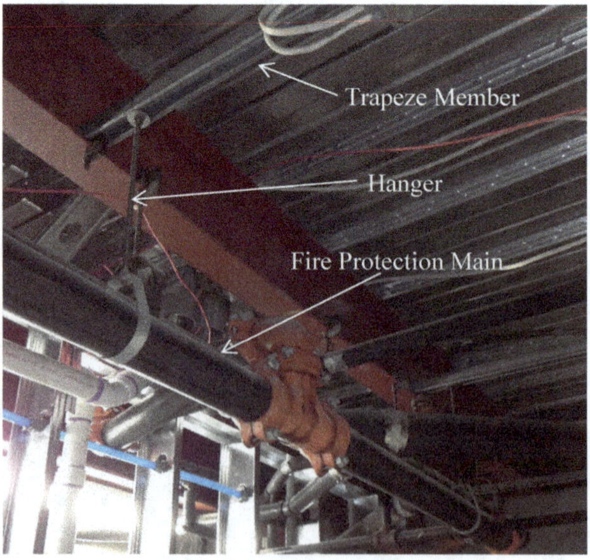

Fig. 11.8 Trapeze hanger with pipe supported very close to one side

was run parallel to the main structural member and about 6 in. away, so a trapeze was placed between this structural member and another one about 6 ft away. Rather than calculate this like the first example assuming that the pipe is in the middle of the span, we can use the formula to calculate an effective span with $a = 0.5$ and $b = 5.5$. The result would be:

$$L = \frac{4ab}{a+b} = \frac{4(0.5)(5.5)}{0.5+5.5} = 1.8\text{ft}$$

This reduces the span from 6 ft to 1.8 ft. Assuming that this pipe is 4 in. Schedule 40 pipe, this reduces the section modulus from 0.59 to 0.2, which opens up additional possibilities for trapeze members including 2 in. Schedule 40 pipe, which would not have been allowed to hold a section modulus of 0.59.

Hanging Horizontal Pipe

Sections 9.2.3 and 9.2.4 of NFPA 13 contain the installation criteria for hangers supporting horizontal pipe. The most important information in these portions of the standard is the table that provides the user with the maximum allowable distances between hangers (Table 9.2.2.1(a)). This table is based on the type and nominal size of the pipe. The following is a summary of this table using the sizes of pipes that are typical to standpipe systems:

Note that Table 11.1 shows "N/A" or "Not Applicable" for the threaded thinwall pipe for sizes 4 in. and larger. This is probably due to the fact that NFPA 13 does not allow the use of threaded thinwall pipe unless it is specifically listed for fire protection system use. If the pipe is specifically listed, the hanger spacing will be a part of the listing, so the rules will not come out of NFPA 13 for that product.

The rest of the rules in NFPA 13 for hangers on horizontal pipe are divided into rules for branch lines and rules for mains. It makes the most sense to follow the rules for mains for the horizontal standpipe piping. Even though there are some portions of standpipe systems that are called "branch lines", they are not the same as branch lines on sprinkler systems and will not be a similar size, so the rules for sprinkler system mains should be used for standpipe system branch lines.

Technically, NFPA 13 does not have any specific rules for the maximum allowable unsupported length of pipe for mains. There is a rule for branch lines that says

Table 11.1 Maximum allowable distance between hangers (in feet)

	2½ in.	3 in.	4 in.	6 in.	8 in.
All steel pipe except threaded thinwall	15	15	15	15	15
Threaded thinwall	12	12	N/A	N/A	N/A
Copper tube	12	12	15	15	15
Lined ductile iron	N/A	15	15	15	15

that the maximum allowable unsupported length of pipe is 5 ft for any steel pipe size greater than 1.5 in. This rule is typically used for mains, even though it is not explicitly stated that way in the standard.

There are some horizontal portions of standpipe systems that are close to the floor rather than being close to the ceiling. For these types of pipe, NFPA 13 has some rules for pipe stands in Section 9.2.6. As a start, NFAP 13 allows pipe stands constructed out of 2 in. pipe to support any large fire protection piping up to 10 in. Schedule 40 as long as the height of the pipe stand does not exceed 4 ft. This rule is typically used for backflow preventers, suction pipe for pumps, and discharge pipe for pumps. If you want to exceed 4 ft in height with the pipe stand, Table 9.2.6.3.1 provides allowable heights up to 26 ft depending on the diameter of the pipe being used to make the stand and the diameter of the pipe being supported.

The pipe stand needs to be anchored to the floor to keep it from tipping over. The pipe being supported by the stand also needs to be connected to the pipe stand. This is typically done with a U-bolt. The piping being supported also needs to be restrained to restrict movement from pressure surges and thrust forces at places where the pipe changes direction. A pipe ring or clamp can be used to accomplish this.

Supporting Vertical Pipe

Section 9.2.5 in NFPA 13 contains the rules for supporting vertical pipe. The user is allowed to use pipe clamps on the vertical section of pipe or a hanger on a horizontal section of the pipe if it is within 24 in. of the centerline of the vertical portion of the pipe. Supports are typically put at the bottom of the vertical section, at the top of the vertical section, and at 25 ft maximum intervals if the vertical portion of pipe exceeds 25 ft.

One method of support that NFPA 13 expressly prohibits for the vertical support of pipe is the use of riser clamps anchored to walls using hanger rods in the horizontal position. While this can be helpful to align the pipe and secure the lateral position of the pipe, it is not acceptable to hold the vertical load of the pipe. Figure 11.9 shows a standpipe system in a parking garage where the horizontal use of a hanger rod at the top of the riser helps to secure the lateral position of the standpipe, but the vertical load is supported by the riser clamp at the floor.

Standpipes are typically installed in multiple story buildings. NFPA 13 has a special section for vertical pipe in such buildings, Section 9.2.5.4. This section requires riser supports in four locations:

1. At the lowest level. Where the standpipe does not start at the ground level, the lowest level is the first ceiling above the location where the standpipe starts.
2. At each alternate level above the lowest level. Note that if the levels are very far apart, the distance between the vertical supports is not allowed to exceed 25 ft.
3. Above and below any offsets.
4. At the top.

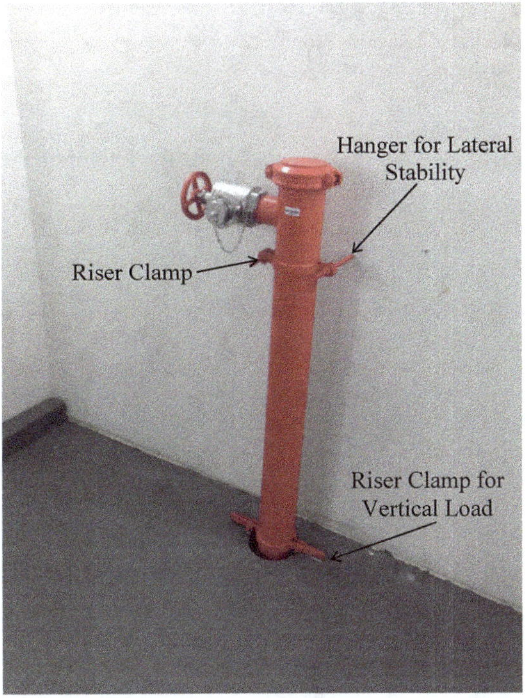

Fig. 11.9 Horizontal hanger rod to secure lateral position of standpipe and riser clamp at floor to support vertical load

It may seem odd, but the pipe also needs to be restrained against upward movement when flexible couplings are used to join the pipe. The challenge is that when the pipe is pressurized, it has a tendency to move upward and small movements can be amplified by the number of joints. In order to keep the pipe from moving upward, a riser clamp can be used on the underside of the ceiling slab. Figure 11.10 ties together the vertical support rules for a standpipe.

In Fig. 11.10, the red horizontal pipe is the interconnection piping between the standpipes. It leads to the standpipe in the stairwell, which is the blue pipe. The standpipe drops to the first hose connection, which is about 4 ft off the floor. Since the standpipe does not go all the way down to the floor, this is not considered the lowest level. As the standpipe goes up through the stairwell, the lowest level of support occurs as the standpipe penetrates the second floor. Since there is a riser support at the second level, there would not normally be a support at the third level since supports are only required every other level, but in this case, there is an offset of the standpipe, so a support is necessary under the offset. A support is required above the offset as will, so there is a support at the fourth level. Supports continue every other level at the sixth and eighth floors. A support is also required at the top. The hanger for the horizontal connection to the sprinkler system counts as the support for the top of the standpipe as long as that hanger is within 24 in. of the center of the standpipe. At each of the supports on the standpipe, a riser clamp is also installed under the ceiling to prevent upward movement of the pipe.

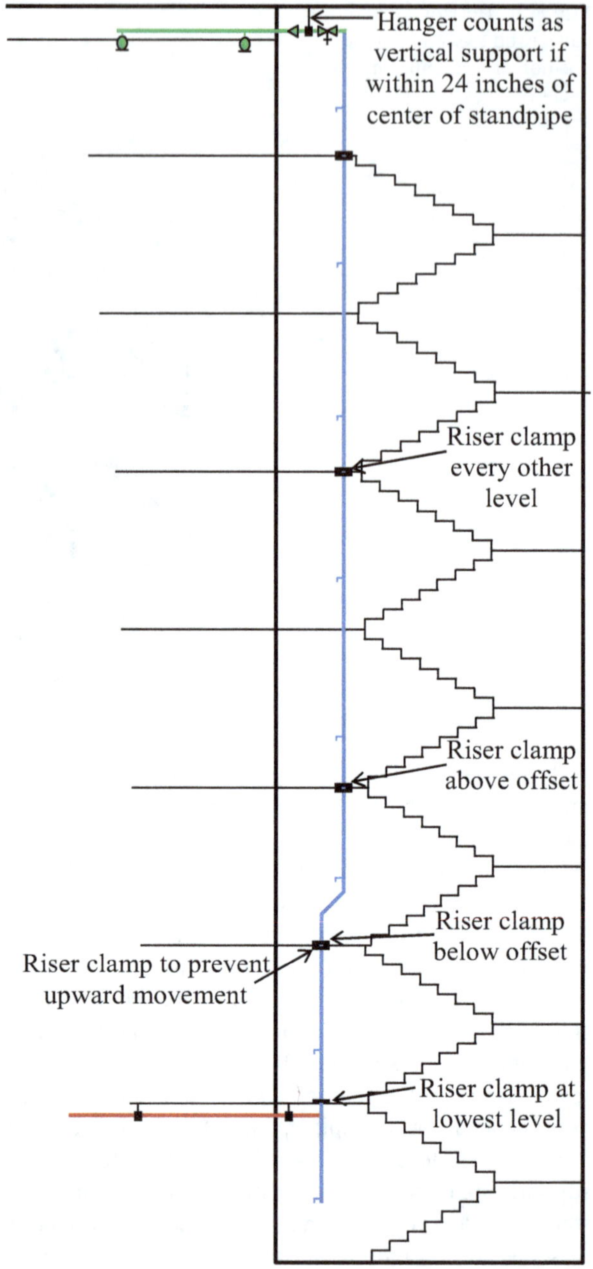

Fig. 11.10 Summary of vertical support rules for a standpipe

Seismic Design Considerations

NFPA 14 also has very little to say about protecting standpipes from damage during an earthquake. Section 6.1.2.5 of NFPA 14 says, "Where standpipe systems are required to be protected against damage from earthquakes, standpipe systems shall be protected in accordance with NFPA 13." Sections 4.4.2.2.3 and 4.4.2.4.7 of NFPA 14 allow tabs to be welded to the pipe in the field for the connection of earthquake braces as long as the rules of NFPA 51B are followed for welding on the job site. This is all NFPA 14 has to say about earthquakes, so we have to go back to NFPA 13 to get the protection rules. However, NFPA 13 does not answer the question of whether seismic considerations are necessary in the design of the system. Instead, the answer to the question will be found in one of the following places:

1. Building or fire code
2. Regulation from some authority (such as the Nuclear Regulatory Commission)
3. Insurance company recommendations
4. Architect specification

The International Building Code references ASCE 7, Minimum Design Loads for Buildings and Other Structures to determine whether seismic considerations are going to be required. ASCE 7 requires seismic consideration for fire protection systems in buildings that are in seismic design categories C, D, E and F, which include portions of about 40 of the states in the United States. Seismic activity, which used to be considered just a "California problem", has become much more of a concern almost everywhere. Section 1613.3.5 of the IBC explains how to determine the seismic design category using three variables:

1. The mapped spectral response acceleration parameter at a 1-second period, which is called S_1 in most structural engineering references.
2. The spectral response acceleration parameters SDS and SD1 calculated using formulas in Section 1613.3.4 in the IBC.
3. The Risk Category of the building. This depends on how the building is being used. Table 1604.5 of the IBC has a list of which buildings fall into each of the four categories with Risk Category I being the least risky and Risk Category IV being the most. Interestingly, a tank used for storing fire protection water and a pump house automatically fall into Risk Category IV, even if the building that they are providing fire protection for is a lower Risk Category.

If you are in an area of the United States and are unsure of what the Seismic Design Category is for a particular structure, the information for placing it into a category like the spectral response acceleration parameters described above can be obtained from the United States Geological Survey at their website www.usgs.gov using the longitude and latitude of the property where the building is going to be constructed. Since the structural engineer on a project will need this information for the design loads of the building, you can sometimes find this information in the structural specifications for the building or you can call the structural engineer and see if they will share the data with you.

Once a document requires seismic consideration in the design, NFPA 13 contains the rules for how to handle those considerations. Many people refer to these considerations as "seismic bracing", but there is much more to the design of the standpipe system. NFPA 13 actually requires flexibility, bracing and restraint of piping. For standpipe systems, the rules for flexibility and bracing will be important. Restraint is an issue for smaller pipes that are not braced, which really will not apply to standpipe systems.

It may sound like a contradiction in terms when NFPA 13 requires both flexibility and bracing, but the reason that both are necessary is that during an earthquake, the floor/ceiling assemblies tend to move parallel to each other, but potentially at different speeds or in different directions. The philosophy of NFPA 13 is to keep the horizontal pipe near the ceiling rigid to the ceiling so that it moves with the ceiling and does not bump into other objects moving with the ceiling (this is where the bracing comes in). At the same time, vertical portions of pipe will need some flexibility where they connect to horizontal pipes, especially if they are connected to horizontal pipes on different floors since the floors move at different speeds or in different directions. Also, horizontal pipes will need some clearance where they go through walls because the walls will not move with the floor/ceiling assemblies during the earthquake. So bracing and flexibility/clearance all go hand-in-hand.

Flexibility and Clearance

NFPA 13 requires that the piping be provided with flexibility in three ways: flexible couplings, seismic separation assemblies, and clearance around pipe. A flexible coupling is defined by Section 3.5.8 of NFPA 13 as a "fitting that allows axial displacement, rotation, and at least 1° of angular movement of the pipe without inducing harm on the pipe. For pipe diameters of 8 in. and larger, the angular movement shall be permitted to be less than 1° but not less than 0.5°." For standpipe systems, where the pipe is always 2.5 in. or larger in diameter, flexible couplings are required in the following locations:

- Within 24 in. of the top and within 24 in. of the bottom of all vertical pipes more than 7 ft in length. If the pipe is between 3 ft and 7 ft in length, a single flexible coupling is allowed. If the pipe is less than 3 ft long, no flexible coupling is required.
- If the building is a multistory building, within 12 in. above every floor.
- If the building is a multistory building, within 24 in. below every floor. If the flexible coupling in this case ends up being above the connection for the sprinkler system at this floor level, an additional flexible coupling is required either in the vertical portion of the sprinkler riser (if there is one) or on the horizontal portion of the sprinkler pipe within 24 in. of the standpipe.
- On both sides of concrete or masonry walls. In order to count, it must be within 12 in. of the wall. If clearance is provided around the pipe as it penetrates the wall, the flexible couplings are not required.

- Within 24 in. of one side of a building expansion joint. A building expansion joint is usually a fiber strip used to separate blocks of concrete to allow expansion as the temperature changes so that pressure does not build up in the concrete and cause cracking.
- Within 24 in. above any intermediate point of support on any vertical pipe.
- Within 24 in. below any intermediate point of support on any vertical pipe.

Figure 11.11 shows where the flexible couplings would go in a typical standpipe. The small bright red lines represent the flexible coupling. Starting at the top of the

Fig. 11.11 Flexible couplings for typical standpipe

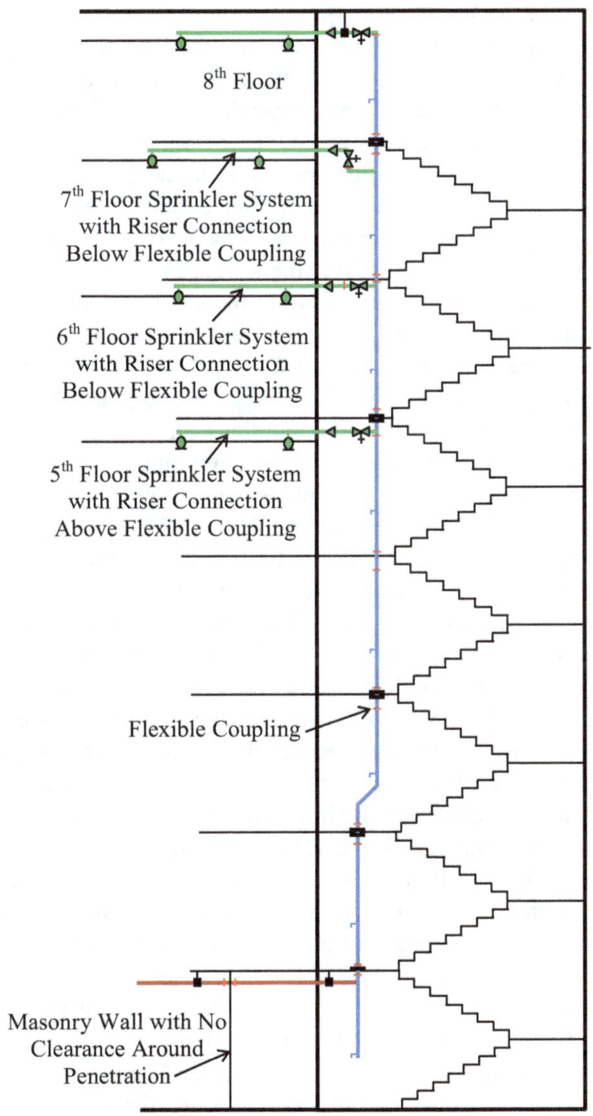

standpipe, there is a flexible coupling right at the connection to the eighth floor sprinkler system, which satisfies the requirement for a flexible coupling at the top of the vertical pipe. Working down the standpipe, the next flexible couplings are above and below the standpipe penetration of the eighth floor. The connection for the seventh floor sprinkler system is below the flexible coupling under the eighth floor, so an additional flexible coupling is placed on the vertical portion of the sprinkler piping. Continuing down the standpipe, the next flexible couplings are above and below the standpipe penetration of the seventh floor.

The connection for the sixth floor sprinkler system is below the flexible coupling, so an additional flexible coupling is placed on the horizontal sprinkler piping. Farther down the standpipe, the next flexible couplings are above and below the standpipe penetration of the sixth floor. The sprinkler piping connection for the fifth floor sprinkler system is above the flexible coupling, so an additional flexible coupling is not required on the sprinkler system. Continuing down the standpipe, there are flexible couplings above and below each penetration of the floor by the standpipe. At the second floor penetration, there is a flexible coupling below the second floor, which counts as the coupling at the bottom of the standpipe even though there are a few more feet of pipe towards the bottom. This counts as the flexible coupling at the bottom because it is the lowest point of support for the piping.

For the horizontal piping in the standpipe system, the dark red interconnection piping needs flexible couplings on both sides of the masonry wall shown on the first floor because there is no clearance provided around the piping. It is probably worth noting that most stairwell walls are masonry and the horizontal sprinkler pipe would probably require flexible couplings that are not shown in Fig. 11.11. The flexible couplings are not shown because the intent of this text is to focus on the standpipe requirements, not the sprinkler requirements.

In addition to flexible couplings, NFPA 13 requires a device called a "seismic separation assembly" in situations where piping crosses "seismic separation joints" aboveground. A seismic separation joint is a feature of a building that allows even more movement of the building than an expansion joint. The seismic separation joint allows two parts of a building to move in completely separate directions, like two separate buildings, during an earthquake. As you can imagine, if the standpipe system piping starts to move in two different directions, as the portions of the building move completely independently, the standpipe piping needs to be extremely flexible. NFPA 13 calls for this flexibility through the installation of a seismic separation assembly as the piping crosses the joint. See section 9.3.3 of NFPA 13 for the requirements of seismic separation assemblies.

There are two ways to make a seismic separation assembly. The first is to put together six elbows with ten flexible couplings and different short lengths of pipe between them that will allow movement in all three dimensions. An example of a seismic separation assembly constructed this way is found in NFPA 13 in Figure A.9.3.3(a). Very few people choose to build seismic separation assemblies this way anymore because it is very labor intensive and it takes a great deal of space, which is difficult to fit in a drop ceiling space. Instead, most people make the seismic separation assembly by buying a specifically listed device with flexible braiding that can

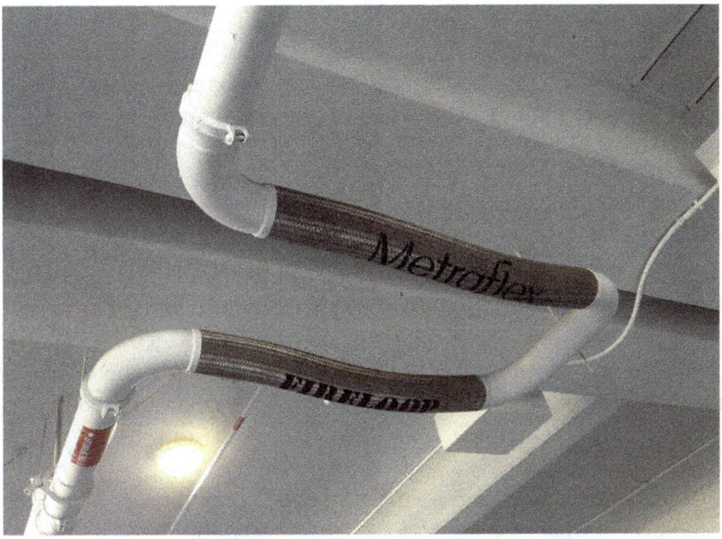

Fig. 11.12 Seismic separation assembly for standpipe system

move in all three directions. Figure 11.12 shows a listed seismic separation assembly for a 6 in. pipe serving a fire protection system in a parking garage where a seismic separation joint allows two portions of the garage to move like separate buildings. The seismic separation joint is the gap right above the assembly.

If standpipe piping crosses a seismic separation joint below ground, NFPA 13 allows the installation of the piping without a seismic separation assembly. Having the pipe below ground limits the amount of motion that can occur in different directions and the other flexibility in the system should be able to take care of this minor differential movement.

The seismic separation assembly is so flexible that immediately upstream and downstream of the assembly, four way braces need to be installed. The four way braces need to be within 6 ft of the seismic separation assembly, but not attached to the assembly itself. More information on braces will be included in a later part of this chapter.

The clearance requirements are found in section 9.3.4 of NFPA 13. The section starts out with the requirement that clearance be provided around all pipes that penetrate walls, floors, platforms, and foundations. These building elements (walls, floors, platforms and foundations) have the ability to move in different directions from the pipe during an earthquake, so there needs to be some space around the pipe to allow it to move in a different direction. The standard goes on to specifically state that this requirement includes the piping for drains, fire department connections, and other auxiliary piping. However, there are a number of exceptions to the rule for clearance. The following situations are not required to have clearance around the pipes that penetrate walls, floors, platforms and foundations:

- Piping passing through gypsum board (or other material that will fall apart before damaging the pipe) as long as the wall is not required to have a fire resistance rating. What this portion of the standard recognizes is that in some cases, the pipe is stronger than the wall and during an earthquake, the pipe will be able to move wherever it needs to move and it will break the wall. While this will keep the fire protection system operational, it will create a hole in the wall, so you are only allowed to do this if the wall is not required to have a fire resistance rating.
- Pipe with flexible couplings on both sides of the wall. The flexible couplings need to be within 1 ft of the wall on both sides. By providing this amount of flexibility, the pipe will be able to move without being damaged by the wall moving slightly differently during the earthquake. This option is sometimes selected for pipe penetrating walls that have a fire resistive rating because the integrity of the walls needs to be maintained, so a hole around the pipe is not an acceptable option.
- Pipe in a multistory building where a flexible connection is placed above and below the floor as discussed above for flexible couplings.
- Pipe that runs perpendicular to the studs in a wall or the joists in a floor/ceiling assembly where the pipe runs through successive studs or joists. In this situation, the pipe will move with the wall.
- Nonmetallic pipe that has demonstrated an inherent flexibility. Note that the flexibility that the committee is looking for is something similar to what steel pipe has when there are flexible couplings within 1 ft on both sides of a wall.
- Pipes supported by holes in the structural members as discussed earlier in this chapter for hanger requirements.
- Where pipes are being held in place laterally by holes in structural members, the member is allowed to be a tight fit with no clearance around the pipe.

Where clearance is required, there are a couple of ways that it can be provided. The pipe can be installed in a sleeve that is larger than the pipe or the pipe can just go through a hole in the wall that is bigger than the pipe. Either way, there are two requirements for the amount of clearance that needs to be provided:

1. For pipes 3.5 in. nominal size and smaller, a hole or sleeve 2 in. (nominal size) larger is required. This does not mean that there needs to be 2 in. all the way around. It just means that the hole needs to be 2 in. larger. So, a 2.5 in. pipe needs to go through a hole or pipe sleeve that is at least 4.5 in. in size. Note that the hole or sleeve size just needs to be bigger than the nominal pipe size, not the actual outside diameter of the pipe. So, even though 2.5 in. pipe has an outside diameter of 2.875 in., the hole only has to be 4.5 in. in diameter, not 4.875 in. Also, there is not requirement for the pipe to go exactly through the center of the hole.
2. For pipes 4 in. in size or larger, the hole or sleeve needs to be 4 in. larger than the nominal size of the pipe. Again, the measurement is just made on the nominal size, not the actual size. So, a 6 in. pipe needs to go through a 10 in. hole or sleeve.

There is one additional rule when it comes to clearance. One of the concerns is that the piping move with the building's structural members, which it will do if the pipe is supported by those structural members. But if the pipe is not supported from the building's structural members, then there is a chance that it will move in a different direction, so a minimum clearance of 2 in. is required.

For situations where the wall has a fire resistance rating, or if the building owner wants to fill the hole for sound control or for aesthetics, the hole is allowed to be filled with flexible material. If the purpose for the fill is to maintain a fire resistance rating for the wall, the fill material has to have a fire resistance rating at least as long as the wall. The material also needs to be compatible with the pipe material so that it does not damage or corrode the pipe.

Bracing of Horizontal Standpipe Piping

Once it has been determined that seismic protection is required, bracing needs to be provided for all standpipe piping. NFPA 13 requires bracing for all piping that is 2.5 in. in size or larger. Since standpipe piping is required to be 2.5 in. in size or larger, all of the pipe in a standpipe system that can carry water to hose connections will need to be braced. Piping to drains is not required to be braced beyond the drain valve since a break in this piping will not cause a problem with the use of the standpipe system during a fire.

There are two types of braces (lateral braces and longitudinal braces) and all piping that is being braced needs both types. Lateral braces prevent the pipe from moving from side to side. Longitudinal braces prevent the pipe from moving in the direction of the run of the pipe. See Fig. 11.13 for a graphical representation of the directions in which the different braces will work. The blue arrows show the direction in which the braces work and the red symbols are the typical ones used for these braces. In some cases, the design professional might want to provide a four-way brace in a specific location, which performs as both a lateral and a longitudinal brace.

In buildings that are required to have earthquake protection for the standpipe system piping, there is an exception to the requirement for lateral braces if the pipe is supported by very short hanger rods and if the maximum expected earthquake load is half the weight of the water filled pipe or less. The philosophy is that if the pipe is hung from very short rods, the amount of movement during an earthquake will be limited, so you don't need to brace the pipe. In order to be considered short enough to use this exception, the distance from the top of the pipe to the point where the pipe is attached to the building structure cannot be more than 6 in. In the 2016 edition of NFPA 13, this section was further modified to limit it to feed mains 6 in. in size or less and cross mains 4 in. in size or less. Since this will apply to horizontal interconnection piping in a standpipe system, which is much more like feed mains than cross mains, it would make sense to allow this rule to be used for pipe sizes up to 6 in. Note that if this rule is used to eliminate the lateral braces, the longitudinal

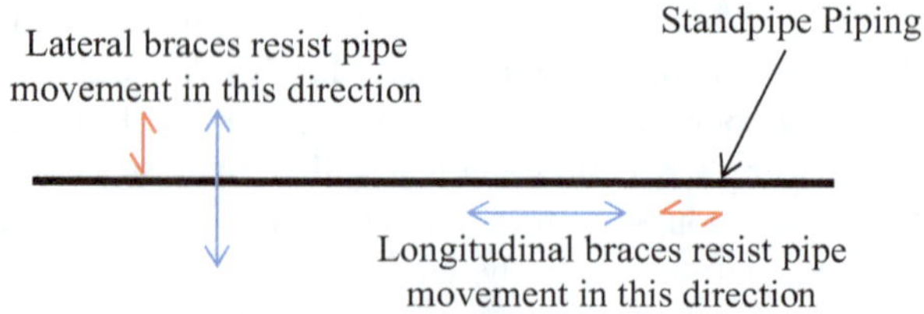

Fig. 11.13 Lateral and longitudinal brace directions

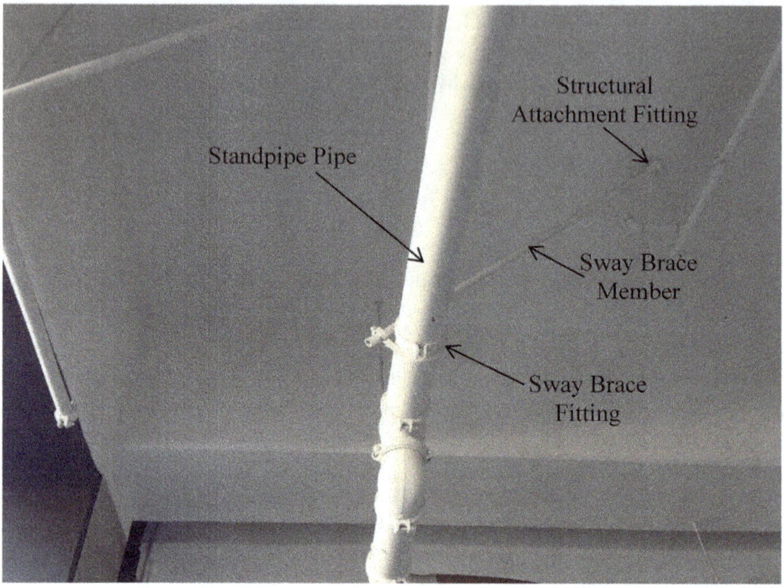

Fig. 11.14 Three parts of an earthquake brace

braces still need to be installed. A few hangers can be a little longer than described above, see section 9.3.5.5.10.2 of NFPA 13 for more information.

Just like hangers, braces have three different parts. There is the part that connects to the standpipe pipe, the part that connects to the building structure, and then there is the part that connects the first two parts. The part that is connected to the building structure is called the "structural attachment fitting" and it includes a fastener that physically keeps the brace attached to the building. The part that is connected to the standpipe piping is called the "sway brace fitting". The part that joins these other two parts is called the "sway brace member." Figure 11.14 shows the three different parts of an earthquake brace. This particular type of brace is a lateral brace, but longitudinal braces have the same parts.

All of the parts of the brace need to be listed except the sway brace members if they are made out of pipe, angles, flats or rods. NFPA 13 contains tables to determine if these common materials can handle specific earthquake loads. If the design professional wants to use different materials for the sway brace member, the material will need to be listed. Whatever material is chosen for the sway brace member, it needs to be able to handle both tension and compression forces since you don't know which direction an earthquake will move the pipe. There are products called "tension only" earthquake braces, which are essentially very strong cables that are listed for use as lateral braces. These tension only systems can be used if they are installed on both sides of the pipe so that they will prevent movement in either direction. They also need to be used with hangers that resist the upward movement of the pipe in order to keep the pipe in place.

Each part of the brace needs to be strong enough to withstand the loads that will be created by the earthquake. Frequently, the weakest part of the brace is the fastener that connects the structural attachment fitting to the structure, so that particular part will need to be looked at carefully. The following is a six step process that can be used to determine the acceptable sizes, materials and locations for sway braces:

1. Tentatively decide on a spacing and configuration that you want to try for the braces.
2. Calculate the load that the earthquake will place on the braces.
3. Make sure that the spacing is appropriate for the load. If the load does not allow for the spacing, go back to Step 1 and put the braces closer together.
4. Calculate whether the brace member can handle the load. If so, go to the next step. If not, select a different brace member that can handle the load or go back to Step 1 and put the braces closer together.
5. Calculate whether the sway brace fitting can handle the load. If not, select a different sway brace fitting that can handle the load or go back to Step 1 and put the braces closer together.
6. Calculate whether the structural attachment fitting and fastener can handle the load. If not, select a different structural attachment fitting or fastener that can handle the load or go back to Step 1 and put the braces closer together.

While these steps seem pretty straightforward, there are a few challenges within each step. It is possible that at any one of the steps, you could find that the plan is not working and that you will have to go back to the start and shorten up the spacing between braces. If that happens, it changes the load being carried by the brace, so you have to go back through the steps to verify that everything still works. The following is a more in-depth discussion of each individual step:

Step 1: Tentative spacing and configuration for the braces. This may be a function of how far apart the major structural members of the building are placed. Smaller structural members may not be able to handle the load, so you may be at the mercy of the spacing of the bigger members. The good news is that the standpipe system does not have branch lines like a sprinkler system, so the loads are just going to be the loads of the standpipe piping itself.

The distance between lateral braces is not allowed to exceed 40 ft. The distance between the last brace and the end of the pipe is not allowed to exceed 6 ft. The last piece of pipe on the main is required to have a lateral brace on it, which is usually taken care of with the brace that is within 6 ft of the end, but occasionally, a smaller piece of pipe might end up at the end of the brace, so it will need a lateral brace on

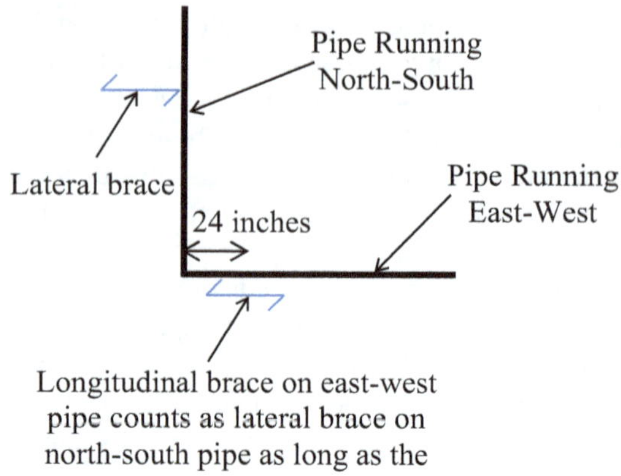

Fig. 11.15 Pipe running East-West has a longitudinal brace that serves as a lateral brace for pipe running North-South

it. If there is a change in direction of the pipe, a longitudinal brace on one pipe is allowed to act as a lateral brace for the other pipe if the longitudinal brace is within 24 in. of the centerline of the piping braced laterally as shown in Fig. 11.15. When there is a change in direction, the piping in each direction does need its own braces unless the pipe is less than 12 ft in length (Section 9.3.5.7.2).

Longitudinal braces are not allowed to be spaced more than 80 ft apart and the maximum allowable distance from the end of the pipe to the last longitudinal brace is 40 ft. If the pipe changes direction, a lateral brace on one pipe is allowed to act as the longitudinal brace on the other pipe if the lateral brace is within 24 in. of the change in direction and if the pipe being braced laterally is on a pipe of the same or larger size as the pipe being braced longitudinally.

In fire sprinkler systems, the main piping always connects to the system riser, and the riser is required to have a four-way brace, so the measurement of the distances between braces typically starts from this four-way brace. But a standpipe system, the horizontal interconnection piping does not start at a system riser, so there will not automatically be a four-way brace to start your brace spacing from. In this situation, the interconnection piping will start and end at a standpipe, which does follow the rules for bracing of vertical pipe, but may not have a four-way brace, so this connection should be considered the end of the pipe for brace spacing.

As we work through this step-by-step process for the first time, we will use lateral braces as an example to illustrate each step. Longitudinal braces need to follow the same process. The example will be a 120 ft long 6 in. Schedule 40 pipe run horizontally being used as a feed main to interconnect standpipes. To comply with Step 1, we will start with the tentative spacing of 40 ft between lateral braces, making a total of four lateral braces (one at the very end of the pipe, the next one 40 ft from the end, the next one 80 ft from the end, and at the other end, the last lateral brace will be 120 ft from the end where we started).

Step 2: Calculate the Load on the Braces. The horizontal loads that may be created by an earthquake are expressed as some number multiplied by the weight of the water filled pipe. Usually, this number is less than one, but there are a few places in the United States where the earthquake might create a horizontal force greater than gravity. Table A.9.3.5.9 of NFPA 13 contains the weight of water filled Schedule 10 and Schedule 40 steel pipe. For other types of pipe, the manufacturer should be able to provide the weight of the pipe when filled with water.

NFPA 13 permits two methods for determining the horizontal force. The first is to use Section 13.3.1 of ASCE 7 to determine the load that this document would require, then multiply that load by 0.7 because ASCE 7 uses a different method of calculating stresses for components than NFPA 13, which puts more of the safety factor in the component evaluation. It would not be fair to put additional safety factors on both the load and the component evaluation, so lowering the load obtained from ASCE 7 is permitted.

The second way to obtain the horizontal load is to start by obtaining the short period response parameter S_S, which is a variable associated with the soil conditions and location of the building site in relation to fault lines or other places of seismic activity. The S_S value can be obtained from the authority having jurisdiction, the structural engineer, or from the U.S. Geological Survey website if the building is in the United States and you have the longitude and latitude of the building site. Once you have the S_S value, Table 9.3.5.9.3 in NFPA 13 converts the S_S value into a C_P value. Table 11.2 is a partial listing of S_S to C_P conversions. Once you have the C_P value, the horizontal force (F_H) can be calculated using the following formula where W_P is the weight of the water filled pipe:

$$F_H = 1.15 C_P W_P$$

If you don't have any way of finding the C_P value, NFPA 13 does allow you to assume that the C_P value is 0.5, which is conservative for most of the United States. The 1.15 multiplier in the formula above is a safety factor that helps to account for the extra weight of valves and fittings on the fire protection system piping.

For our example, assume we are building this system in a place where the S_S value is 0.9, which would make the CP value 0.48 according to Table 11.2. The

Table 11.2 Partial list of S_S to C_P conversions (see Table 9.3.5.9.3 in NFPA 13 for more detail)

S_S	C_P
0.33 or less	0.35
0.5	0.4
0.7	0.42
0.9	0.48
1.0	0.51
1.2	0.57
1.4	0.65
1.6	0.75
1.8	0.84
2.0	0.93
2.2	1.03
2.4	1.12

weight of the 6 in. water filled pipe will be 1268 lb (31.69 lb per ft from Table A.9.3.5.9 times 40 ft of pipe supported by any one brace; 31.69 × 40 = 1268). The maximum horizontal load that the earthquake might cause would be 700 lb (1.15 × 0.48 × 1268 = 700).

Step 3: Make Sure the Spacing is Appropriate for the Load. NFPA 13 contains 12 different tables (Tables 9.3.5.5.2(a) through 9.3.5.5.2(l)) that match the maximum allowable load to the spacing considering the type of pipe being supported. Each of the 12 different tables corresponds to a different type of pipe that is being supported. Since some of these pipes, such as CPVC won't be used for standpipes, some of these tables can be ignored when making this all work for a standpipe system. The user in this case will mostly use Tables 9.3.5.5.2(a) through 9.3.5.5.2(d). Some of these tables provide values in the metric system while others in the traditional foot-pound units. You only need to worry about meeting one table, so pick whichever one is convenient given the units you are working with in your system.

For our example, Table 9.3.5.5.2(c) is the appropriate table to use (Schedule 40 steel pipe in the standpipe system). According to the table, placing the lateral braces at 40 ft spacing is permitted as long as the load on the brace does not exceed 3713 lb. Since our example has a load of 700 lb calculated in Step 2, it is well below the limit and we are allowed to space our lateral braces at a distance of 40 ft apart.

Step 4: Calculate Whether the Brace Member can Handle the Load. NFPA 13 has a series of tables (Table 9.3.5.11.8(a) through Table 9.3.5.11.8(c)) that provide that maximum allowable load for braces that are constructed out of Schedule 40 steel pipe, angle iron, all thread rod, rod only threaded at the ends, and flat steel. These are the products that are commonly used for braces. The reason that there are three different tables is that the user is provided with an opportunity to trade off the length of the brace for the load that the brace can carry. Longer braces cannot carry as much of a load. By shortening a brace, the load that any material can handle can be increased.

The length of a brace, by itself, is actually insufficient for describing the load that the brace can handle. The load is also affected by the thickness of the brace. The comparison of the length to the thickness is called the "slenderness ratio", which is calculated by taking the length of the pipe (l) and dividing by the "least radius of gyration" (r). Don't worry about exactly what the "least radius of gyration" for a material is, just understand that it has something to do with the thickness of the object and that the good folks who write NFPA 13 have done the calculation for you. The "slenderness ratio" is expressed in NFPA 13 using the symbol "l/r".

The reason that there are three different tables in NFPA 13 that provide the maximum acceptable loads for brace members is that the calculations were performed with slenderness ratios of l/r = 100, 1/r = 200, and l/r = 300. You are allowed to use any of the tables. You just need to make sure that the length of the brace and the load that the brace can handle are both acceptable on the table that you select.

As you look at the tables in NFPA 13, you will notice that the load that the brace member can handle is dependent on the angle of the brace. The earthquake loads that we are concerned with are horizontal in nature, so the closer to horizontal a

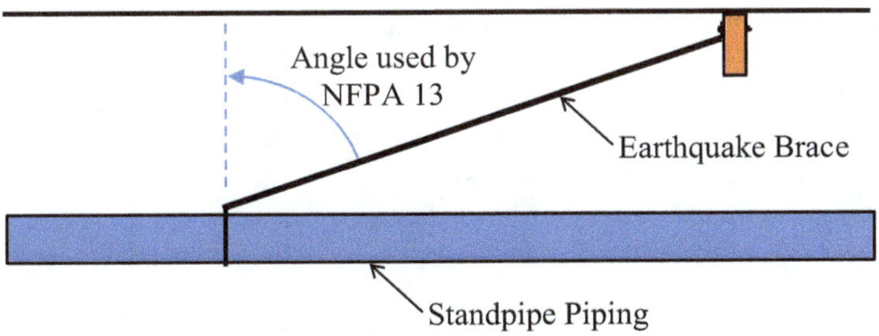

Fig. 11.16 Measuring the angle for use in the seismic rules of NFPA 13

brace is, the more of a load it can handle. However, NFPA 13 does not express the angle of the brace in terms of a comparison to the horizontal. Instead, all of the angles in NFPA 13 for all of the seismic issues are expressed in terms of a comparison to the vertical as shown in Fig. 11.16.

In the example that we have been working through so far, we have not discussed the length that the brace needs to be, the angle in which it will be installed, or the material that the brace will be made out of. Let us assume for the sake of this discussion that the brace will need to be at least 5 ft long and that it will be installed at an angle of 60° to the vertical. From Step 2, we know that the load that the earthquake might cause is 700 lb. If we go to Table 9.3.5.11.8(a) in NFPA 13, which is for slenderness ratios of l/r = 100, we see that we could only use 1.5 in. (maximum allowable load of 8825 lb) or 2 in. (maximum allowable load of 11,818 lb) Schedule 40 pipes for a brace member. While many of the other objects in this table can handle a load greater than 700 lb, the braces need to be shorter than the 5 ft we need. For example, a 3/8 in. all thread rod can handle a load of 773 lb at an angle of 60° from the vertical, the brace could only be 7 in. long with this slenderness ratio and we need it to be 5 ft long, so this just won't work.

Another option would be to look at Table 9.3.5.11.8(b) using a slenderness ration of l/r = 200. This opens up the option of 1 in. Schedule 40 pipe, which can go up to 7 ft long and can handle a load of 1604 lb. Both of these values exceed what we want to do with the brace, so it would be acceptable. Using this table also opens up all of the angle iron options except the smallest one.

If we wanted to make the brace out of a rod, we could go to Table 9.3.5.11.8(c), which is for a slenderness ratio of l/r = 300. The 7/8 in. rod that is only threaded at the end can be used at a length of 5 ft—5 in. and can handle a load of 867 lb. Since both of these values exceed our plan of using a brace that is 5 ft long and supporting a load of 700 lb, we will be able to use any of the materials here including the 7/8 in. rod.

Step 5: Calculate Whether the Sway Brace Fitting can Handle the Load. The information on the sway brace fitting does not come from NFPA 13. This comes from the manufacturer, who gets the information from the listing laboratory when their product gets listed. However, NFPA 13 does require that an adjustment be made to the manufacturer's information (Section 9.3.5.2.3) because the load that the fitting can handle is always determined by the listing laboratory in the horizontal position. The

Table 11.3 Adjustment to sway brace fitting based on angle from vertical

Angle from vertical (°)	Adjustment
30–44	Divide listed load rating by 2.0
45–59	Divide listed load rating by 1.414
60–89	Divide listed load rating by 1.155
90 (horizontal)	Don't adjust the load rating. Use at full load rating

earthquake brace is almost never used in the horizontal position, so the load that the fitting can handle needs to be adjusted based on the actual angle from vertical at which that the brace is being used. Table 11.3 shows the adjustment that needs to be made based on the angle.

For our example, assume that we have found a sway brace fitting that is listed for 1000 lb. We are using the brace at an angle of 60°, so we need to divide the listed load by 1.155 to see if it can still handle our load. This brace fitting at this angle can handle a load of 866 (1000/1.155 = 866). Since our load is only 700 lb and the brace fitting can handle 866 lb, we can use this brace fitting at this angle.

Step 6: Calculate Whether the Structural Attachment Fitting and Fastener can Handle the Load. As far as the structure attachment fitting goes, the same situation applies as the sway brace fitting. The manufacturer will have a fitting listed at a specific load and you need to divide by the number based on the angle from the vertical to see if the fitting can handle your load at the angle in which you want to use it. We go to tables in NFPA 13 to determine whether the attachment to the building structure will work. Tables 9.3.5.12.2(a) through (j) provide different loads that different fasteners can handle when used in different structural members. The following is a list of which tables apply to which fasteners:

- Table 9.3.5.12.2(a): Applies to Wedge Anchors in 3000 psi Lightweight Cracked Concrete on a Metal Deck
- Table 9.3.5.12.2(b): Applies to Wedge Anchors in 3000 psi Lightweight Cracked Concrete (not on a Metal Deck)
- Table 9.3.5.12.2(c): Applies to Wedge Anchors in 3000 psi Normal Weight Cracked Concrete
- Table 9.3.5.12.2(d): Applies to Wedge Anchors in 4000 psi Normal Weight Cracked Concrete
- Table 9.3.5.12.2(e): Applies to Wedge Anchors in 6000 psi Normal Weight Cracked Concrete
- Table 9.3.5.12.2(f): Applies to Undercut Anchors in 3000 psi Normal Weight Cracked Concrete
- Table 9.3.5.12.2(g): Applies to Connections to Steel Using Unfinished Steel Bolts
- Table 9.3.5.12.2(h): Applies to Through-Bolts in Sawn Lumber or Glue-Laminated Timbers

- Table 9.3.5.12.2(i): Applies to Lag Screws and Lag Bolts in Wood
- Table 9.3.5.12.2(j): Contains some adjustment factors for wood with different specific gravity than the assumptions that went into Table 9.3.5.12.2(h) or Table 9.3.5.12.2(i).

Each of the tables up above provides the user with the acceptable load that a fastener can handle when the brace is at a certain angle and when the fastener is placed in a specific direction. In order to use the tables discussed above, the user needs to know the angle at which the brace will be used and which of the fastener condition will be present. There are three possible fastener positions:

1. The fastener will be inserted vertically up into the structural member as shown in Part A, Part B and Part C of Fig. 11.17. As the earthquake occurs, a sheer stress will be developed on the fastener from the horizontal forces and any vertical force will work in the opposite direction of the fastener (down).
2. The fastener will be installed horizontally into the structural member in the same direction as the brace as shown in Part D, Part E and Part F of Fig. 11.17. The horizontal force of the earthquake will be trying to pull the fastener away from the structural member.

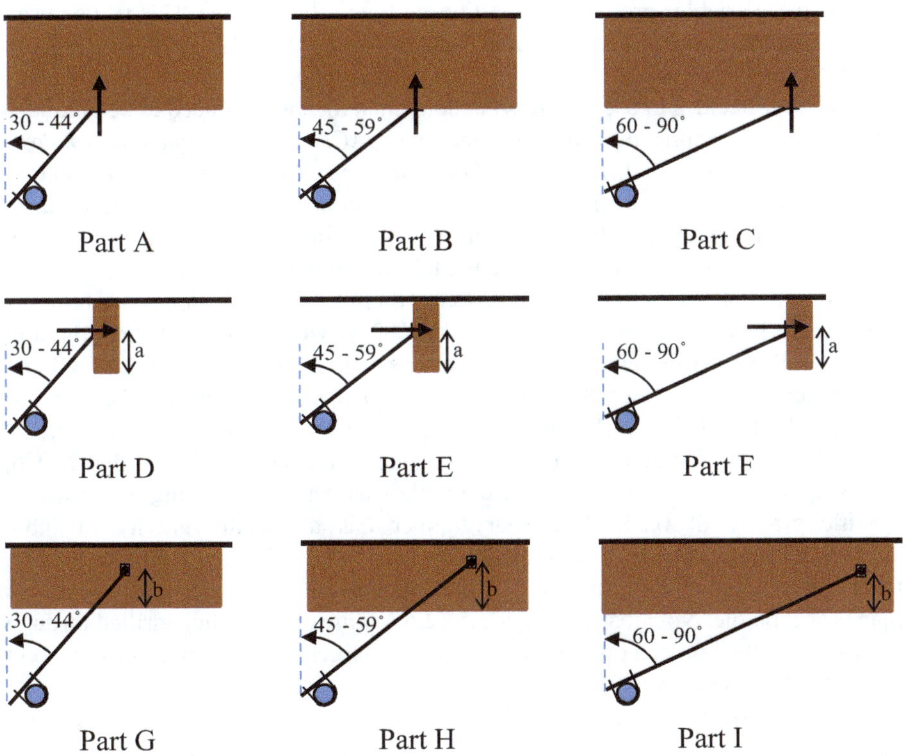

Fig. 11.17 Fastener direction options for a lateral brace

3. The fastener will be installed horizontally into the structural member in a direction perpendicular to the brace as shown in Part G, Part H and Part I or Fig. 11.17. The horizontal force of the earthquake will be creating a sheer stress on the fastener.

Figure 11.17 shows the fastener direction options for a lateral brace. There are nine different options with three different brace angles and three different fastener directions shown as Parts A through I. The same options apply to longitudinal braces, but only lateral braces are shown in the figure in order to simplify the drawings. The brace angles are shown from 30° to 90°. Angles less than 30° are not allowed by NFPA 13 because they are too close to the vertical and therefore will not resist significant horizontal forces.

In Parts A through F of the figure, the fastener is shown by the straight black arrow. In Parts D, E and F, the distance “a” from the fastener to the bottom of the structural member needs to be at least half of the depth of the structural member. For wood beams, “a” is also not allowed to be less than 3 in. In Parts G, H and I, the distance “b” is required to be at least four times the width of the fastener and at least 1/3 the depth of the structural member. The distance “b” is also not allowed to be less than 3 in. for wood beams.

Once the user knows the brace angle and the fastener direction, the combination of these two variables provides the part of the figure that applies. This is then used to determine the maximum allowable load in Tables 9.3.5.12.2(a) through (j) in NFPA 13.

If the user decides to use concrete anchors, then the anchors need to be prequalified for seismic application in accordance with *ACI 355.2 Qualification of Post-Installed Mechanical Anchors in Concrete and Commentary.* See section 9.3.5.12.8 in NFPA 13 for more information on concrete anchors for earthquake brace fasteners including information on an alternate method of finding the load that the anchor can handle instead of using the tables mentioned above.

In the example that we have been working on in this step-by-step procedure, the brace angle has already been determined at 60°. If we assume that the structural member will be a 4 in. wide beam made of oak with a specific gravity of 0.63 and that the fastener will be a 5/8 in. bolt that will go all the way through the wood as shown in Part F of Fig. 11.17, then the fastener load shown in Table 9.2.5.12.2(h) will be 685 lb. At first glance, this does not appear to meet our design load of 700 lb, but as the note to the table explains, the table was determined using wood with a specific gravity of 0.35. Since our lumber has a specific gravity of 0.63, Table 9.2.5.12.2(j) allows us to increase the load by 1.25, so the load that the fastener can handle is 856 lb (685 × 1.25 = 856). This meets our design load, so this plan is acceptable. Note that section 9.3.5.12.6 requires that the hole drilled through the wood for the bolt can only be 1/16 of an inch larger than the bolt. Since the bolt is 5/8, the hole needs to be 11/16 of an inch.

Having followed all six steps in the process and meeting all of the load requirements, all three of the brace components that have been proposed here will be sufficient for the potential earthquake that might occur in the area of this building.

Bracing of Vertical Standpipe Piping

The requirements for bracing vertical piping are much simpler than the requirements for horizontal pipe. All that is required is a four-way brace at the top of any vertical section of pipe more than 3 ft long and possibly additional braces at 25 ft intervals down the standpipe. The additional braces at 25 ft intervals are not required where the standpipe goes through at least one floor every 25 ft with the clearance required, but no additional clearance. In most buildings, this is satisfied with a typical standpipe installation since the standpipe will be at a main floor landing or an intermediate floor landing and these will be 10 or 12 ft apart.

For the four-way brace at the top of the riser, the designer is allowed to connect the brace to horizontal piping at the top of the standpipe if the brace is connected within 24 in. of the centerline of the standpipe. If this is done, the four-way brace needs to take into account the load of the vertical pipe as well as the load of the horizontal pipe. See section 9.3.5.8 of NFPA 13 for more information on bracing of vertical piping.

Protection of Piping from Mechanical Damage and Fire

Section 6.1.2 of NFPA 14 requires that the piping be protected from both fire and mechanical damage. In many situations, the piping is protected from fire when it is placed in a location that also protects it from mechanical damage. For example, when the pipe is placed in a noncombustible exit stair with walls having a 2-h fire resistance rating, the pipe is protected from mechanical damage while at the same time, it is also protected from fire.

Table 6.1.2.2 of NFPA 14 tells the user which types of pipes in Class I and Class III standpipe systems need protection. Rather than repeat the table here, a list will be provided of the types of pipe that need protection and the types of pipe that do not:

- Class I and Class III Systems in High Rise Buildings
  - In a building of Type I construction, vertical standpipe piping requires protection whether it is sprinklered or not.
  - In a building of Type I construction not sprinklered in accordance with NFPA 13, the horizontal pipes (horizontal standpipes or interconnection piping) and branch lines require protection.
  - In a building of Type II construction, vertical standpipe piping requires protection whether it is sprinklered or not.
  - In a building of Type II construction not sprinklered in accordance with NFPA 13, the horizontal pipes (horizontal standpipes or interconnection piping) and branch lines require protection.
  - In a building of Type III, Type IV and Type V construction, all pipes are required to be protected regardless of whether the building is sprinklered or not.

- Class I and Class III Systems in Non High Rise Buildings
  - All vertical standpipes are required to be protected regardless of construction type or whether the building is sprinklered.
  - Horizontal piping (horizontal standpipes and interconnection piping) and branch lines do not require protection.
  - Piping to extra hose connections that have to be added due to travel distance requirements is not required to be protected.

Once NFPA 14 requires protection for the piping, there are three options for how the protection can be provided:

1. Enclosed in a fire rated exit stairway.
2. Enclosed in construction equivalent to the rating of the exit stairways in the building.
3. Listed fire wrap or other insulating material applied directly to the pipe with a rating equal to that of the exit stairways in the building.

If the exit stairways in a building are not required to have a fire resistance rating, then the pipes in Class I and Class III systems are not required to be protected. The pipes in Class II systems are never required to be protected regardless of whether the building is high rise or not and regardless of whether the building is sprinklered or not.

For the piping that is not required to be protected in accordance with the rules above, there still needs to be some protection if the piping is in a location where it could be hit or struck by anything happening in the area. For example, a branch line to a hose connection in a non-high rise parking garage does not need to be in a stairwell or protection equivalent to a stairwell, but it still needs to be protected from a car running into it. Such protection could be from steel posts, concrete barriers, or some other approved means of protection.

For standpipe systems that also provide water to sprinkler systems, care should be taken to make sure that the standpipe piping is not run through combustible areas with significant fire hazards such as areas where flammable and combustible liquids are stored. You would not want an incident damaging the standpipe that supplies water to the sprinklers that are supposed to take care of the incident.

The International Building Code (IBC) has a different set of rules for protecting the pipe from damage and fire. Sections 905.4.1 and 905.6.1 require that all piping outside of stairwells for Class I and Class III standpipe systems be either in a sprinklered building or surrounded by construction that meets the same level of fire resistance as the stairwells. Whether the sprinkler system meets NFPA 13, NFPA 13R, or some other set of requirements would be up to the authority having jurisdiction since the section just says that the sprinkler system needs to be approved. This rule applies to all pipe, both horizontal and vertical, regardless of whether the building is high rise or not.

The requirement to protect the pipe is different than the rules in NFPA 14. Since the IBC is the primary legal document, it takes precedent over the rules in NFPA 14 and needs to be followed in any jurisdiction that adopts both documents.

Section 905.5.2 of the IBC agrees with NFPA 14 that Class II standpipe system piping does not need any specific protection.

Protection of Piping from Freezing

Section 6.1.2.3 of NFPA 14 requires that piping for manual wet and automatic wet standpipe systems be protected from freezing. There are a number of options for protecting piping from freezing. Whichever of the options is selected, the piping needs to be kept between the temperatures of 40°F and 120°F. The options for protecting the pipe from freezing include:

- Heating the space around the pipe with a reliable heating system.
- Tenting insulation over piping in an unheated space with a heated space underneath. This method is successfully used with sprinkler piping in an attic with insulation tented over the pipe trapping the heat rising from the heated room below. The technique can be adapted to standpipe piping.
- Heat tracing. This is the process of using specially designed wires with a current running through the wires creating heat. Some of the heat tracing products include insulation over the heat tracing to keep the warmth around the pipe. If heat tracing is used, it needs to be specifically listed for fire suppression systems and it needs to be installed in accordance with its listing including the manufacturer's instructions. The heat tracing system also needs to be supervised so that if it stops working, a signal is sent to a constantly attended location.

One of the methods for the prevention of freezing that is not acceptable for standpipe systems is the use of antifreeze. Even before the antifreeze concerns of NFPA 13 regarding the combustibility of antifreeze products, NFPA 14 prohibited the use of antifreeze. The volume of the piping in standpipe systems is too large to have antifreeze be a practical solution. The expansion of the liquid at different temperatures would make the expansion chamber difficult if not impossible to design and maintain.

If the requirement to keep the piping heated to the proper temperature cannot be met, then the system will need to be redesigned to be a manual dry system, a semi-automatic dry system, or an automatic dry system.

Protection of Piping from Corrosion

Section 6.1.2.4 of NFPA 14 requires that piping in standpipe systems be protected from corrosion. Potential sources of corrosion include:

- Weather in situations where the piping is exposed to the outside air.
- Salt air where the piping is exposed to the outside air near the ocean.
- Exposure to corrosive fumes from industrial processes or the use of corrosive products near the exposed pipe.

One way to deal with corrosion is to use corrosion resistant pipe, fittings and hangers. Another way to deal with corrosion is to use a corrosion resistant coating over the pipe, fittings and hangers. It also might be possible to put a type of construction around the piping and/or increase the ventilation conditions around the piping so that corrosive fumes don't build up.

Another type of corrosion that might be possible is internal corrosion from the combination of air and water in the pipe. Such corrosion most frequently occurs in dry systems. One way to prevent or slow down this type of internal corrosion is to pitch the piping so that the water can be drained after each time the system is tripped or used. Horizontal piping is required to be pitched toward a drain at a slope of ¼ in. per every 10 ft (section 6.1.3 of NFPA 14). If the piping is in a constantly refrigerated location such as a walk-in refrigerator or walk-in freezer, the piping is required to be pitched toward a drain at a slope of ½ in. per every 10 ft (section 6.1.4 of NFPA 14).

Test Yourself

11.1. If 6 in. Schedule 40 pipe weighs 31.69 lb per ft and the hangers are spaced at 15 ft intervals, what load does each hanger need to be able to handle?

11.2. A trapeze member needs to support 4 in. Schedule 40 pipe in the center of an 8 ft span. What is the smallest size Schedule 40 pipe that could serve as a trapeze member for this situation?

11.3. If the standpipe system piping described in Question 11.2 is 1 ft from one of the supports instead of being in the center of the span, what is the smallest size Schedule 40 pipe that could serve as a trapeze member for this situation?

11.4. Given the part of the building in Fig. 11.18 with the major structural members in the floor/ceiling assemblies being steel I-beams 8 in. deep and spaced 25 ft apart, answer the following questions assuming that earthquake protection is required for the piping and no expansion joints or seismic separation joints are in this part of the building:

(a) If the S_S value is 0.5, where on the 6 in. Schedule 10 horizontal feed main (or interconnection piping) are flexible couplings and braces required?
(b) Assuming that the sprinkler piping on each floor is tied into the standpipe 18 in. below each floor slab, where on the 4 in. Schedule 10 vertical standpipes are flexible couplings and braces required?

11.5. Take the worst-case lateral and longitudinal braces on the horizontal piping in question 11.4 and answer the following questions assuming that each brace needs to be 30 in. long and installed at an angle of 34° from horizontal:

(a) Will a sway brace fitting rated for 800 lb will be sufficient.
(b) What is the smallest all thread rod that can be used as the brace member?
(c) What is the smallest size unfinished bolt that could be used to fasten the brace to the steel structural member?

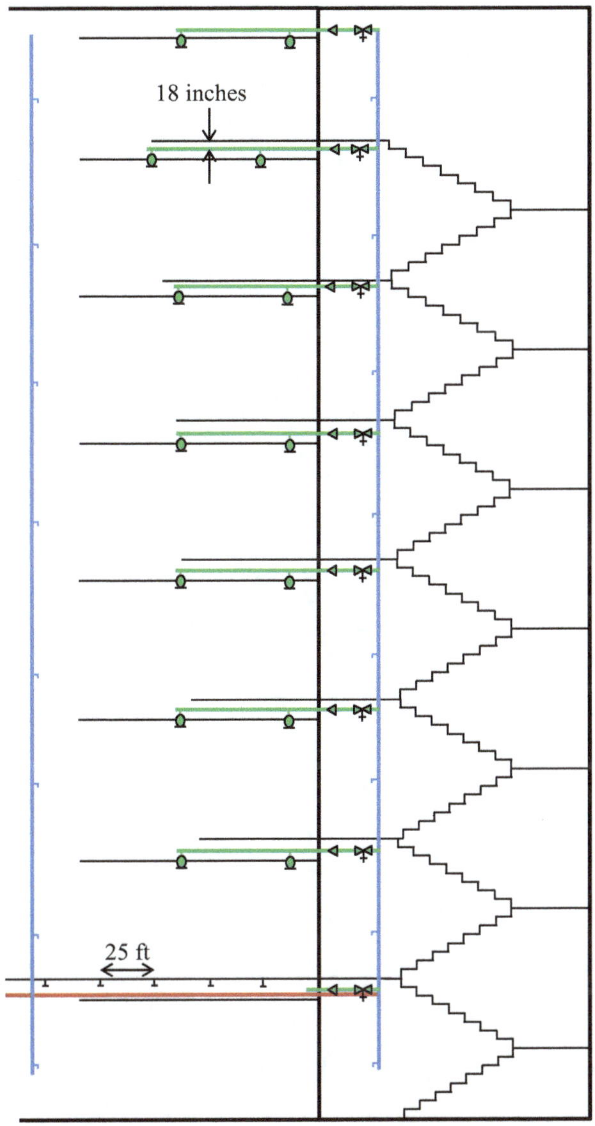

Fig. 11.18 Standpipe system for question 11.4

Chapter 12
Horizontal Standpipes and Lateral Piping

There are a couple of reasons that horizontal piping will be used in a standpipe other than the feed mains that interconnect vertical standpipes. The first type of horizontal pipe is a branch line that connects a single hose connection to a standpipe. The second type of horizontal pipe is a horizontal standpipe, which is defined as any horizontal pipe that connects two or more hose connections on the same level of a building. This chapter will cover both situations.

While it is certainly possible for Class II systems to have horizontal standpipes and branch lines, it would make no difference in these systems whether the piping was considered part of a horizontal standpipe or part of the standpipe system. For Class II systems, the rules for pipe sizes and hydraulic calculations would all be the same. But for Class I and Class III systems, the concept of a horizontal standpipe or a branch line has serious consequences, so they need to be differentiated from each other and from other types of standpipes. This chapter will focus only on the rules for Class I and Class III systems for horizontal standpipes and branch lines. Rather than say "for Class I and Class III systems" each time throughout the chapter, the assumption will just be made that everything in this chapter applies only to Class I and Class III systems.

Branch Lines

By definition (Section 3.3.2 of NFPA 14) a branch line is a run of piping that is generally in the horizontal plane that connects a single hose connection to a standpipe. Figure 12.1 shows a branch line (the red line) that has been added to the third floor of a building because the arrangements of rooms and corridors on this particular floor does not meet the travel distance requirements to the hose connections in the stairwell whereas on the other floors of the building, the travel distance rules can be met from the hose connections in the stairwell. Putting a separate vertical standpipe into the building in this situation would not be cost effective, so the branch line

K.E. Isman, *Standpipe Systems for Fire Protection*,
DOI 10.1007/978-3-319-47750-3_12

Fig. 12.1 Branch line on a single building floor

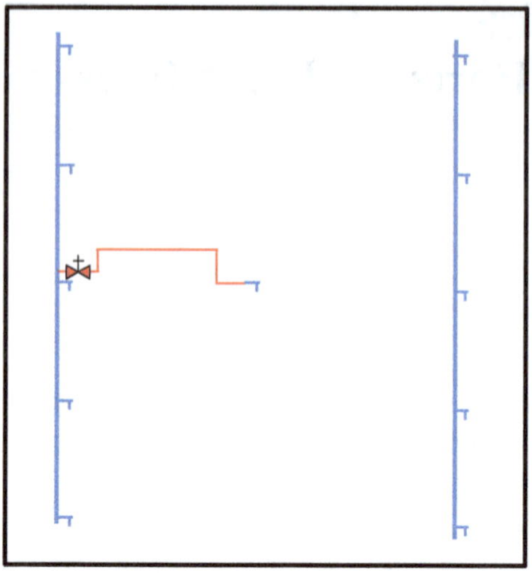

makes sense in this case. While the general direction of the piping is horizontal, the branch line is allowed to have some vertical pipe in it as shown.

In Fig. 12.1, a control valve is shown at the connection of the branch line to the standpipe. This control valve is required by Section 6.3.3 of NFPA 14 in situations where the branch line exceeds 40 ft in length. The 40 ft distance needs to be measured along the run of the pipe, not just the straight distance between the hose connection and the standpipe. So, the length of the vertical pipe needs to be considered. The purpose for this control valve is to be able to isolate the branch line from the standpipe so that the standpipe does not need to be shut down if there is a problem with the branch line.

A second reason for using branch lines in a standpipe system is to handle the situation where a vertical standpipe will be difficult to install. This might happen on every floor in a building such as where the hose connections at a horizontal exit are fed from a standpipe in a stairwell rather than a standpipe of their own. The arrangement is shown in Fig. 12.2 shows an example of this use of branch lines. The arrangement shown in Fig. 12.2 also helps in the situation where a pressure reducing valve is required. The valve can be located in the stairwell where it is easier to install a drain riser and then the branch line can take water from the pressure reducing valve to the hose connection in the middle of the building.

Branch lines have to be sized in order to provide the required flow and pressure to the hose connection. Section 7.6.3 of NFPA 14 requires that the branch line be at least 2.5 in. in size for Class I and Class III standpipe systems. Hydraulic calculations will show whether the 2.5 in. size will work or not. The pressure loss of 250 gpm flowing through 2.5 in. pipe is as follows:

- 0.215 psi per ft for Schedule 40 pipe in a wet system (C = 120)
- 0.157 psi per ft for Schedule 10 pipe in a wet system (C = 120)
- 0.302 psi per ft for Schedule 40 pipe in a dry system (C = 100)
- 0.220 psi per ft for Schedule 10 pipe in a dry system (C = 100)

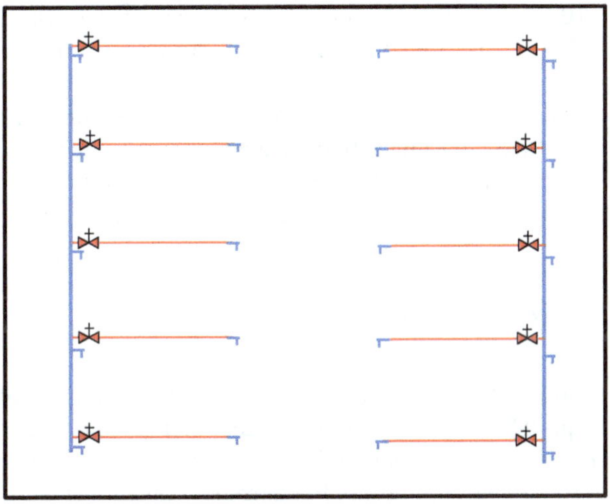

Fig. 12.2 Branch lines on each floor of a building

Whether this friction loss will be too much for the system or not will depend on the length of the pipe and the water pressure available at the connection to the standpipe. In addition to the actual length of pipe, the equivalent length of the fittings and valves need to be considered. For example, in the branch line shown in Fig. 12.1, the pipe may only be 50 ft long, but there is also a tee at the connection to the standpipe, a control valve and two elbows. The equivalent lengths for all of these fittings and valves are: 12 (tee), 7 (butterfly valve), 12 (two elbows), making the total equivalent length of the valve and fittings 31 ft. The total length of the pipe and fittings is then 81 ft (50 + 31 = 81). The total friction loss would then be somewhere between 12.7 psi and 24.5 psi depending on whether the system is wet or dry and whether the pipe is Schedule 10 or 40. If the system does not have this pressure available, then the pipe size would need to be increased to 3 in., which would drop the friction loss about 65 %.

For the situation shown in Fig. 12.1, the branch line creates an extra hose connection in the building that has no effect the total flow demand for the system. The total flow demand is based on the number of standpipes in the building, not the number of hose connections. The branch line is considered an extension of the standpipe on the left and therefore does not increase the flow demand for calculation purposes.

It is possible that the situation shown in Fig. 12.1 would have a small effect on the hydraulic calculations if the extra hose connection was at a different location. If the extra hose connection was at a remote location connected to the most remote standpipe, then it might be the one where the first 250 gpm flow demand would be calculated with the second most demanding hose connection on that same standpipe. This does not change the total flow required to be calculated in the standpipe, but might slightly alter the hose connections from which that flow would be expected.

For the situation shown in Fig. 12.2, the placement of the hose connections in a vertical line at the end of each branch line coming from one of the standpipes replicates those of a vertical standpipe, even if a vertical standpipe is not installed. Therefore, NFPA 14 specifically says that when this condition occurs, the hydraulic calculations need to be treated as if an additional standpipe was present (see Section

7.10.1.1.6 and the accompanying annex text in Section A.7.10.1.1.6 and Figure A.7.10.1.1.6). Likewise, the row of hose connections at the end of the branch lines coming from the other standpipe also create another vertical standpipe for hydraulic calculation purposes. So, even though the situation shown in Fig. 12.2 only has two standpipes and many branch lines, it would need to be calculated as though the building had four standpipes, which would require 1000 gpm in a sprinklered building and 1250 gpm in an unsprinklered building.

Horizontal Standpipes

A horizontal standpipe is defined in Section 3.3.16.1 of NFPA 14 as, "The horizontal portion of the system piping that delivers the water supply for two or more hose connections, and for sprinklers on combined systems, on a single level." This distinguishes a horizontal standpipe from a branch line so that they are two distinct types of pipe. If the horizontal pipe serves only a single hose connection, then it is a branch line. If the horizontal pipe serves more than one hose connection, it is a horizontal standpipe. The horizontal standpipe serves the same function as a vertical standpipe, to carry water to hose connections, while at the same time, the horizontal standpipe is permitted to supply sprinkler systems. Horizontal standpipes are typically installed in one or two story buildings that have a very large floor plan with multiple hose connections on each floor, making one or two horizontal runs of pipe more efficient than many vertical runs.

Figure 12.3 is an isometric view of a two-story building with two horizontal standpipes in blue. The vertical red line represents the feed main or interconnection piping between the two horizontal standpipes. The feed main runs vertically because the horizontal standpipes on each floor need to be connected to the water supply and fed from the fire department connection (FDC). Since each of the horizontal standpipes is (by definition) a separate standpipe, they each need their own control valve as shown in the figure.

Horizontal standpipes are just like vertical standpipes in concept, just laying on their sides. Therefore, horizontal standpipes have the same installation requirements as vertical standpipes. The minimum pipe size is 4 in. Each standpipe needs its own control valve. In the case where the horizontal standpipe also serves as a water supply for the sprinkler system, a separate control valve and check valve will be required for the sprinkler system at the connection to the horizontal standpipe.

The situation where horizontal standpipes are different than vertical standpipes is in the hydraulic calculations. Sections 7.10.1.1.2 and 7.10.1.2.2 of NFPA 14 combine to require that in the situation where there are three or more hose connections on a horizontal standpipe, the hydraulic calculations must show that 250 gpm can be achieved through the three most hydraulically demanding hose connections on the highest horizontal standpipe with another 250 gpm from other standpipes (horizontal or vertical) in the building until a total flow of 1000 gpm is achieved in sprinklered buildings or 1250 gpm in unsprinklered buildings. This assumption has to do with the total number of hose connections that are likely to be used simultaneously by a fire department during a single fire.

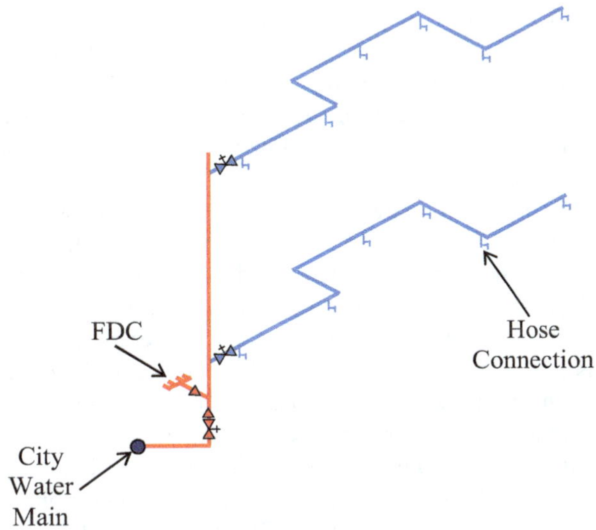

Fig. 12.3 Isometric view of horizontal standpipes

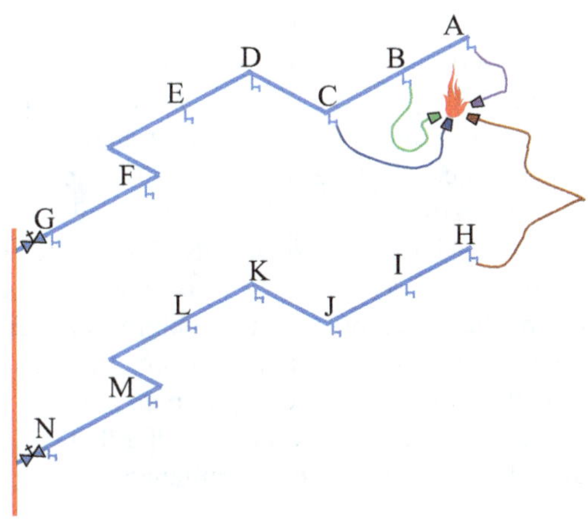

Fig. 12.4 Hose connections on horizontal standpipe used for fighting a fire near the remote portion of the system

For example, in Fig. 12.4, the fire is on the second floor of the building between hose connections A and B. It is possible that the fire department will connection hose to fight the fire at hose connections A, B, C, and H. In a sprinklered building, this would be the maximum 1000 gpm and no additional hose connections would be calculated. In an unsprinklered building, no additional flow would be added because there are no additional standpipes on this system and you are only required to pick up three hose connections on one horizontal standpipe and one hose connection on the other, so no additional flow would be added. If there was a third standpipe (vertical or horizontal) in the building and the building was not sprinklered, an additional hose connection on the third standpipe would need to be calculated, even though the chances of a fire department traveling two additional floor levels to fight a fire would be pretty slim.

The pressure requirements for hose connections on horizontal standpipes are the same as those for vertical standpipes. The minimum pressure that needs to be provided is 100 psi. The maximum pressure that is allowed at the outlet is 175 psi. If the pressure in the horizontal standpipe is greater than 175 psi, a pressure reducing valve is going to be required between the horizontal standpipe and the hose outlet, which can be built right into the hose valve. With the overwhelming majority of horizontal standpipe systems being in one and two story buildings, it is most common to see these systems by manual wet systems, so the maximum pressure is usually not a problem. The fire department can control what pressure they put into the system in the fire department connection.

The hydraulics of the situation can become complicated if the horizontal piping is split somewhere in the building and each direction serves more than one hose connection. In these situations, the argument could be made that you have two horizontal standpipes on a single floor and therefore, you need more flow from the water supply. For example, consider the one story building shown in plan view in Fig. 12.5. Part A of the figure shows a single run of piping to six hose connections. This would require a total flow of 750 gpm with 250 gpm from the hose connections at A, B and C. Part B of the figure shows the piping being split and serving three hose connections each. In this case, the flow demand would be 1000 gpm because there are two horizontal standpipes in the building and the requirements of NFPA 14 are to take 750 gpm from the first standpipe and 250 from additional standpipes.

Note that in Fig. 12.5, the additional 250 gpm flow is shown from hose connection E instead of hose connection D, which is further down the pipe. This is because there is no difference in the hydraulic calculations between 250 gpm coming off the system at Node D or Node E. NFPA 14 does not require that the flows be balanced to the higher pressure when you connect a second standpipe. So, once you have established the most remote standpipe and calculated the maximum number of hose connections on that standpipe, every connection on the other standpipes would provide the same hydraulic requirements. In this case, the hose is shown connected at Node E because it is closer to the fire that would most likely cause hose connections at A, B, and C to be used simultaneously.

Protection of Horizontal Standpipes and Branch Lines

Horizontal standpipes and branch lines need to be protected from mechanical damage, fire, earthquake, corrosion and freezing just like the other standpipe piping in the system. The rules were discussed completely in Chap. 11 of this text. The only area that could use a little more explanation is the earthquake protection rules.

NFPA 14 references back to NFPA 13 to pick up the installation and design criteria for earthquake protection of standpipe system piping. NFPA 13 requires earthquake bracing for all pipes serving as mains and all other pipes 2.5 in. in size or larger. This means that all horizontal standpipes and all branch lines will require both lateral and longitudinal bracing in accordance with NFPA 13. See Chap. 11 of this text for specific information on how to protect such pipe.

Part A – One Horizontal Standpipe

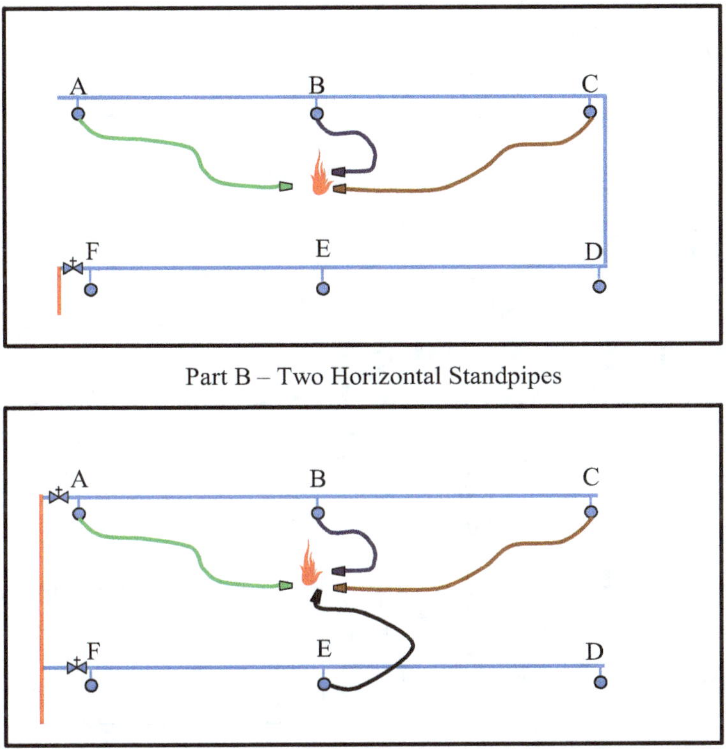

Fig. 12.5 Plan view of building showing consequences of splitting horizontal standpipe piping

Test Yourself

12.1. What is the difference between a branch line and a horizontal standpipe?

12.2. Does a branch line that runs a total of 30 ft require a control valve where it is connected to the standpipe?

12.3. What would be the friction loss in a 2.5 in. Schedule 40 steel pipe branch line for a manual wet standpipe system if the branch line piping includes 60 ft of pipe, four elbows, a tee and an OS&Y gate valve?

12.4. What would be the friction loss in the 2.5 in. pipe in question 12.3 if it was Schedule 10 pipe instead of Schedule 40?

12.5. The manual wet horizontal standpipe system shown in the isometric view in Fig. 12.6, will be for a sprinklered two story shopping mall. All pipe will be Schedule 10 steel. All valves are OS&Y gate valves. The lengths of pipe are shown in the figure, but do not include equivalent lengths of fittings or valves. What will be the flow and pressure requirements at the fire department connection? A blank hydraulic calculation form has been provided as Fig. 12.7 in order to make the calculations easier.

Fig. 12.6 Manual standpipe system for two-story shopping mall in question 12.5

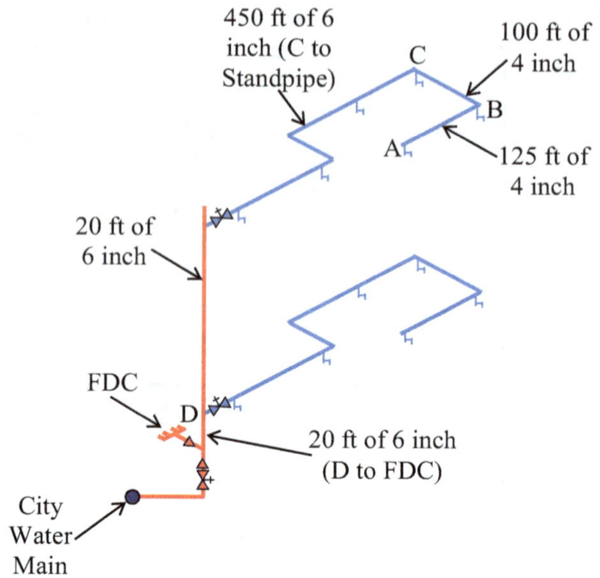

Node 1 / Node 2	Elev 1 (ft) / Elev 2 (ft)	K-Factor	Flow added – this step (q) / Total flow (Q)	Nominal ID / Actual ID	Fittings – quantity and length	L ft / F ft / T ft	C Factor / P_f per foot (psi)	total (P_t) / elev (P_e) / frict (P_f)	Notes

Fig. 12.7 Blank hydraulic calculation form for question 12.5

Chapter 13
Acceptance Testing of Standpipes

Once the standpipe system is finished, a series of tests need to be performed before turning over the system to the building owner. There are two reasons to run these tests. The first is to prove that the equipment was installed correctly. The second is to establish baseline performance so that long-term pass/fail criteria can be established for the periodic testing and maintenance that will be done on the system in the future. The testing for the underground system needs to be done before the aboveground piping is connected to the underground so that dirt and debris are not transferred into the aboveground piping.

Chapter 11 of NFPA 14 contains the requirements for acceptance testing of standpipe systems. This chapter will also reference NFPA 24 for the acceptance testing of underground pipe leading to standpipe systems. The acceptance tests of NFPA 24 are copied into NFPA 13 in section 10.10 of that document, so if you don't have a copy of NFPA 24, you can use NFPA 13 for the acceptance test requirements of the underground pipe. The installing contractor is generally required to do three things during the acceptance testing:

1. Notify all of the authorities having jurisdiction (AHJ's) and the building owner's representative of the date and time of the tests.
2. Perform all of the acceptance tests.
3. Complete and sign the acceptance test certificates.

If the underground contractor and the aboveground contractor are different companies, which they frequently are, then completely separate acceptance testing will be performed. The aboveground contractor needs to be informed as to when the underground tests are being made so that they can witness the underground testing. Nobody wants to connect the aboveground pipe to the underground pipe until they are sure that the underground was installed and flushed correctly. Witnessing the tests is the only way that some aboveground contractors will be satisfied that the underground tests were performed to their satisfaction.

It is important to note that the AHJ's are not required to approve the date and time of the tests (as the rules are written by NFPA 13, NFPA 14 and NFPA 24). The

K.E. Isman, *Standpipe Systems for Fire Protection*,
DOI 10.1007/978-3-319-47750-3_13

AHJ's need to be notified, but the NFPA committees did not want to give them veto authority for when the tests were going to be run. The concern of the committees is that the completion of the standpipe system can hold up the completion of the entire construction project and it is not fair to hold up the whole construction project due to scheduling problems of the AHJ. That being said, many AHJ's will not accept the results of tests unless they are there to witness them, so many local jurisdictions have passed their own rules that require the acceptance tests to be scheduled with the local AHJ and to have them on hand to witness tests as they are performed. A few jurisdictions have permitted contractors to perform acceptance tests with third party firms witnessing the tests or video tape evidence that specific tests have been performed.

Acceptance Tests for Underground Pipe and Equipment

Sections 11.2.1 and 11.4.4 of NFPA 14 reference NFPA 24 for the flush test and the hydrostatic test of the underground pipe. These are the two most important tests to perform before accepting the installation of the underground pipe, and they should be run in that order with the flush test first and then the hydrostatic test. You don't want to run the hydrostatic test with dirt or debris in the pipe because a failure during the high pressure of the hydrostatic test might cause dirt or debris to become projectiles and do more damage.

Both of these tests are usually run with the underground pipe still somewhat exposed in the trench. NFPA 13 (and NFPA 24) requires the sections of pipe between the joints to be filled in, but the joints still need to be exposed (see Section 10.10.2.2.4). The advantage to running the test this way is that when a problem is discovered during the test, the pipe joints are exposed and the problem can be addressed immediately. The disadvantage to running the test this way is that the trench is exposed. It may take some time to schedule the test and during that time, the exposed trench is a potential hazard on the job site. Care needs to be taken so that people don't fall into the exposed trench. If, for safety reasons, the trench can't remain open all the way, one option is to fill the trench completely, but this comes with the risk that if a problem occurs during the test, you may need to dig up the pipe; and it may not be clear exactly where you need to dig, so you may need to take a significant amount of the dirt back out of the trench to find the problem. Clearly, the best option is to keep track of the schedule closely so that you can keep the AHJ informed and run the test as soon as the pipe is finished while the trench is still open at the joints.

Flush Test of Underground

Section 10.10.2.1 of NFPA 13 and NFPA 24 contains the requirements for running a flush test of the underground piping. All of the underground piping, including the piping from the main into the building that connects to the system riser (called

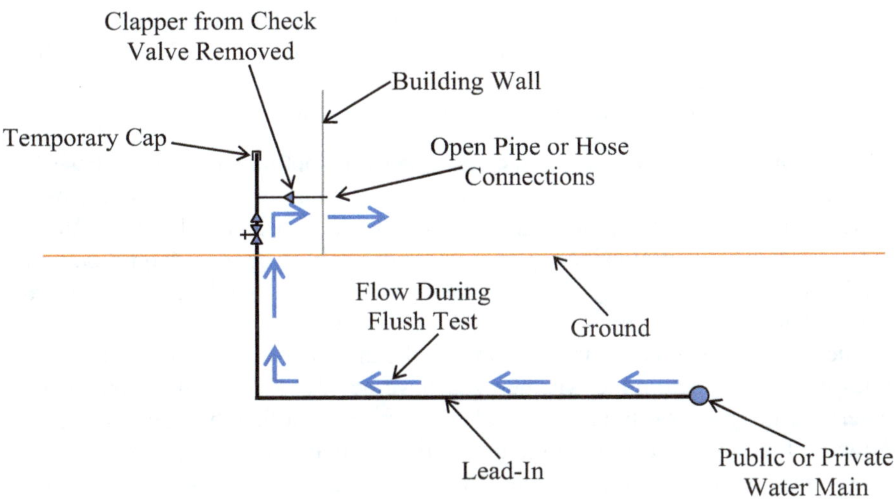

Fig. 13.1 Flushing arrangement for lead-in

the "lead-in"), needs to be completely flushed before connecting to the aboveground piping. In order for this test to be conducted correctly, a significant amount of water will need to flow through the piping, so the test plan needs to include a consideration for where the water can safely be discharged. There are a number of options including the use of the fire department connection piping as shown in Fig. 13.1.

In order to flush the pipe using the arrangement in Fig. 13.1, a temporary cap needs to be installed on the riser so that all of the water coming into the building will go back out through the fire department connection. Since it is extremely common to have the fire department connection attached to the riser above the system control valve and check valve, these may need to be in place prior to flushing the underground lead-in. The check valve in the fire department connection piping is also typically in place before flushing the underground lead-in, so the clapper will need to be removed or bypass piping will need to be installed around the check valve so that the flushing can be conducted. Once the water gets to the outside of the building, the bare end of the piping can be used if it is safe to discharge the water right there at the building wall. But if the water needs to be directed away from the building, the fire department connection can be completed with hose couplings and hose can be connected to the couplings to take water to a safe discharge location.

In order to conduct the flush test, enough water needs to flow through the piping to move any dirt, debris or obstruction that might be in the piping. The person running the test has an option to use one of two methods to determine the amount of water that will flow during the test. The options are spelled out in Section 10.10.2.1.3 of NFPA 13 and NFPA 24. The two options are:

1. Create a flow equivalent to a velocity of 10 ft/s in the underground pipe. For 4 in. pipe, this would be 390 gpm. For 6 in. pipe, this would be 880 gpm. For 8 in. pipe, this would be 1560 gpm.
2. Run the test at the maximum flow available from the water supply.

Both options have their pros and cons. The 10 ft/s option has been the basic test that has been done since at least the 1960's. The velocity of 10 ft/s was determined (in tests conducted by engineers at Industrial Risk Insurers) to be the minimum necessary to move certain size objects that were likely to be in underground pipes. So, the advantage to running the test at this flow is that you will be sure to move any objects that are in the piping.

The reason that some contractors do not like the 10 ft/s option is that it is sometimes difficult to determine whether you have achieved this flow. Accurate flow meters are not necessarily always available during a test like this. The flow option shown in Fig. 13.1 might allow the use of pitot gauge to determine the flow if the hose connections are already installed on the FDC. But if the flow is coming out of the open 4 in. pipe, a pitot gauge might not provide an accurate representation of the flow because a pitot gauge works well for a 2.5 in. outlet, but not as well for large outlets like 4 in. pipe. The large outlet typically does not have water flowing out around the full opening of the outlet, so the formula that converts velocity pressure in psi to flow in gpm would not be correct.

The option to just open the system and flow what you can from the water supply was added to NFPA 13 and NFPA 24 much more recently. This option was added for contractors that did not have the ability to measure the flow. The advantage to doing the test this way is that you don't need to take any measurements of flow. The disadvantage is that you may not achieve enough flow to more debris or obstructions that might be in the pipe. The logic behind this is that the maximum flow available from the water supply on any given day should be greater than the amount of water that would be used during a fire. So, if there are debris or obstructions in the pipe and they do not move during the test, they should not move during the use of the system, so they should stay where they are in the pipe and never become a problem in the aboveground portion of the system.

The flush test is not too complicated. You just need to determine where you are going to put the water, start the flow, and continue to flow water until the water runs clear. It can be difficult sometimes to determine whether the water is flowing clear. Many contractors place a burlap sack over the location of the discharging water. The sac will collect any debris coming from the pipe and let the water flow through. The test continues until the burlap sack is no longer collecting any debris.

It is important to note that the second option (flowing the maximum available from the water supply) needs to be done with openings that create at least as much cross sectional area as the pipe. This is important because you need to create a significant amount of flow, even if you are not required to measure it. Some contractors are used to installing pipe for plumbing systems, which only require a flush test from a small opening (sometimes as small as 1 in.). A test for a fire protection system from such a small opening would not create sufficient flow and would therefore not be acceptable. Many aboveground contractors want to witness the flush test before they connect their aboveground pipe just to make sure that the test was conducted through appropriate size openings so that a reasonable flow was achieved.

Hydrostatic Test of Underground

The goal of the hydrostatic test is to make sure that the pipe can hold pressure, proving that the pipe has been joined properly and that the pipe and fitting materials are free of flaws that would cause the piping to burst before its useful life is over. The basic procedure of the test is to fill the pipe with water, then increase the pressure in the piping until the test pressure is reached. The pipe then needs to hold pressure for the duration of the test. Since the test is run with water in the non-flowing condition, the test has been given the name "hydrostatic" using the Greek prefix for water (hydro) and the Greek word for a lack of movement (static).

For aboveground pipe, no loss of pressure is allowed during the hydrostatic test, but for underground pipe, some small loss of pressure is allowed. There are two reasons that have been given for why underground pipe is allowed to lose a small amount of pressure. The first is that it is hard to seal up the drains on dry barrel hydrants, so during the hydrostatic test on underground mains connected to hydrants, some water will go out the hydrant drains. The second reason that has been given for allowing pressure loss in the underground is that the joints in underground pipe are hard to tighten and with a long run of pipe, it is difficult to get every single joint so tight that it will not weep a little bit.

The fact that water is used instead of air increases the safety of the test. Water is an incompressible fluid, so if a catastrophic failure occurs during the test, the water absorbs most of the energy by transferring it through the fluid and allowing that energy to disperse throughout the system piping. When a failure does occur during the hydrostatic test, water will escape from the system, but it will not be in a form that will do harm to anyone. If the piping were pressurized with air during the test and a failure were to occur, the air would be compressed during the reactionary force at the failure point, increasing the energy right at that point. The extra energy right at the failure point could cause pieces of the pipe wall to fracture and become projectiles, moving away from the point of failure like shrapnel from an explosion. These flying pieces of pipe wall could severely injure any person in the vicinity. So, a test with water under pressure is much safer than a test will air under pressure.

Some people like to conduct air pressure tests before water pressure tests because of the concern for water damage. In these cases, they want to prove that the pipe joints are holding together before they put water into the pipe. An air pressure test can be safely conducted on a piping system at much lower pressures than 200 psi. NFPA 13 calls for some air pressure testing of pipes with a pressure of 40 psi. With this much lower air pressure, if a failure does occur, there will not be as much energy in the system and there should not be enough energy to cause the pipe wall to become flying shrapnel. Air pressure tests should never be conducted on any pipe with a pressure in excess of 40 psi.

The pressure that will be required during the test will depend on the maximum static pressure expected in the piping. If the maximum static pressure expected in the piping is less than 150 psi, then the test pressure will be 200 psi. If the maximum static pressure expected in the piping will be over 150 psi, then the test pressure will need to be 50 psi over the maximum expected static pressure. For example, if the maximum expected static pressure in the underground pipe is expected to be 165 psi, then the test pressure will need to be 215 psi (165 + 50 = 215).

One of the frequent situations where the pressure in the underground main will be above 150 psi is where a fire pump feeds an underground private fire service main. The pressure in the main will be the suction pressure (maximum static pressure adjusted to the elevation of the pump suction flange) plus the maximum net pressure created by pump (the churn pressure), plus the elevation difference between the discharge flange of the pump and the underground fire service main. For example, if the static pressure from a public water supply (adjusted to the elevation of the pump suction flange) is 50 psi and the pump's churn pressure is 120 psi, and the elevation change between the pump discharge flange and the underground private main is 7 ft, then the maximum expected pressure in the main will be 173 psi (50 + 120 + (7 × 0.433) = 173), which means that the hydrostatic test would need to be run at 223 psi (173 + 50 = 223).

The pressure that needs to be achieved during the test is measured at the lowest elevation of the pipe being tested. The actual location where the gauge needs to be placed is described in Section 10.10.2.2.3 of NFPA 13 (or NFPA 24) as an outlet of a hydrant on the system or at the lowest point in the system if there is no hydrant. The important item to note is that this is at the lowest point in the system, so the whole pipe is not hydrostatically tested at 200 psi (or 50 psi over the maximum expected pressure). Other points in the system will be tested at a lower pressure due to the elevation difference.

The duration of the test is a minimum of 2 h. In this period of time, it will become apparent if there are any problems with the way that the pipe was put together. Even though the pipe may not be rated to hold this pressure (many underground pipes are only rated for 150 psi), the higher pressure of the hydrostatic test will not be a problem for the pipe during this relatively short duration.

For many years, the amount of water that was allowed to leave the piping (sometimes called weepage) during the hydrostatic test was expressed in gallons. The formula to calculate the amount has changed over the years, but for the last few editions of NFPA 13 (and NFPA 24), the formula has been:

$$L = \frac{SD\sqrt{P}}{148{,}000}$$

Where: L = Allowable loss of water in gallons per hour
S = The length of pipe being tested in feet
D = The nominal diameter of the pipe being tested in inches
P = The average gauge pressure during the test in psi

Using the formula above, a person can calculate how much water is allowed to weep out of the fittings during the 2 h of the hydrostatic test. Note that the formula above provides the amount of weepage per hour. Since the test is run for 2 h, the value obtained from the formula is multiplied by two to get the total weepage allowed in gallons.

The term "weepage" is being used to describe the water that can leave the system even though the variable in the formula is "L". Many people believe that the "L" stands for leakage, which would be an accurate term for water leaving the system. But because the amount allowed by NFPA 13 (or NFPA 24) is so small, many of the committee members get upset at the use of the term "leakage" because it implies a constant stream of water. They would much rather use the term "weepage" because it implies a small loss of water droplets slowly over time rather than a constant stream. The term "weepage" may not be an official word in many English dictionaries, but it is used in the water flow industry to indicate a very small amount of water discharging from a portion of a system. Here is an example of a calculation for the maximum allowable amount of weepage in an underground private main with the following characteristics:

- The underground pipe being installed in 450 ft long.
- The maximum expected pressure in the system is going to be 173 psi, so the pressure during the test will be 223 psi.
- The pipe will be 8 in. Class 52 lined ductile iron.

Using the three conditions stated above, the amount of weepage can be calculated as 0.36 gal/h ($L = 450 \times 8 \times (223)^{0.5}/148{,}000 = 0.36$). Since the test will be run for 2 h, the total allowable amount of weepage would be 0.72 gallons ($0.36 \times 2 = 0.72$). This would be approximately 92 fluid ounces or a little less than three quarts.

The challenge with presenting an acceptable loss in fluid in terms of gallons is that in order to determine whether you passed the test or not, you would need to somehow either have a way of conducting the test so that you could measure how much fluid you lost during the test, or you would need to be able to collect all of the water droplets that weep from each joint in the system to determine whether you exceeded the maximum allowable amount or not. Collecting all of the water droplets is clearly not an option, so you are left with trying to conduct the test in a way that would indicate how much water left the system during the test. One way to do this would be to pump into the system from a calibrated container. With the pump constantly on during the test, you could check the water level at the beginning and end of the test to determine if too much water discharged during the test. The challenge with this sort of testing arrangement is that the pump needs to be cooled during such a long duration of running (2 h). Most water pumps use coolant water from the suction side of the pump, so it would be difficult to determine whether the water that left the calibrated container was due to coolant or water weepage in the pipe. In order to deal with this problem, the coolant water could be returned to the calibrated container, but if the container was relatively small, this water would heat up during the test and towards the end of the two hour duration might not be sufficiently cool to provide any benefit to the pump as coolant water.

By the 2013 edition of NFPA 13 and NFPA 24, the committee responsible for these documents decided that a better method of determining the pass/fail criteria of the hydrostatic test was needed. They settled on section 10.10.2.2, which now allows a pressure loss of 5 psi during the test. This is much easier to determine without worrying about measuring the loss of water. The test can even be run without keeping the pump running for 2 h. Instead, the pump can be started to get the pressure up to the test pressure and then the pump can be shut down. The test pressure needs to stay in the system for 2 h. If, at the end of 2 h, the system pressure has not dropped any more than 5 psi below the test pressure, then the system passes the test without worrying about how much water discharged from the joints. This 5 psi pass/fail criteria becomes the value used by almost everyone running the test now and the weepage calculation, while it remains in the standard, does not need to be calculated for most system acceptance tests anymore since the 5 psi rule handles the situation so much better.

Other Underground Acceptance Tests

In addition to the two major acceptance tests for the underground pipe (flush test and hydrostatic test), there are a few other tests that need to be run before the underground pipe can be placed in service. The following is a summary of these other tests:

- *Hydrants*: All private fire hydrants connected to underground mains need to be fully opened while one of the outlets is open to flow water and flush out the barrel. This test needs to be conducted at system pressure, so if a pump is upstream of the hydrant, the test needs to be run with the pump in service and turned on. The hydrant needs to be closed under system pressure as well to make sure that the valve seats properly. After this test is over for dry barrel hydrants, the person doing the testing needs to make sure that the barrel of the hydrant drains properly. One way to do this is to put your hand over the opening in the hydrant. You should feel the suction as air is being drawn into the barrel as the water drains down on the inside.
- *Control Valves*: All control valves need to be fully closed and then fully opened under system pressure to make sure that they work properly. If fire pumps are upstream of the valves, this test needs to be conducted with the pump running to make sure that the valves work under the maximum expected system pressure. Also, after closing and opening any valve, a flow test should be conducted downstream to make sure that the valve really opened at the end of the test.
- *Backflow Preventers*: All backflow preventers that are incorporated into the underground mains need to be forward flow tested to make sure that they are working. The goal is to open hose connections or hydrants downstream of the backflow devices in places where it is safe to flow large quantities of water. All that you need to do is establish a flow equivalent to the demand of the fire

protection system through the backflow device. In a standpipe system, this flow demand might be as high as 1250 gpm depending on the type of standpipe system, the number of risers, and whether or not the building is sprinklered. This test is just to exercise the internally loaded check valves in the backflow device and to prove that it can be done in the future, so it can be combined with the aboveground tests that are going to be discussed next in this text.

Underground Test Certificate

In order to provide documentation that the proper tests were conducted, and to provide a record of the specific measurements made during the test, a test certificate needs to be completed at the end of the test and turned over to the building owner, with copies going to any authority having jurisdiction that is interested. NFPA 13 and NFPA 24 mandate the use of a specific form, which is Figure 10.10.1 within those documents. This form is problematic for underground contractors for two reasons. First, the form is two pages long, which can be seen as excessive since it is providing a location for the data from a maximum of only five different tests. Second, the form is copyrighted by the NFPA, making it illegal to copy it. The NFPA does not sell pads of the form or make it available for people to use, so the NFPA puts people in the position of mandating a specific form and then making it illegal for people to copy or use that form.

There are two ways that people end up complying with the requirement to provide a certificate at the end of the test. The first is that they go ahead and copy the NFPA form and use it with their client. To date, this author is unaware of the NFPA going after someone who does this, even though it is technically a violation of the NFPA's copyright. The other way is for people to use a slightly modified form, such as those produced by the National Fire Sprinkler Association (and available through their website at www.nfsa.org). Organizations such as the National Fire Sprinkler Association produce a form that is slightly modified from the NFPA's form so they are not in violation of the NFPA copyright, but their form still has all of the information requested by the NFPA form, so it usually makes everyone happy.

A modification of this second option would be to simplify the form even more, potentially getting it down to a single page. There is a great deal of information on the NFPA form that is not really needed for documenting the results of the test, such as explaining the test procedure and providing the formulas for certain calculations. The fact that these procedures and formulas are in the standard should be sufficient for documentation purposes. A simplified form has been included in this text as Fig. 13.2 and can be used by the rules of NFPA 13 under the equivalency clause (section 1.5) or the alternate arrangement clause (section 1.7) as long as the authority having jurisdiction agrees that this form is equivalent (which they should since it provides the same information). Similar clauses are in NFPA 24. Note that

Underground Acceptance Test Certificate

ISMAN ASSOCIATES

All questions that are answered "No" are required to be explained in the Comments Section at the end of this certificate or on a separate attachment.

Description of Property

Property Name:________________

Property Address:________________

Systems Served by Private Main:________________

Plans

Name and address of Authority that Approved the Plans:________________

The installation conforms to the plans? ❑ Yes ❑ No

Pipe and Joints

Pipe type:________________

Standard to which pipe was made:________________

Joint type:________________

Standard to which joints are made:________________

Are joints properly restrained? ❑ Yes ❑ No

Flush Test

All new underground and lead-in pipe properly flushed? ❑ Yes ❑ No

Company that did flushing:________________

How was flushing water obtained?
❑ Public Main ❑ Tank or Reservoir ❑ Fire Pump

What was the outlet through which the water flowed?
❑ Hydrant Butt ❑ Open Pipe ❑ Hoses ❑ Other

Provide size of opening:________________

Explain "Other" Outlet:________________

How much flow was obtained?
❑ Maximum from water supply ❑______ gpm

Hydrants

How many connected to underground? ________

Type and make:________________

Fully opened under system pressure? ❑ Yes ❑ No

Barrel flushed out? ❑ Yes ❑ No

Closed under system pressure? ❑ Yes ❑ No

If dry barrel, drained on its own? ❑ Yes ❑ No ❑ N/A

Threads compatible with fire dept.? ❑ Yes ❑ No

Control Valves

Fully opened under system pressure? ❑ Yes ❑ No

Fully closed under system pressure? ❑ Yes ❑ No

Left in fully open position? ❑ Yes ❑ No

Hydrostatic Test

All new underground and lead-in piping properly hydrostatically tested? ❑ Yes ❑ No

Pressure at start of test:______psi and end:______psi

Duration of test:__________ hours

Total leakage (weepage) measured:________ gallons

Total leakage (weepage) allowed:________ gallons

Is leakage (weepage) acceptable? ❑ Yes ❑ No

Is pressure loss acceptable? ❑ Yes ❑ No

Were joints exposed during the test? ❑ Yes ❑ No

Backflow Preventers

Forward flow test performed? ❑ Yes ❑ No

Satisfactory flow achieved? ❑ Yes ❑ No

Describe means used to achieve forward flow:

After Test Requirements

Owner's representative shown location of control valves? ❑ Yes ❑ No

Owner's representative given information on care and maintenance of system? ❑ Yes ❑ No

System left in service? ❑ Yes ❑ No

Comments – Check box if attachments included ❑

Witnesses and Signatures

Company Name of Contractor:________________

Name of witness for Contractor:________________

Title of witness for Contractor:________________

Signature of witness for Contractor:________________

Date of witness for Contractor:________________

Company Name of Property Owner:________________

Name of witness for Property Owner:________________

Title of witness for Property Owner:________________

Signature of witness for Property Owner:________________

Date of witness for Property Owner:________________

Fig. 13.2 Alternative underground test certification form

Acceptance Tests for Aboveground Pipe and Equipment

The rules for the acceptance testing of the aboveground portion of the standpipe system will be found in Chap. 11 of NFPA 14, which starts out with the requirement that the acceptance tests be performed prior to the occupancy of the building. The standpipe system is an important part of the fire protection for a building and people should not be occupying the building until all of the fire protection systems are functional. Occasionally, it might be necessary to occupy a building before all of the fire protection systems are finished. In such a case, the authority having jurisdiction is permitted to allow the building to be occupied if the Impairment Procedures of NFPA 25 are followed. These Impairment Procedures will be discussed in Chap. 14 of this text.

If the reader has started here in this text and skipped the earlier portion of this chapter regarding the importance of conducting the test of the underground prior to connecting the aboveground piping, the reader is encouraged to go back and read the earlier portion of this chapter. Before conducting the aboveground acceptance testing, the contractor is required by NFPA 14 to verify that the proper testing has been completed on the underground pipe and equipment.

The first test that NFPA 14 describes for the aboveground piping is a flush test of the fire department connection piping. There is no specific flow that needs to be achieved for this test. The contractor just needs to create flow in the fire department connection piping to make sure that any dirt or debris that got into the piping during the construction process is removed.

The second test that is described by NFPA 14 is a hose thread compatibility test. Someone needs to go to all of the hose threads on all of the fire department connection inlets and all of the hose connection outlets and make sure that the hose that the local fire department will use can be threaded onto every one of these connections. The contractor should have a hose coupling, cap or plug that is compatible with the local fire department threads with which to conduct this test. It is important to check every thread since you never know if the correct products were shipped by the manufacturer. A good contractor will check these threads prior to installing the product so that a problem can be discovered before the device is installed, but it should be checked again during the acceptance test as a good practice unless the original check was well documented during installation.

Hydrostatic Test of Aboveground Piping

All of the aboveground piping and equipment, including the fire department connection piping, needs to be subjected to a hydrostatic test. The test is very similar to the test described for underground piping, with a few minor variations. The pressure that needs to be achieved during the test is the same as the underground: a minimum of 200 psi or 50 psi in excess of the maximum system pressure, whichever is greater.

Just like the underground test, this pressure is measured at the lowest level of the system. However, the difference between the aboveground test and the underground test is that the aboveground pipe is going to be much higher above the lowest point in the system. So, if the test pressure that needs to be achieved is 200 psi, that is the pressure that will need to be achieved that the lowest point in the system. Five or six stories higher, the pressure might only be 174 psi (200 – (60 × 0.433) = 174), which is perfectly acceptable given how NFPA 14 is written. If the building is divided into multiple vertical zones, then the lowest point in the system is the lowest point in the zone, not the lowest point in the building. The test pressure needs to be maintained for a period of at least 2 h.

The pass/fail criteria for the aboveground testing are a little different than what was acceptable for the underground piping. For the aboveground piping, no leakage or weepage is allowed. The minimum test pressure is required to be maintained during the test. However, the authority having jurisdiction is encouraged to take into account the laws of physics. Water and air do not maintain a constant volume as temperatures change. While the system is being filled with water, some trapped air will exist in the system. If the temperature changes during the 2 h of the test, the water may contract, allowing compressed air trapped in the system to expand, which may result in a small drop in pressure without any leakage or weepage occurring. The extent to which the pressure might change is impossible to say on a general basis because it is dependent on the volume of the system, the amount of trapped air, the temperature of the water going into the system and the temperature of the pipe over the period of the test. In general, the pressure would go down if the temperature went down during the test. So, the contractor and authority having jurisdiction should be aware of temperature changes during the test and should look for leakage or weepage if the pressure goes down during the test. If all of the following conditions existed during the test: no leakage or weepage can be found, the temperature went down during the test, there is no other explanation for the pressure going down, the drop in pressure was reasonable given the size of the system and the size of the temperature drop, the pressure is holding steady after the pressure drop, then the conclusion can be drawn that the drop in pressure was not due to a leak or weepage in the system and the system can be accepted.

If the building is being completed during cold weather and the standpipe system is exposed to the cold temperatures, NFPA 14 allows an air pressure test to substitute for a hydrostatic test until the weather gets warmer and a hydrostatic test can be completed at that time. Due to the safety concerns discussed in the underground section of this text, the air pressure test can only be conducted at a maximum pressure of 40 psi. This pressure needs to be held in the system for 24 h and pressure losses over that period are not allowed to exceed 1.5 psi. The same comment on temperature changes applies here so the starting temperature and ending temperature should be noted to make sure that the change in pressure was not due to any change in temperature.

When repairs and replacements are made to existing systems, the new portion of the system is required to be hydrostatically tested. If the new work can be isolated, then the pressure to which the new work needs to be exposed will be 200 psi or

50 psi over the maximum working pressure, whichever is greater. But if the new work cannot be isolated (such as a new hose connection being put in the system or a new control valve on a standpipe) then the pressure during the hydrostatic test will be the system pressure when the valve is opened. The reason that this lower pressure is acceptable is that there is a concern over exposing older pipe, fittings and equipment to pressures greater than the pressure at which they are rated. This can be done safely for new systems, but for older pipe that has begun to corrode, safety margins have been taken away and it is best not to over-pressurize this pipe or equipment.

During the hydrostatic test, gauges that have already been installed at the top of each standpipe need to be inspected to make sure that they are reading a reasonable pressure given what the test pressure is at the bottom of the system (remember that is will be lower due to the elevation difference). This pressure needs to be recorded from each gauge.

There are contractors that deal with other water based systems (not fire protection systems) that like to use additives like sodium silicate or other chemicals during a hydrostatic test. Sodium silicate is a chemical that is a liquid in contact with water, but hardens to a solid in contact with air. Other chemicals, like brine, actually cause a little bit of corrosion in contact with air, which may seal a small leak. These contractors like such products because they seal leaks automatically and help them pass the test. The problem is that in a fire protection system, there may be trapped air somewhere in the system and these chemicals might cause a blockage where the come into contact with the air. Worse yet, these chemicals might cause corrosion or blockage in the system at a sprinkler downstream of the standpipe system, which would prevent the control or suppression of a fire. The use of sodium silicate or any other chemical during the hydrostatic test is specifically forbidden by NFPA 14 (and all other NFPA codes and standards).

Flow Tests

The next acceptance test that needs to be performed is a flow test of the standpipe system to make sure that the design pressure (typically 100 psi for Class I and Class III systems and 65 psi for Class II systems) can be achieved at the most remote hose connection while the demand flow has been created through the piping system. For an automatic standpipe system, the flow for this test should be easy to achieve since the water supply is permanently connected to the system. For a manual standpipe system, the flow needs to be created through the fire department connection, which can be done in a variety of ways including:

1. Use of a portable pump or fire department pumper putting the necessary flow and pressure into the fire department connection from a hydrant or tank as shown in Fig. 13.3.
2. Use of a portable pump or fire department pumper recirculating the necessary flow through the system piping and back to the pump using hoses and/or temporary pipe as shown in Fig. 13.4.

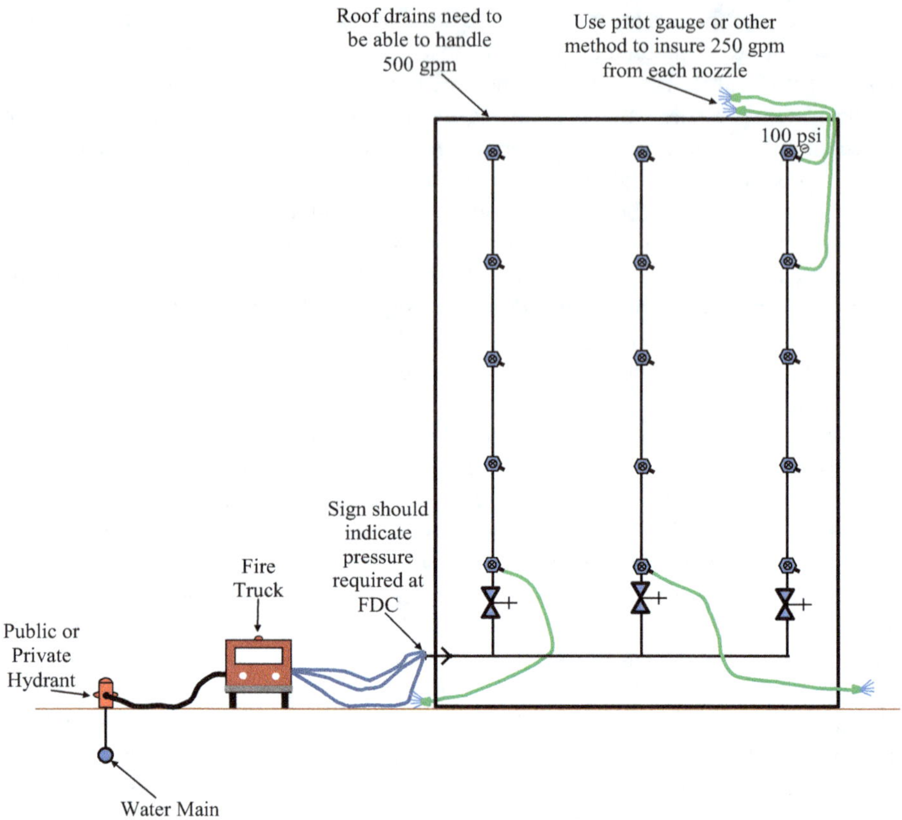

Fig. 13.3 Use of fire truck pumping into FDC with flow to drains for Class I flow test

3. Use of a temporary water supply such as a public or private main that can provide the system flow as shown in Fig. 13.5.

For the test method shown in Fig. 13.3, the hoses need to be connected to the two most remote connections away from the water supply. These hoses are shown going up to the roof since this is frequently the most convenient place to discharge the water. If this is the technique used during the testing, the roof drains will need to be able to handle the 500 gpm total from these hoses. The hoses could also be run down the stairs to places at the ground level that could handle the flow. The other two hoses can be connected anywhere that is convenient to discharge water from the other two standpipes. These are not required to be at the top of the standpipe since there is no need to verify any pressure in the standpipes that are not the most remote.

When testing in accordance with the methods shown in Fig. 13.3 or Fig. 13.4, the portable pump or fire truck will need to be set up to pump the required flow into the fire department connection at the required pressure. Section 6.4.5.2.2 of NFPA 14 requires that the pressure be indicated on the sign at the FDC. However, this is rarely seen in practice. Typically, the sign only shows the pressure that the fire

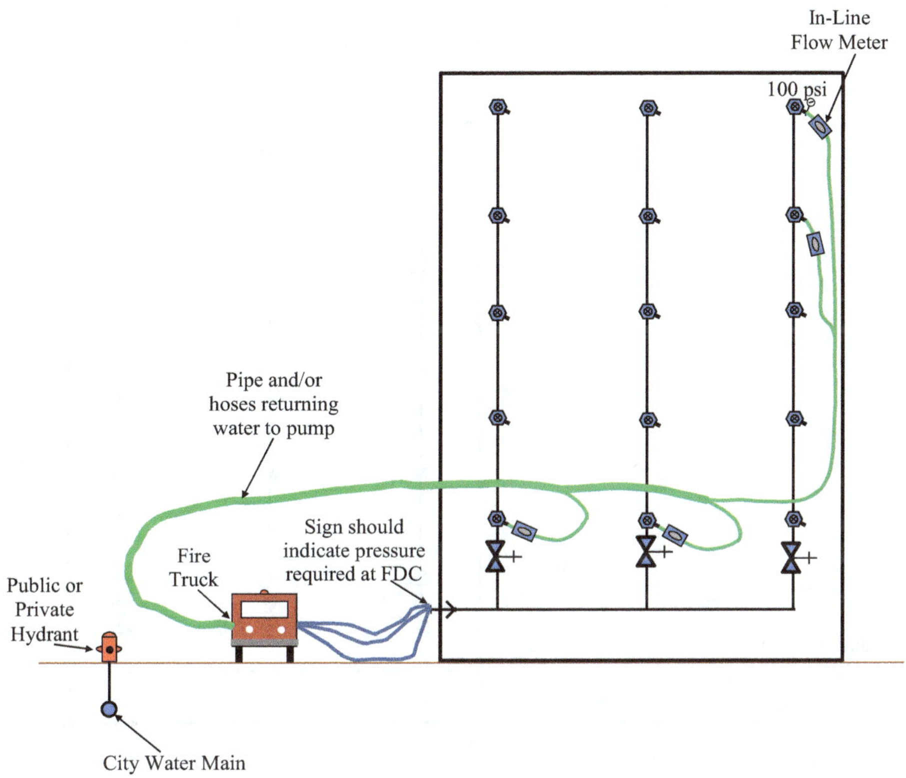

Fig. 13.4 Recirculating test water to fire truck during Class I flow test

department needs to pump in if that pressure exceeds 150 psi. If the firefighters do not see a pressure on the sign, it is standard operating procedure to just pump into the system at 150 psi. Figure 13.6 shows a fire department connection for a building where the fire department will need to pump into the system at 222 psi. This is the pressure that would be required during the acceptance test.

When testing in accordance with the method shown in Fig. 13.4, the pump puts water into the fire department connection at the proper pressure. The water flows through the piping until it comes to the four hose connections that are open (the two most remote and then one on each of the other two standpipes as is convenient). At each of these hose connections there needs to be an in-line flow meter, which is adjusted to create as close to 250 gpm as possible. The hose from these connections carries the water back to where it is connected to larger pipes or hoses that bring the water back to the pump.

With the test method shown in Fig. 13.5, the water supply does not need to meet the pressure demand of the standpipe system. As long as the flow demand can be met, a pressure gauge can be placed at the fire department connection and another pressure gauge can be placed at the most remote hose connection outlet. The difference between the readings on the gauges will be the effect of friction loss and

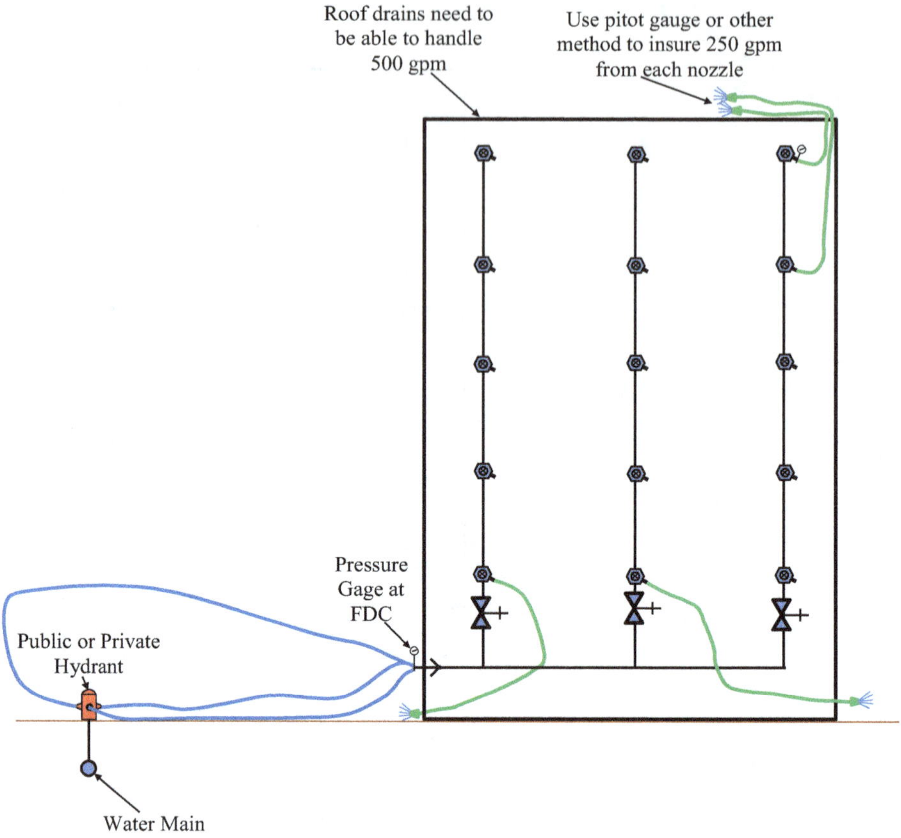

Fig. 13.5 Flow test using public or private main that can provide demand flow and pressure

elevation loss in the system with the minimum flow established. This difference can be used to determine whether the system passes or fails the test. For example, if the pressure at the FDC is 80 psi and the pressure at the most remote outlet is 32 psi (when all of the proper flows are coming from each of the hose connections), then the difference between these values will be 48 psi (80 – 32 = 48). This means that when 1000 gpm is pumped into the FDC at 150 psi, the pressure loss will be 48 psi and the water will discharge from the outlet at 102 psi (150 – 48 = 102). Since this value exceeds the minimum of 100 psi at the hose connection, this will be an acceptable performance of the standpipe system, so the system should be accepted.

Pressure Regulating Devices

Every pressure regulating device on the system needs to be tested to make sure that it is appropriately dropping the pressure. The test is conducted by flowing 250 gpm through the Class I or Class III outlet with a pressure gauge on the upstream side

Fig. 13.6 Fire department connection sign showing the minimum pressure the the fire department needs to pump in as 222 psi

and a pressure gauge on the downstream side of the device. The difference between the readings on the pressure gauges will indicate whether the device is working properly. In order to interpret the values correctly, you need to know what the settings on the device is supposed to be and how the device is intended to operate. This information should be available on the system calculations, which need to be available during the acceptance test.

During the test, the static pressure on both the upstream and downstream sides of the pressure reducing valve needs to be recorded. Then the residual pressures upstream and downstream with the 250 gpm flowing need to be recorded. After recording these values, they need to be evaluated to make sure that they are appropriate given the setting of the valves. The flow during the test will need to be placed in an appropriate drain that can handle the flow, which is why NFPA 14 requires a drain riser adjacent to the pressure reducing valve.

If the pressure reducing valves are in series (such as with a master pressure reducing valve assembly), then three tests will need to be performed. The first test will be the first pressure reducing valve (the one closest to the water supply), which will be the same test as with any other pressure reducing valve. The second test will be with the second pressure reducing valve tested with the first valve in service, which needs to make sure that the second valve swings open and does not reduce the pressure. The third test needs to be performed using the second pressure reducing valve without the first valve in service using the bypass line around the first valve. This last test needs to show that the second valve in series can take over and reduce the pressure if the first valve does not operate properly.

Other Acceptance Tests

There are a few other tests that need to be run during the acceptance testing of the aboveground system. Each of these tests is performed to make sure that each part of the system is working correctly. The additional tests are:

- *Backflow Preventers*: If a backflow preventer is installed as a part of the aboveground system, a forward flow test needs to be conducted by opening devices downstream to create the system demand flow through the backflow device. In many cases, this test can be combined with the flow test discussed above as long as the flow is coming through the backflow preventer.
- *Main Drain Test*: For systems connected to a water supply, a main drain test needs to be performed. There are two purposed to a main drain test. The first is to make sure that the valves between the drain and the water supply are open. The second is to establish a baseline pressure for use in comparison to the periodic tests that will be performed for years after the acceptance test. The procedure for a main drain test is simple. You record the static pressure from the water supply, open the main drain all the way, and then record the residual pressure. When shutting the drain down, watch the pressure gauge and make sure it returns to the static pressure immediately. If the gauge is slow to return to the static pressure, a valve is probably partially closed or there is an obstruction in the supply piping.
- *Automatic Fill for Suction Tanks*: If there is an automatic fill valve on a suction tank, the fill valve needs to be tested by draining a little water out of the tank with the fill line shut. Make sure that the water level is below where the automatic fill is supposed to start. Open the line and make sure that the valve automatically senses the loss of water and opens to fill the tank. The flow should be determined either by using a pitot gauge or flow meter, or by timing how long it takes to raise the water level a specific distance and using the volume of the water to calculate the flow. For example, a tank with an 18 ft diameter contains a volume of 63.6 ft^3 ($3.14159 \times 9 \times 9 \times 0.25 = 63.6$) when the water level goes up 3 in. This is 475.7 gallons ($63.6 \times 7.48 = 475.7$) of water. If the automatic fill takes 23 s ($23/60 = 0.383$ min) to raise the water level in this tank a distance of 3 in., then the fill was providing a flow of 1242 gpm ($475.7/0.383 = 1242$). If the system has a demand of 1000 gpm, this would meet the requirement to fill the system as quickly as the water is being used, which is a requirement of NFPA 20 for very tall buildings. At the end of the test, the automatic fill valve should sense the correct water level when the tank is full and automatically shut down.
- *Pumps*: Fire pumps need to be acceptance tested in accordance with NFPA 20. Those test procedures are too numerous to recount here.
- *Automatic and Semiautomatic Dry Systems*: There are a number of tests that need to be run on automatic and semiautomatic dry standpipe systems. These tests are:
  - *Water Delivery Time*: For systems with more than 750 gallons of volume within the piping system above the dry-pipe or preaction valve, water must be

delivered to the most remote hose connection within 3 min of opening the hose valve. For systems with fire pumps, this test can be conducted with the pump operational.

  - *Remote Activation Device*: For semiautomatic dry systems, every single remote activation device (pull station or button) that is designed to trip the preaction valve needs to be tested to make sure that it actually does trip the valve. The most efficient way to do this test is with at least two people. One to travel around the building and test the remote activation device. The other one to stay at the preaction valve to make sure that it trips, then reset it for the test of the next device.
  - *Air Pressure Test*: In order to make sure that the piping will hold air pressure, the system must be pressurized to 40 psi with air, which it needs to hold for 24 h. If the air pressure drops more than 1.5 psi during the 24 h of the test, the system fails the test. See the earlier discussion in this chapter on changes in temperature when interpreting the results of this test.

- *Valves*: All control valves need to be run full their full range all the way open and then all the way closed. Each of the valves on the hose connections need to be run through this same test with the caps tightly on so that only the caps fill up with water when the hose connection is opened. After the test, the caps need to be opened up and the water drained.
- *Alarm and Supervisory Devices*: All waterflow alarms and supervisory devices need to be tested in accordance with NFPA 72. This basically consists of making sure that the waterflow alarms sense the amount of flow that they are supposed to sense and send the correct signal to the alarm panel. Likewise, the supervisory devices such as the control valve supervisory switches and the low water level indicators in tanks sense what they are supposed to and send the correct signal to the panel. Finally, the panel needs to be checked to make sure that it is communicating correctly with the constantly attended location.
- *Signs*: A number of signs are required by NFPA 14. These were outlined in Chap. 5 of this text. During the acceptance test, the person doing the testing needs to make sure that the correct signs are in the correct place with the correct information.

Plans, Drawings and Reports

The building owner is supposed to get a copy of the standpipe system plans. This includes the calculations that went into proving that the system piping was sized correctly. Typically, the plans are turned over the building owner at the time of acceptance testing, although they can be provided at any time. The acceptance test form will include a question regarding whether the plans have been turned over, so many contractors do it while they are filling out the certificate to make sure that they answer the question correctly.

The building owner also needs to receive a copy of all of the information generated during the acceptance test including the acceptance test certificate. This information is supposed to be kept by the building owner for the life of the system. The acceptance test certificate includes important baseline information that needs to be used to determine whether the system is still performing in an acceptable way years down the road.

Just as with the underground acceptance test, a form needs to be filled out for the aboveground acceptance test. NFPA 14 contains a form for the acceptance test that is three pages long. Like the underground form, it can be shortened and still contain the important information. Figure 13.7 is an alternate to the NFPA 14 form spread over two pages. It could be used under Section 1.4 of NFPA 14 as an equivalent method to the form mandated by that standard. The form shown in Fig. 13.7 is copyrighted by Isman Associates and is not permitted to be used without express written permission by Isman Associates.

The installing contractor is supposed to turn over manuals for each of the major components of the system so that the building owner knows how to use, inspect, test and maintain the system and all of its components. This includes a requirement to turn over a copy of NFPA 25 to the building owner so that they understand the requirements for inspection, testing and maintenance and the frequency with which the work needs to be done.

Test Yourself

13.1. For an underground system with a maximum expected static pressure of 130 psi, what pressure would be needed at the lowest point of the system for the hydrostatic test?

13.2. For an 8 in. underground pipe that is 700 ft long, how much leakage (or weepage) would be allowed during the hydrostatic test if the water supply consists of an elevated tank with a water level 150 ft above the pipe?

13.3. For an automatic wet standpipe system where the maximum static pressure from the water supply is 50 psi and the churn pressure of the pump is 140 psi, what will be the pressure expected on the gauge at the top of the vertical standpipe 90 ft above the pump during the hydrostatic test?

13.4. For a standpipe system with six vertical standpipes, what flow will need to be achieved during the flow test and where will the flow need to be created if the building is also fully sprinklered in accordance with NFPA 13?

13.5. During an acceptance test of a standpipe system, the pressure reducing valve at a hose connection is being tested. The pressure reducing valve is a Model D valve in accordance with Fig. 13.8. While 250 gpm is flowing through the valve, the inlet pressure is 230 psi and the outlet pressure is 148 psi. Is this result acceptable?

Standpipe System Acceptance Test Certificate

ISMAN ASSOCIATES

All questions that are answered "No" are required to be explained in the Comments Section at the end of this certificate or on a separate attachment. Make sure you have proof of underground tests before connecting aboveground pipe.

Description of Property and System

Property Name:____________________________

Property Address:__________________________

____

____

System Type: ❑ Automatic Wet ❑ Manual Wet
❑ Automatic Dry ❑ Semiautomatic Dry ❑ Manual Dry
❑ Check if standpipe is combined with sprinklers
Number of Vertical Standpipes: ______________
Number of Horizontal Standpipes:______________

Plans

Name and address of Authority that Approved the Plans:__________________________________

____

____

The installation conforms to the plans? ❑ Yes ❑ No

Water Supply

Type: ❑ Water Utility ❑ Gravity Tank ❑ Suction Tank
❑ Open Reservoir ❑ Other ____________________

Capacity of Supply:____________________ gallons

If Utility: Static Pressure: ____________________

Residual Pressure:_________ Residual Flow:_________

If fire pump, manufacturer: ____________________

Model:______________ ❑ Electric ❑ Diesel ❑ Other

Rated Flow:________ Rated Pressure:____________

Churn (Shutoff) Pressure: ________________

Pipe and Joints

Pipe type:__________________________________

Joint type:__________________________________

Control Valves

Main control valve location:______________________
Supervision: ❑ Locked ❑ Sealed ❑ Supervisory Switch
❑ Other____________________________________
Do all standpipes have a shutoff valve? ❑ Yes ❑ No
All hose valve operated properly? ❑ Yes ❑ No

Backflow Preventers

Size:____ Make:____________ Model:______________
❑ Double Check ❑ Reduced Pressure Zone
Forward flow test performed? ❑ Yes ❑ No
Satisfactory flow achieved? ❑ Yes ❑ No
Describe means used to achieve forward flow:

____

Hose Threads

Threads on FDC and hose connections verified to match local fire dept. ❑ Yes ❑ No

Dry-Pipe Valve

Size: _______ Make:_____________ Model:___________

Water Pressure:___________ Air Pressure: ___________

Trip time after remote hose con. opened:____________

Air pressure at trip:___________

Water Delivery Time: ___________

Deluge or Preaction Valve

Size: _______ Make:_____________ Model:___________
Activation: ❑ Hydraulic ❑ Electric ❑ Pneumatic

Supervisory Air Pressure:___________

Type of Actuation Device:____________________________
Each actuation device tested? ❑ Yes ❑ No
Each actuation device operated properly ❑ Yes ❑ No

Water delivery time to most remote connection:________

Air Pressure Test for Dry Systems

All pipe air pressure tested at 40 psi? ❑ Yes ❑ No
System passed test, less than 1.5 psi loss? ❑ Yes ❑ No

Pressure Regulating Devices

If more than 9 devices, use a separate page.

	Valve 1	Valve 2	Valve 3
Location			
Model			
Static In			
Static Out			
Residual In			
Residual Out			
Flow			

	Valve 4	Valve 5	Valve 6
Location			
Model			
Static In			
Static Out			
Residual In			
Residual Out			
Flow			

	Valve 7	Valve 8	Valve 9
Location			
Model			
Static In			
Static Out			
Residual In			
Residual Out			
Flow			

Waterflow and Supervisory Signals

Waterflow alarm sent appropriate signal? ❑ Yes ❑ No
All appropriate supervisory signals sent? ❑ Yes ❑ No

 Page 1 of 2

Fig. 13.7 (**a**) Page 1 of standpipe acceptance test form; (**b**) Page 2 of standpipe acceptance test form

Standpipe System Acceptance Test Certificate

All questions that are answered "No" are required to be explained in the Comments Section at the end of this certificate or on a separate attachment.

Hydrostatic Test
All piping properly hydrostatically tested? ❑ Yes ❑ No

Pressure at lowest point in system during test:_______psi

Duration of test:_______________ hours

If cold weather, all piping air tested at 40 psi? ❑ Yes ❑ No

Pipe passed test? ❑ Yes ❑ No

Do full hydrostatic test in warm weather

Pressure at Top of Standpipes			
Location	Pressure	Location	Pressure

Are test results acceptable? ❑ Yes ❑ No

Main Drain Test
Static Pressure:____________

Residual Pressure:____________

Are test results acceptable? ❑ Yes ❑ No

Flow Test
System flow tested at demand flow? ❑ Yes ❑ No

Acceptable pressure at remote outlet? ❑ Yes ❑ No

How was flow determined? ❑ Flow Meter ❑ Pitot Gauge

Method of testing: ❑ Automatic Supply ❑ FDC

Pressure at water supply:____________

Pressure at remote hose connection:____________

Total Flow:____________

If pitot: nozzle size:_________ Pitot Pressure:_________

Test Blanks
If test blanks were used, how many?____________

Location of test blanks:____________________

How many blanks were removed?______________

Welded Piping
Was any piping welded? ❑ Yes ❑ No

If yes, Certify to the Following:

Welding follows AWS D10.9 AR-3 ❑ Yes ❑ No

Welders qualified by AWS D10.9 AR-3 ❑ Yes ❑ No

Documented quality control procedure? ❑ Yes ❑ No

All cutout discs retrieved? ❑ Yes ❑ No

Opening smooth? ❑ Yes ❑ No

Diameter not penetrated? ❑ Yes ❑ No

Slag and residue removed? ❑ Yes ❑ No

After Test Requirements
Do you certify that no chemicals such as sodium silicate, brine or other corrosive chemicals have not been used? ❑ Yes ❑ No

Have the following been left with the owner's rep.:

System components manuals? ❑ Yes ❑ No

Care and maintenance instructions? ❑ Yes ❑ No

Copy of accepted plans and calc's? ❑ Yes ❑ No

NFPA 25? ❑ Yes ❑ No

Hydraulic nameplate provided? ❑ Yes ❑ No

System left in service? ❑ Yes ❑ No

Comments – Check box if attachments included ❑

____

____

____

____

____

____

____

____

____

____

____

____

____

____

____

____

Witnesses and Signatures
Company Name of Contractor:____________________

Name of witness for Contractor:____________________

Title of witness for Contractor:____________________

Signature of witness for Contractor:____________________

Date of witness for Contractor:____________________

Company Name of Property Owner:____________________

Name of witness for Property Owner:____________________

Title of witness for Property Owner:____________________

Signature of witness for Property Owner:____________________

Date of witness for Property Owner:____________________

Page 2 of 2

Fig. 13.7 (continued)

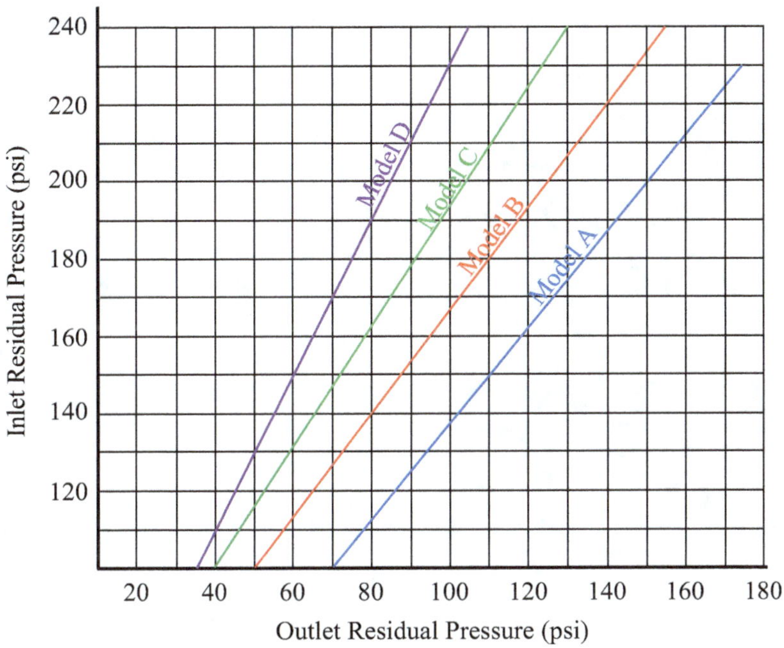

Fig. 13.8 Residual pressure performance for pressure reducing valve in questions 13.5

Chapter 14
Periodic Inspection, Testing and Maintenance

Unless they are properly maintained, even the best designed standpipe systems can fail during a fire emergency, which can lead to serious damage and potentially life loss as well. Proper maintenance includes inspections, tests, preventative work to prolong the life of the equipment, and repair of the equipment when inspections or tests reveal that there is a problem. The minimum inspection, testing and maintenance requirements that are adopted as law for most fire protection systems are found in NFPA 25, Inspection, Testing, and Maintenance of Water-Based Fire Protection Systems.

According to NFPA 25, the definition for "inspection" is the process of performing a visual examination of an object to determine whether it appears to be operational and free of damage. There are two important parts to this definition. The first is that the inspection is done by looking at an object. There is no requirement to touch it or move it. In order to conduct an inspection, you don't need to take it apart (with one exception that we'll discuss later in this chapter), and you don't need to try it out to see if it works.

The second important part of the definition of "inspection" is that the person performing the inspection only needs to make sure that the object "appears" to be functional. This means that the object needs to appear normal to a reasonable and knowledgeable person without necessarily seeing all of the parts. An inspection does not ensure that a piece of equipment will work, but it does help establish that its condition has not changed over a period of time since the last inspection. A chain of inspections helps to provide a long-term picture of a fire protection system.

The definition of a "test" involves the operation of a device to determine if it is functional. Another definition of "test" is a measurement of some characteristic to determine if it meets requirements. Either way, a person performing a test needs to touch the device being tested. At the end of the test, there needs to be some sort of criteria to determine whether or not the object passes the test or not.

The term "maintenance" is defined in a number of ways. First, maintenance is the preventative work that is performed to keep a device functional. For example,

K.E. Isman, *Standpipe Systems for Fire Protection*,
DOI 10.1007/978-3-319-47750-3_14

when you change the oil in your car, you are performing preventative maintenance. The term “maintenance” is also used to refer to the repairs that are done when equipment breaks down or fails to operate properly.

A building owner needs to have a complete program of inspection, testing and maintenance to make sure that their standpipe system remains in good shape. The owners do not want to wait until there is a fire to see if their standpipe system will work. Likewise, firefighters don’t want to wait until there is a fire in a building to see if the standpipe system will work. The rest of this chapter will discuss the inspection, testing and maintenance requirements for a standpipe system. For other fire protection systems, there are similar requirements, but NFPA 25 should be consulted for more details.

While NFPA 25 is a standard that is written in language that is intended to be adopted as law, and has been adopted as law in most states in the United States, it has not been published for as long as the NFPA installation standards like NFPA 14. The first edition of NFPA 25 was published in 1992. Prior to that, there was a Recommended Practice for Care and Maintenance of Standpipe Systems (NFPA 14A). As a Recommended Practice, NFPA 14A could not be adopted as law and therefore, many of its requirements were ignored by building owners. Therefore, there are many older standpipe systems that have been ignored for years and are just now beginning to be properly inspected, tested and maintained as NFPA 25 is being adopted and enforced.

NFPA 25 requires inspection, testing and maintenance to be performed on a specific schedule, but each item has its own frequency associated with it. Some items need to be performed as often as weekly while other items only need to be performed once every five years. While building owners might be willing to do the weekly items on their own (or more realistically speaking, they might not be willing to pay someone to come to their building on a weekly basis), but many building owners are likely to hire a professional firm to perform the quarterly, annual and less frequent items.

It is important to note that the responsibility for performing the requirements of NFPA 25 resides with the building owner. The building owner can hire a contracting firm to perform some of the work, but ultimately, it is the building owner’s responsibility to make sure the work gets done.

The requirements that will be discussed here come from Chap. 6 (Standpipe and Hose Systems) and Chap. 13 (Valves, Valve Components and Trim) of the 2014 edition of NFPA 25, which is the most recent edition when this text was being written. There are other requirements that may apply to a standpipe system, especially one that has a fire pump or one that also feeds a fire sprinkler system. This text will only include the requirements for standpipe systems. Requirements for inspection, testing and maintenance of fire pumps, sprinkler systems, or other fire protection systems are outside the scope of this text.

Inspection Requirements for Standpipe Systems

The inspection requirements will be organized here by pieces of equipment and the frequency of the most common inspection item. Not every piece of equipment will be found in every standpipe system. If your standpipe system does not have one of these pieces of equipment, you can skip that item. The following is a list of inspection requirements for standpipe systems:

1. The enclosure around dry-pipe, preaction and deluge valves needs to be inspected during cold weather to make sure that the minimum temperature of 40 °F is being maintained. The inspection needs to be done daily unless a low temperature alarm is installed in the enclosure and monitored at a constantly attended location in which case the inspection can be performed weekly.
2. Master pressure reducing valve assemblies need to be inspected weekly to verify the following:
   (a) The downstream pressures are in accordance with the system design.
   (b) The supply pressure is in accordance with the design.
   (c) The devices are not leaking.
   (d) The devices and trim are in good condition.
3. In general, gauges are supposed to be inspected on a weekly basis unless there is a more specific rule further down in this list. For preaction and deluge valves, the gauge on the water supply side needs to show a reasonable water pressure.
4. For dry-pipe systems that do not have low pressure alarms, the gauge needs to be inspected weekly to make sure that the water pressure gauge is showing a reasonable water supply pressure, that the air/nitrogen gauge is showing a correct gas pressure (given the ratio to the water pressure) and that the gauge on the quick opening device is showing the same pressure as the air/nitrogen gauge.
5. The gauges on preaction systems that show the air pressure in the system need to be inspected on a monthly basis to insure that the correct pressure is being maintained.
6. The gauges on dry-pipe systems that have low pressure alarms are permitted to be inspected monthly. The inspection needs to make sure that the water pressure gauge is showing a reasonable water supply pressure, that the air/nitrogen gauge is showing a correct gas pressure (given the ratio to the water pressure) and that the gauge on the quick opening device is showing the same pressure as the air/nitrogen gauge.
7. The relief port on reduced pressure backflow preventers needs to be inspected weekly to make sure that it is not discharging water, which would indicate that one of the check valves inside the device has failed.
8. Control valves, including control valves on backflow preventers need to be inspected monthly if locked or electrically supervised. If the valve is sealed with plastic or lead wire tamper evident seals, the inspection needs to be weekly. The inspection needs to include:

(a) The valve is in the normal open or closed position. Almost all valves need to be kept in the open position. Pump test line valves and manual tank fill valves are about the only control valves that are kept closed on fire protection systems.
(b) The supervision (seal, lock, or tamper switch) is in place.
(c) The valve is accessible.
(d) The wrench is present for post indicator valves.
(e) The valve is not leaking.
(f) The valve is properly identified.

9. Preaction and deluge valves need to be externally inspected monthly. The inspection needs to include:

   (a) The valve is free of physical damage.
   (b) The trim valves are in the appropriate open or closed position.
   (c) The valve seat is not leaking.
   (d) The electrical components are in service.

10. Dry-pipe valves need to be externally inspected monthly. The inspection needs to include:

   (a) The valve is free of physical damage.
   (b) The trim valves are in the appropriate open or closed position.
   (c) The intermediate chamber is not leaking.

11. Alarm valves need to be externally inspected on a monthly basis. The inspection needs to include:

   (a) The gauges indicate that normal water supply pressure is being maintained.
   (b) The valve is free of physical damage.
   (c) The control valve is in the correct position and all trim valves are in the correct open or closed position.
   (d) The retarding chamber and alarm drains are free of leaks.

12. Gauges on automatic wet and semiautomatic dry standpipes need to be inspected quarterly. The gauges are required to be in good condition and need to be showing a reasonable water pressure (which is specifically the requirement of Section 6.2.2.1 of NFPA 25, which is very odd since a semiautomatic dry standpipe will not be indicating water pressure on a quarterly basis, but that is exactly what NFPA 25 requires).
13. Standpipe hose connection valves need to be inspected quarterly. The inspection needs to include:

   (a) Hose caps are in place and not damaged.
   (b) Hose threads are not damaged.
   (c) Valve handles are present and not damaged. Figure 14.1 shows a standpipe system that did not meet this inspection item.
   (d) Gaskets are not damaged or showing signs of deterioration.

Fig. 14.1 Standpipe hose connection valve missing a valve handle

(e) The valve is not leaking.
(f) Valves are not obstructed and appear capable of normal operation.

14. Fire department connections need to be inspected quarterly. The inspection needs to include:

(a) The FDC is visible and accessible.
(b) Couplings a swivels are not damaged and rotate smoothly.
(c) Plugs and caps are in place and are not damaged. Figure 14.2 shows a standpipe FDC that did not do well on this inspection item.
(d) Gaskets are in place and are in good condition.
(e) Identification signs are in place.
(f) The check valve is not leaking.
(g) The automatic drain valve is in place and is operating correctly.
(h) The FDC clapper is in place and is operating properly. Figure 14.3 shows an inspector checking the clapper. Never stick your hand inside the FDC until you make sure that there are no foreign obstructions inside. You may need a flashlight to see into the dark opening to make sure that it is safe.
(i) There are no obstructions inside the hose connections.

15. Supervisory switches need to be inspected quarterly to make sure that they are free from physical damage. Note that this is not a test. This is just a visual observation that the switch appears to be in good condition.
16. Low temperature alarms in dry-pipe, preaction and deluge valve enclosures need to be inspected annually at the start of cold weather to verify that they are free of damage.

Fig. 14.2 Missing cap on standpipe FDC

Fig. 14.3 Checking the clapper in an FDC

17. For preaction or deluge valves that cannot be reset from the outside, an internal inspection needs to be conducted annually when the valve is being reset after it has been trip tested.
18. Dry-pipe valves need to be internally inspected annually when the valve is being reset after it has been trip tested. This requirement applies to all dry-pipe valves, even those that can be externally reset.

Fig. 14.4 Corrosion on a vertical standpipe support

19. Hose connection and hose rack pressure regulating devices need to be inspected annually. The inspection needs to include:
    (a) The handwheel is not broken or missing.
    (b) The outlet hose threads are not damaged.
    (c) The valve is not leaking.
    (d) The hose adapter and cap are not missing.
20. Piping needs to be inspected on an annual basis. NFPA 25 is not specific as to what the inspection needs to include, but Table 6.1.2 says that damaged pipe needs to be replaced and that missing or damaged pipe supports need to be repaired or replaced. There is no information in NFPA 25 on what to do with standpipe supports that are corroded, but perhaps there should be because vertical standpipe supports sit on the ground where water also collects, making corrosion more of a concern than it is with sprinkler pipe that is mostly up in the air. Corrosion is not necessarily damage, so it is not clearly covered by the language in NFPA 25. It is debatable how much corrosion should be allowed. Figure 14.4 shows a vertical standpipe support with some corrosion on it. Should this be replaced? The answer is a difficult judgement call. This support will probably hold the vertical load for now, but how much longer will it last as the corrosion continues?
21. Hose, hose storage devices, hose cabinets, and hose nozzles, need to be inspected annually. The details of the inspection requirements are in NFPA 1962, but the basics are to make sure that the hose and equipment used with it are in good shape and appear to be able to get the water from the hose connection to the nozzle.

22. The hydraulic nameplate needs to be inspected annually. The nameplate needs to have the design information for the standpipe system. The plate needs to be legible and securely attached. If the sign is missing or no longer legible, then a new sign needs to be placed on the system. Many years ago, standpipe systems were allowed to have the pipe size decided by a pipe schedule rather than hydraulic calculations. Such systems originally did not have a hydraulic nameplate. In order to distinguish these from hydraulically designed standpipe systems, the pipe schedule standpipe systems are now required to have a sign that says "Pipe Schedule System" attached so that people know what type of system it is and that if a sign is missing, it needs to be replaced.
23. Alarm valves need to be internally inspected once every five years. During the inspection, all strainers, filters and orifices need to be inspected as well.
24. Check valves and backflow preventers need to be internally inspected once every five years. The inspection needs to show that all components operate correctly, move freely, and are in good condition.
25. For preaction or deluge valves that can be reset from the outside, an internal inspection needs to be conducted once every five years.
26. Strainers, filters, restricted orifices, and diaphragm chambers in dry-pipe, preaction and deluge systems need to be inspected once every five years.

Testing Requirements for Standpipe Systems

The testing requirements will be organized here by pieces of equipment and the frequency of the most common test. Not every piece of equipment will be found in every standpipe system. If your standpipe system does not have one of these pieces of equipment, you can skip that item. The following is a list of inspection requirements for standpipe systems:

Main Drain Test

Every standpipe system is required to have a main drain, which is an opening through which most of the water in the system can be taken out and disposed of safely. The main drain will be located somewhere on the hose connection side of the main control valve for the system so that the control valve can be closed and the water drained for repairs or maintenance. In some cases, the main drain is a simple 2 in. connection as shown in Fig. 14.5, but in many cases, the main drain will be a hose connection at the bottom of the system. Regardless of how the main drain is arranged, it needs to be tested on a regular basis.

There are three reasons to conduct the test. First, the test is conducted to make sure that the valve(s) between the water supply and the system are open and that the path is unobstructed. Main drain tests are sometimes the mechanism used to find partially closed valves or obstructions in the pipe.

Fig. 14.5 Two-inch main drain connection

The second reason to conduct the main drain test is to exercise a backflow preventer, if one is installed between the water supply and the main drain. Backflow preventers have check valves that are constantly being pushed closed by internal springs. If the springs stay pushed in the closed position too long, they tend to get stuck in that position, and then might not open fully during a fire when full flow needs to occur. Opening the main drain on a regular basis helps to exercise the springs.

The third reason to conduct the test is to keep track of the water supply. Pressure readings are taken during the test and compared to previous readings. By themselves, they don't tell you much about the water supply because there is no direct way to tell how much flow is moving through the drain during the test. The fact that the tests are frequently done at different times of day during different seasons also makes it difficult to tell anything from the results of a single test due to the random nature of water usage in public water mains. But together over a long period of time, the pressure readings provide a picture that helps you spot potential concerns regarding the water supply.

NFPA 25 requires the main drain test to be performed on an annual basis unless there is a backflow preventer between the water supply and the drain. If there is a backflow preventer, the test is required to be run at least quarterly. Since the purpose of the main drain test is to check the water supply, the valves between the system and the water supply and the pipes between the water supply and the system, NFPA 25 allows a single main drain test to be performed in the situation where more than one fire protection system is manifolded together sharing the same water supply pipes and water supply control valves. After one of these system valves is closed and opened again, the main drain test will need to be performed to make sure that the valve really did open.

Table 14.1 Summary of main drain tests

Date	Static pressure (psi)	Residual pressure (psi)
15 September 2016	60	42
20 December 2016	73	53

Interpreting the results of the main drain test takes some consideration. There are two pressure readings, the static and the residual pressure. These need to be compared to the static and residual pressures from the previous readings. If the static pressures are similar, but the residual pressures are different, then there is a reason to be concerned. If the static pressures and residual pressures are different, but the differences between the readings are similar between tests, then the test readings are probably okay.

Take for example the following two main drain tests. On September 15, 2016, the static pressure at the main drain was 60 psi and the residual pressure with the main drain valve fully open was 42 psi. On December 20, 2016, the test was run again with the static pressure being 73 psi and the residual pressure being 53 psi with the main drain test valve fully open. The results of the tests are summarized in Table 14.1.

For a building owner, a table like Table 14.1 can be helpful in summarizing the results from main drain tests. Over time, more rows can be added as additional tests are performed. In this case, it appears that the tests are being performed on a quarterly basis, so it would be helpful to compare a test done in the fall to a test done the prior fall as well as the most recent previous test. The very first test should have been conducted during the acceptance test (see Chap. 13 of this text), and the owner should have the record of this test to use as the baseline for interpreting the results because the acceptance test records need to be kept for the life of the system.

For the situation in Table 14.1, the difference between the static pressures is not a cause for concern. Static pressures in a water supply frequently go up and down more than 13 psi (the difference between 73 and 60 psi) and it is common to see more water usage in the early fall than in the early winter, so a higher pressure in December is to be expected. The important consideration is that the difference between the static and residual pressure in each case is very similar. In the case of the September test, the difference between the static and residual pressure is 18 psi (60 – 42 = 18) while the difference between the static and the residual pressure in December was 20 psi (73 – 53 = 20). This shows that the conditions have not changed all that much during the 3 month time period between September and December. While there was a little more loss during the December test, this could be explained by the additional flow that was created when the drain was opened at a slightly higher pressure, which caused a little more friction loss in the pipe between the water supply and the drain connection.

NFPA 25 requires that when there is a 10 % or more reduction in the residual pressure reading from any previous reading (initial acceptance test or previous periodic test that you still have in the records), the drop in pressure needs to be investigated and explained. If the explanation is that there is a problem with the system, then the problem needs to be corrected. It is quite possible that the drop in pressure is nothing more than the regular fluctuation of the pressure in the water supply (as was discussed in the example above) in which case no repair of the system is required. But someone does need to look at this information and explain it with logical reasons before determining whether further repairs are needed or not.

Waterflow Alarms

There are three types of waterflow alarms; mechanical, vane-type, and pressure switch-type. A mechanical waterflow alarm is also called a water motor gong. When water starts flowing in a fire protection system, a small amount of water is diverted through a pipe that turns a water activated paddle wheel. The wheel is connected to a small hammer that rotates as the wheel turns and strikes a small bell every time it comes around the circle. With a continuous flow of water, the hammer hits the bell once or twice each second, and creates an alarm for everyone within earshot of the bell. The advantage to this type of alarm is that it is not dependent on electricity. The water motor gong was extremely popular in older sprinkler systems that predate reliable electrical distribution networks. They can still be found on fire protection systems in new buildings in situations where a designer wants a backup to the electrical waterflow alarms.

A vane-type switch is an electrical device with a plastic paddle that gets inserted into the pipe. The paddle is connected to an arm that penetrates the pipe wall and is connected to a switch. When water begins to flow in the pipe, the paddle gets pushed a little downstream, moving the arm and causing the switch to change position. The switch can be set to immediately send an alarm signal when it moves or a delay can be set at the switch so that it needs to be constantly held in the downstream position for a period of time before it goes into alarm. The advantage to the delay feature is that small pressure surges that push the paddle downstream briefly and then allow the arm to reset in its normal position do not cause alarms. The longest delay allowed by NFPA 72 is 90 s. Vane-type flow switches are only allowed on wet pipe systems because the concern is that the turbulence caused by dry and preaction systems when they trip might pull the paddle off of the arm, which would cause the paddle to become an obstruction in the piping.

Pressure switch-type waterflow alarms are electrical devices that sense the pressure in a small piece of pipe. The pipe is normally kept at atmospheric pressure. When water starts to flow in a fire protection system, a small amount of water is diverted into the pipe connected to the pressure switch. The pressure in this pipe goes up from atmospheric pressure to whatever the water pressure is, causing the switch to go into the alarm mode. Just like the vane-type switch, a delay of up to 90 s can be set so that the switch does not go into alarm from pressure fluctuations. Pressure switch-type waterflow alarms are popular on dry-pipe and preaction systems because there is no concern about damage downstream when the valve trips sending water into air-filled pipe. Pressure switch-type waterflow alarms can also be found on wet systems where the designer did not want to insert a paddle into the main system piping.

The type of switch determines the frequency with which the test of the device is performed. Mechanical waterflow alarms (water motor gongs) need to be tested on a quarterly basis. Electrical switches like vane-type and pressure switch-type only need to be tested once every 6 months (NFPA 25 uses the term semiannually in this case, meaning twice each year). The reason for the difference is that water motor gongs are less reliable. They tend to get dirt, debris and other objects (like bird's nests and bee's nests) inside the bell, which prevents them from operating. More frequent testing is necessary to make sure that the device still works.

Whether the waterflow alarm is mechanical or electrical, the test procedure is the same. The person performing the test needs to open a test connection that simulates the flow of a single outlet (sprinkler or hose connection) in the system to see if the alarm detects the flow and goes into its alarm mode in a reasonable period of time. As stated previously, the acceptable delay for electrical waterflow alarms is 90 s. The acceptable delay in mechanical alarms is 5 min, which allows for the diverted water to get to the water motor gong and start turning it. For electrical systems, the person running the test should go over to the alarm panel and make sure that the panel is indicating the correct alarm during this test rather than some other situation in the building. This verifies the programming of the device correctly within the alarm and detection system.

The type of connection that gets opened for this test will depend on whether the standpipe system is wet or dry. For wet standpipe systems, a test connection will be downstream of the waterflow alarm. This test connection is frequently called an "inspector's test connection" by people in the field, but NFPA standards more frequently refer to it as an "alarm test connection", which is more descriptive of its function. On sprinkler systems, the alarm test connection is typically a ½ in. outlet that simulates the flow of a single sprinkler in the system. For standpipe systems, the connection can be much larger since it only needs to simulate the flow of water from a single hose connection. NFPA 14 allows any size connection up to the size of the hose connection to serve as an alarm test connection. For wet systems, any hose valve that discharges water to an acceptable location, with or without hose, can be used for an alarm test connection.

For dry standpipe systems, automatic and semiautomatic systems will need an alarm test bypass at the dry or preaction valve that controls the flow of water into the system. The alarm test bypass takes some water from the underside of the valve and runs it through the alarm sensing line to the pressure switch without tripping the valve. You would not want to put an alarm test connection downstream of the dry or preaction valve to test the waterflow alarm because you would potentially be filling the system with water during freezing weather just to test the waterflow alarm. The alarm test bypass is a much better mechanism for testing waterflow alarms on dry standpipes. Manual dry standpipes are not required to have a waterflow alarm, so this test can be skipped on those types of systems.

Control Valves and Supervisory Switches

Once each year, all control valves need to be run through their entire range of operation. If the valve is normally open (which almost all control valves in fire protection systems are), then it needs to be fully closed and then fully opened again. If the valve is a normally closed valve, it needs to be opened all the way, and then closed again. In the maintenance section of this text, there will be a discussion of an annual requirement to lubricate the stem of OS&Y valves, fully closing the valve and then opening it again to distribute the lubricant. Many people do this maintenance once each year, which satisfies both the maintenance requirement and the test requirement in one activity.

The procedure for opening an OS&Y valve is to continue turning the valve in the open direction until it cannot be turned anymore and then closing it back one-quarter turn. The purpose of the quarter turn towards the closed position at the end of the opening sequence relieves some of the stress on the valve. For a butterfly valve, the valve can be fully opened and left in that position without worrying about turning it back a quarter turn. In the fully open position, there is no stress on a butterfly valve.

For a post indicating valve (PIV), the procedure is similar to the OS&Y valve. The valve nut needs to be rotated in the open position using the official wrench for the valve. Use of a generic wrench could damage the valve nut. The valve nut needs to be rotated in the open position using a reasonable amount of force (but not too much) until spring or torsion is felt in the wrench. This is the sign that the valve gate is pulled up completely out of the flow stream and is pressing against the top of the casing and the valve stem is twisting a little bit (which is what is creating the feeling of spring or torsion in the wrench). The valve needs to be closed one-quarter turn to relieve the tension in the valve rod. If the person operating the valve cannot open the valve using a reasonable amount of force, use of a longer wrench or cheater bar should be discouraged. This would be a sign that the valve is damaged and needs to be disassembled and repaired rather than forcing it open. If the person opening the valve does not feel torsion when the valve is fully opened or the wrench does not spring back, then there might be an obstruction at the top of the valve casing preventing the gate from fully retracting into the casing. Finally, if the valve nut continues to turn in the open direction without ever stopping, the gate has become disconnected from the stem and the valve needs to be repaired.

Every time that a control valve has been closed and then opened again, a main drain test needs to be performed. The reason for this is that control valves have multiple parts that are connected to each other. We think that an OS&Y valve is open when we see a stem. We think that a butterfly valve is open when the gear indicator is pointing at the word "open". But the reality is that the parts of the valve may have become disconnected. On an OS&Y valve, the gate may have become disconnected from the stem so that when the stem was turned outside the valve, the gate stayed inside, blocking the waterflow. Similarly, the closing mechanism in the butterfly valve could become disconnected from the gear indicator, so that it is closed when the valve reads "open". Conducting a main drain test or some sort of flow test downstream of the valve that was just opened helps to verify that the valve really did open up.

Many control valves will have a supervisory switch (also called a tamper switch) connected to the valve that will send a signal to a constantly attended location if someone starts to move the valve. These switches are required by NFPA 25 to be tested at least once every 6 months. The procedure for testing the supervisory switch is to start to move the valve from its normal position (closing a normally open valve or opening a normally closed valve). The supervisory switch needs to send a signal within the first two revolutions of the handwheel or wrench, or when the stem of the valve has moved more than one-fifth of its length. The signal that goes to the panel needs to be distinctive, meaning that it needs to indicate which of the valves in the system is being tampered with. The signal needs to continue until the valve is returned to its normal (open or closed) position.

In the rare circumstance where this test is being performed on a normally closed valve (like a valve on a test header line for a fire pump), the person conducting the supervisory switch test needs to be aware that they are letting water into pipe that is normally not filled with water. While the valve should not fully open during this test, it will open enough to let some water into the piping downstream. Care should be taken to make sure that devices downstream, such as pump test headers, are closed and that the water in the piping drains properly after the valve is restored to the closed position.

The control valves that are a part of the Class I and Class III standpipe system hose connections are required to be operated fully through their range of operation from closed to open and then back to closed on an annual basis. For the control valves that are on Class II hose connections and hose stations, the control valve needs to be fully operated, but only once every three years (see Section 13.5.6.2.2 of NFPA 25).

Automatic and Semiautomatic Dry Systems

Automatic dry standpipe systems use dry-pipe valves just like dry sprinkler systems, so the rules for testing automatic dry standpipe systems are the same as dry sprinkler systems. Semiautomatic dry standpipe systems use the same valves as preaction sprinkler systems, so the testing rules in NFPA 25 that apply to preaction systems need to be applied to semiautomatic dry standpipe systems. Most of the rules are the same for dry-pipe and preaction systems in NFPA 25 regarding the tests, so they will be covered here together. Where there are slight differences, the differences will be explained here.

Priming Water

On a quarterly basis, the priming water level needs to be tested if priming water is incorporated into the design of the dry or preaction valve. On most valves using priming water, there will be a priming water test port at the proper level in the valve.

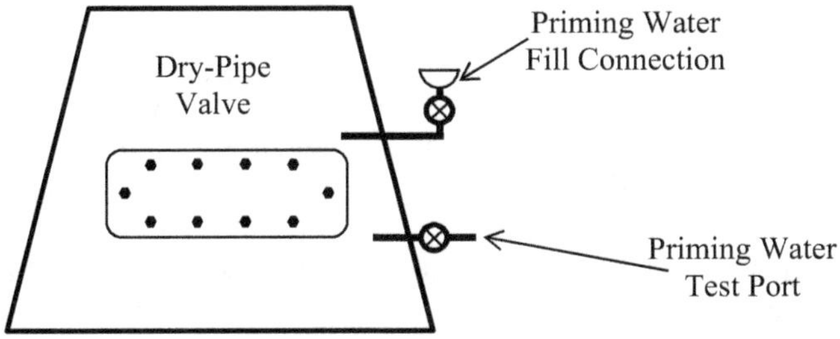

Fig. 14.6 Priming water fill connection that can only be used with no air pressure in the system

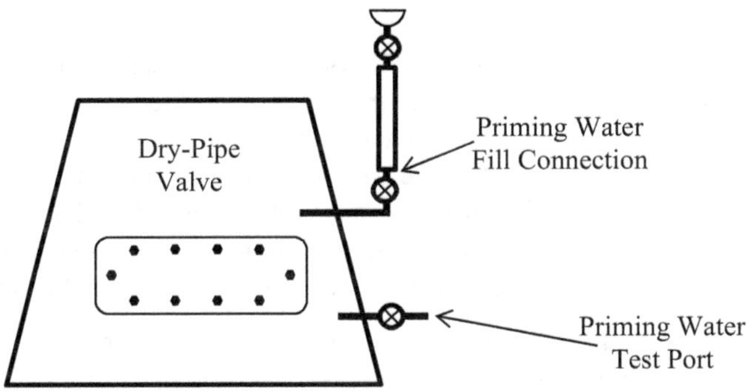

Fig. 14.7 Priming water fill connection that can be used with air pressure on the system, using the two control valves and pipe between as an air-lock

The test is conducted by opening the port and seeing what comes out. If air comes out, water will need to be added. If water comes out, the proper amount of priming water is sitting on the top of the valve. In the situation where water needs to be added, the valve may have to be taken out of service depending on how the priming water fill cup is arranged. Figure 14.6 shows a priming water fill connection that is typical to many dry-pipe valve installations that needs to be used without air under pressure in the dry-pipe valve. If this type of priming water fill connection is provided, then the air will have to be taken off of the system, the priming water needs to be put into the system, and then the air pressure will have to be returned to the system.

Figure 14.7 shows a different kind of priming water fill connection that can be used without taking the air pressure off of the dry-pipe system. The fill connection has two valves that can be operated like an air-lock. As long as one of the valves is always closed, the air pressure remains in the system. To fill the priming water chamber, the bottom valve is closed and the top valve is opened. Water can be put into the large section of pipe between the valves. The top valve can then be closed and the bottom valve opened, allowing the water to drain into the dry-pipe valve. This process can be repeated, closing the bottom valve and opening the top valve until the dry-pipe valve has the right amount of priming water above the clapper.

Trip Tests

The dry or preaction valve needs to be trip tested in some way every year, but the exact frequency and procedure of the test depends on the system and what sort of space it is protecting. For dry and preaction valves protecting freezers, the test is conducted annually in such a manner that the water does not get into the freezer. This is typically done by closing the control valve at the freezer itself and opening a bypass test line that discharges the water in a space prior to the freezer. The valve is tripped by either releasing the air pressure (for a dry-pipe valve) or activating the alarm system (preaction valve). Once this test is complete and the dry or preaction valve is reset, the control valve for the piping in the freezer needs to be opened again.

For dry and preaction systems that do not protect freezers, a full flow trip test needs to be conducted once every three years. It is a good idea to run this test in the spring if possible so that the water does not freeze in the pipes during the test and so that any water that condenses out of the air when the system is refilled after the test has time to drip back to a drain before the onset of cold weather. This test generally needs to be conducted by two people, who need to be in communication with each other. The procedure for the dry-pipe valve full flow trip test is as follows:

1. Open the main drain to flush out any accumulated debris or pipe scale from the system prior to letting the water flow into the standpipe piping.
2. Close the main drain valve.
3. Record the system water pressure and the air pressure.
4. One person goes to the most remote portion of the system and opens a test connection, letting the second person know to start a timer when they do. This test connection is sometimes called an inspector's test connection in the field, but this frequently confuses it with the alarm test connection. NFPA standards tend to refer to this connection as a "trip test connection". In sprinkler systems, the trip test connection simulates the size of the opening of a single sprinkler. But in standpipe systems, the trip test connection is permitted to be as large as one of the hose connections and is actually permitted to be the most remote hose connection on the system rather than installing a separate connection.
5. The second person will be stationed at the dry-pipe valve and will start a timer when the trip test connection or hose connection is opened.
6. The second person then needs to watch the air pressure gauge and note the time and the pressure when the dry-pipe valve trips.
7. The first person waits at the remote hose connection or trip test connection for the water to arrive. When it arrives they tell the second person, who notes the time.
8. After the water flows for a few minutes, and flows cleanly, the test can be terminated and the main control valve can be shut down.
9. The main drain and all low point auxiliary drains need to be opened.
10. Once all the water has drained from the system, the dry-pipe valve can be opened and reset, checking the clapper latch to make sure that the clapper

locked in the open position when the valve tripped. This is important because the main drain for the dry-pipe valve is below the clapper, so if the clapper does not latch in the open position, it will be much more difficult to drain the system in the future.

11. Once the valve has been reset, the priming water (if required) can be put on top of the clapper.
12. All drains can be closed.
13. The air pressure can be turned back on, filling the system to the proper pressure. Dry-pipe systems are supposed to be filled with air to their normal pressure within 30 min. If the dry-pipe system is protecting a freezer that keeps the temperature below 5 °F, the air is allowed to return to normal system pressure in 60 min. This extra allowance for time is in recognition of the fact that as the air gets colder, it takes more volume of air to get the pressure up to the normal point.

Once the dry-pipe valve is reset, the determination needs to be made as to whether the system passed the full flow trip test or not. NFPA 14 has two separate requirements for water delivery time. Section 5.2.1.2.1 allows systems with a total volume in the piping up to 750 gallons to pass the test with no specific time for water delivery. The time still needs to be recorded during the test so that it can be compared to previous tests. If the water delivery to the most remote hose connection is 4 min at the acceptance test and 10 min a few years later with similar water and air pressure, then there would be concern that there are some obstructions in the piping somewhere. The second requirement is for system with more than 750 gallons in piping volume, which requires that the water be delivered to the most remote hose connection within 3 min. While NFPA 25 is silent on pass/fail criteria for this test, it would seem logical to expect the system to continue to perform as it was originally intended.

The person doing the full flow trip test may not have the information as to the volume of the standpipe system, which is critical to the determination as to whether there is a water delivery time requirement or not. If the volume of the system is not available, the person performing the test may need to estimate the volume. Table 14.2 shows the volume inside a foot of pipe of common sizes. By estimating how many feet of pipe of each size are in the system, the person doing the tested can figure out whether the 3 min delivery rule applies or not.

For a preaction valve (on a semiautomatic dry standpipe system), the procedure is similar. Rather than trip the valve with the drop in air pressure by opening a test

Table 14.2 Volume of pipe in gallons per foot

Nominal pipe size (in.)	Schedule 10 steel pipe (gal/ft)	Schedule 40 steel pipe (gal/ft)
2½	0.283	0.248
4	0.740	0.660
6	1.649	1.501
8	2.776	2.66 (Sched. 30)

connection, the valve needs to be tripped from one of the alarm signals. Typically for a semiautomatic dry standpipe system, the alarm signal will come from the manual pull station or button at the hose connections that sends a signal to trip the preaction valve.

Full flow trip tests do not need to be performed every year. In fact, it is best if they are not performed every year because excess corrosion can occur from filling the system with water if all of the water is not drained out after every test. For those years where a full flow trip test is not being performed, a partial flow trip test needs to be performed. This test is performed with the main control valve almost all the way closed. As soon as the preaction or dry-pipe valve trips, the main drain valve is closed all of the way, limiting how far the water gets into the system (limiting the potential for corrosion to occur). All drains still need to be opened to make sure that any water that got into the system gets back out.

Air Leakage

All preaction and dry systems need to be tested once every three years for gas (air or nitrogen) leakage. There are two ways that the test can be performed. The first is to run the gas pressure up to 40 psi and turn off the gas pressure maintenance system (compressor, shop air or other gas maintenance system). If the piping system holds the gas pressure for 2 h and does not drop below 37 psi, then the system passes the test. If the gas pressure drops below 37 psi in the 2 h time period, then the pressure loss within the system needs to be addressed by finding the gas leaks and/or tightening the pipe connections.

The second way to run the gas test is to turn off the gas pressure maintenance system when the pressure is at the normal system pressure. Over the next 4 h, monitor the system pressure. If the low pressure alarm does not go off (which is usually set for 10 psi under the normal air pressure), then the system passes the test. If the low air pressure alarm goes off, then the pressure loss within the system needs to be addressed by finding the gas leaks and/or tightening the pipe connections. Of course, this method can only be used if there is a low pressure alarm on the system.

Low Pressure Alarms

If low pressure alarms are provided on preaction or dry-pipe systems, they need to be tested annually. This needs to be done in accordance with the manufacturer's instructions, which may require simulating a low pressure condition at the device. This can be done if the device is installed on its own little branch of piping with an isolation valve. The isolation valve can be closed and the pressure released from this little section of pipe without dropping the pressure in the rest of the system.

Low Temperature Alarms

If a low temperature alarm is included in the dry or preaction valve enclosure, it needs to be tested annually, just prior to the start of freezing weather. One way to test the alarm would be to allow the temperature in the room to drop below the alarm set point. This may not be easy, especially before the start of cold weather. The manufacturers of these alarms recommend testing these alarms be spraying them with tetrafluoroethane, a material sold at electronics stores under the name "Circuit Cooler" or "Component Cooler,". It lowers the temperature of material and is an inexpensive way to test a low temperature alarm without exposing other objects in the room to cold temperatures. After you have sprayed the sensor with the tetrafluoroethane, the low temperature alarm should sound. Then allow the switch to warm up and the signal should stop as the switch resets.

Pressure Maintenance

Any automatic gas pressure maintenance device (for air or nitrogen) needs to be tested annually, preferably prior to the start of freezing weather. In order to be most efficient, this test should be performed at the same time as the trip test of the dry or preaction valve. The procedures for the test will vary based on the type of equipment, so refer to the manufacturer's instructions for more detail on exactly how to run the test.

Pressure Reducing Valves

There are two tests that need to be performed on pressure reducing valves that are incorporated into hose connections. The first is the full flow test of the device, which needs to be performed once every five years. The test needs to be performed at the highest flow expected through the valve. For pressure reducing valves at Class I and Class III hose connections, this will be 250 gpm.

The exact procedure for the test will depend on exactly what equipment is installed in the standpipe system. In order for the test to be completed correctly, a pressure reading will need to be taken both upstream and downstream of the valve while water is flowing through the valve. The water can be discharged into the drain riser, which is required to have inlets every other floor. Figure 14.8 shows one way to set up the test. In the figure, the equipment shown in red has to be added for the test. The pressure gauge on the upstream side of the pressure reducing valve does not need to be there all the time, but the valved outlet for it needs to be there so that the gauge can be added when it is time to do the test.

The second test that needs to be conducted on pressure reducing valves that are incorporated into hose connections happens during the four years in between the full flow tests. This test is called a partial flow test and it needs to be performed on

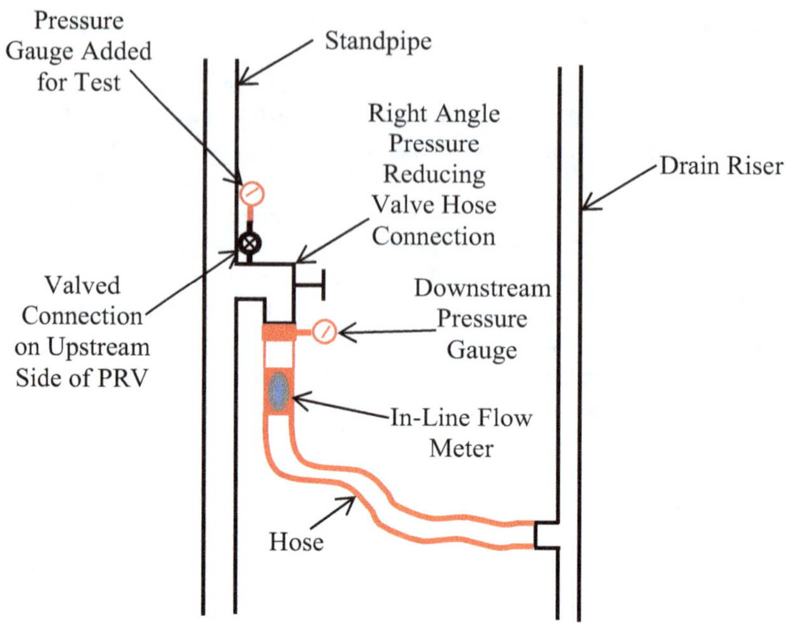

Fig. 14.8 Flow test of pressure reducing valve

each pressure reducing valve annually during these four years. The procedure for a partial flow test is to tighten the cap on the hose connection and open the valve. Water will flow for just a split second until the cap is filled. This is enough to lift the seat of the pressure reducing valve and give it a little exercise. The valve can then be closed and the cap removed to drain the little bit of water that got into the valve during the test. Lastly, put the cap back on and tighten by hand.

For master pressure reducing valve assemblies, the consequence of a failure is more important, so the same types of tests need to be performed, but on a more frequent basis. The partial flow test needs to be conducted quarterly and the full flow test needs to be conducted annually. The tests are a little different to conduct since there is no hose outlet right near the valve. The partial flow test can be conducted using the closest hose connection to the assembly and closing some of the control valves to isolate each pressure reducing valve. The full flow test needs to be conducted with as many hose connections as it takes to get the flow demand moving through the valves. While the first pressure reducing valve can be tested with the assembly set normally, the second pressure reducing valve will need to be tested with the control valves around the first pressure reducing valve closed and the water coming through the bypass to the second pressure reducing valve. Only in this way will the water coming into the second pressure reducing valve actually activate the pilot operated pressure reducing mechanism.

Backflow Preventers

All backflow preventers are required to be forward flow tested at least once every year. The goal of the test is to exercise the internal check valves that are part of the backflow preventer. In order to exercise the check valves completely, a flow equivalent to the fire protection system demand flow needs to be created downstream of the backflow preventer. The flow can be created by opening any number of openings downstream. If there are hydrants or hose connections downstream of the backflow device, which there will be on standpipe systems, these can be used to create the flow for the test.

For a standpipe system, the flow that needs to be created could be as high as 1250 gpm. While this can be done through hose connections, it is still a great deal of water to use during the test. If the test is run for only 5 min, a total of 6250 gallons of water will be used. In communities facing serious drought conditions, this is not a good use of water. NFPA 13 allows the forward flow test to be replaced with an internal inspection of the backflow preventer in communities where a drought or water rationing has lasted more than a year (see Section 13.6.2.1.1).

Hydrostatic Tests

There are a couple of hydrostatic tests that need to be performed on standpipe systems. The piping from the fire department connection to the system needs to be hydrostatically tested at least once every five years at 150 psi for 2 h (Section 13.7.4). NFPA 25 does not explicitly contain any pass/fail criteria for this test, but it would be logical to use the criteria out of NFPA 14, which would not allow a drop in pressure during the 2 h of the test, remembering the discussion earlier in this book regarding the reality of the relationship between pressure and temperature.

The requirement for the hydrostatic test of the fire department connection piping applies to every fire protection system because it is in Chap. 13 if NFPA 25. However, the more specific requirements of Chap. 6 override this provision in some standpipe systems. Chapter 6 still requires that the test be performed, but it requires a higher pressure for some of the systems.

Section 6.3.2.1 requires a hydrostatic test for all of the piping (including fire department connection piping) in manual standpipe systems (wet and dry) and all semiautomatic dry systems at least once every five years. There is an exception that exempts all standpipe systems that are combined with a sprinkler system from this test. Since many manual wet standpipe systems are combined with sprinkler systems, these systems will not need to be hydrostatically tested. For those systems that do need to be tested, the test needs to be conducted at 200 psi or 50 psi in excess of the maximum system pressure, whichever is higher. The pressure needs to be measured at the lowest level in the system or standpipe zone, so the pressure will not be 200 psi (or 50 psi in excess of the maximum system pressure) at the top of the system or zone. The system has to hold the pressure for 2 h without leakage in the piping (although the annex note does say that minor leakage under test pressure is not a reason for failure if the leakage goes away after the test pressure is removed.

The purpose of the hydrostatic test is to make sure that the system will work properly during a fire. The fire department does not want to show up and pump into a system during a fire and find that the extra pressure that they have put into the system causes the pipe to burst. The concerns of the fire department are well founded after years of experience with manual dry standpipe systems. This requirement did not always exist in NFPA 25 and many firefighters have had serious blowouts of manual dry systems. Manual wet systems that are combined with the sprinkler system are continuously subjected to system pressure from the water supply, so they do not need to perform the hydrostatic test because they are constantly being tested.

Flow Tests

A full flow test is required on every Class I and Class III standpipe system at least once every five years. The test needs to be conducted at the standpipe system demand (500 gpm from the most remote standpipe and 250 for the next standpipes until a maximum of 1000 gpm for sprinklered buildings or 1250 gpm for unsprinklered buildings is achieved), which is easy enough to perform on automatic systems because they have a water supply connected that can be used during the test. Manual standpipes also have to be tested under this requirement, which is harder since they don't have connected water supplies that can meet the system demand. In order to conduct this test in a manual standpipe system, a portable pump or fire truck and a water supply will need to be connected through the fire department connection. Several methods of doing this were discussed in the acceptance test portion of this text.

The pass/fail criteria for this flow test are to achieve the system design pressure while the waterflow demand is flowing through the system. While the pressure demand at the most remote outlet today is 100 psi in NFPA 14, this has not always been the case. Many standpipe systems are still in service that were designed for a pressure of 65 psi at the most remote hose connection while the system flow is moving through the system. You will need to do some investigation before performing the test to determine what the pressure requirements will be. If you cannot determine what the pressure requirements were for the design of the system, speak with the authority having jurisdiction to determine what they will be willing to accept. It may be necessary to determine what year the system was installed and then look at the corresponding edition of NFPA 14 or the local codes for that year.

There are typically questions regarding the flow from the additional standpipes during this test. Under fire conditions, the flow might be needed from the most remote standpipes, but that is not necessary during the test. While the flow in the most remote standpipe needs to be created from the most remote location, the flow at the additional standpipes does not need to be from the most remote portions of those standpipes. From a hydraulic perspective, it is only necessary to create the flow somewhere in that standpipe so that the flow occurs in the feed main piping. The flow can be created wherever it is convenient in the additional standpipes, usually at the lowest level in the standpipe as shown in figures in Chap. 13 for the acceptance test.

Gauges

Section 6.3.4 of NFPA 25 requires that all gauges be tested against a calibrated gauge or replaced every five years. If you chose to test the gauge against a calibrated gauge, the gauge needs to be within 3 % of the full scale reading of the gauge. Most gauges on fire protection systems can read up to 300 psi, so the gauge needs to be correct to within 9 psi in order to pass this test. However, most people do not bother to test the gauges, they just end up replacing the gauges once every five years, which is much cheaper.

In order to know when the gauge needs to be tested or replaced, the building owner needs to keep track of when the gauges have been installed. There are a number of ways to do that, including putting a label on the gauge or putting a bar code on the gauge and storing the information in a computer data base that associates the information regarding the gauge with the testing requirements. Another method that is used by some contractors is to keep track of the age of gauges is to write the date of installation of the gauge in grease pencil on the gauge as shown in Fig. 14.9. This

Fig. 14.9 Gauge with installation date in grease pencil making it obvious when five year test or replacement is required

makes it obvious when the gauges are more than five years old and need to be tested or replaced. In the case of the gauges in Fig. 14.9, they were installed in 2013 and will need to be tested or replaced in 2018.

Hose

NFPA 25 refers the user to NFPA 1962 for requirements to test hose and hose stations. That standard requires that hose storage devices be tested annually. The hose itself needs to be tested five years after it is installed, and then every three years after that. The test consists of pressurizing the hose to its service pressure (which should be stamped on the hose) using a procedure that is in Section 4.8 of NFPA 1962.

Maintenance Requirements for Standpipe Systems

In general, many of the maintenance requirements can be summed up by saying that if something is found during the inspection and testing to be broken or not working correctly, then it needs to be repaired or replaced. If something is missing, then it needs to be replaced. Section 6.5 of NFPA 25 contains a list of inspections and tests that need to be performed after parts of standpipe systems are replaced or repaired. There are a few other maintenance requirements for standpipe systems:

1. *Control Valves*: Once each year, the stem of all OS&Y valves need to be lubricated. The valve needs to be closed and then opened again to distribute the lubricant. As discussed in the testing portion of this text, the valves need to be closed and opened again anyway, so performing this maintenance will actually serve as a test as well.
2. *Automatic Dry and Semiautomatic Dry Systems*: There are a number of maintenance rules in NFPA 25 for dry-pipe and preaction systems, which correspond to automatic and semiautomatic standpipe systems. The following are the maintenance requirements:

   (a) If the air or nitrogen in the system is leaking enough to cause low pressure alarms, the pipe needs to be tightened or repaired.
   (b) When the dry-pipe valve or preaction valve is opened to reset it after a trip test, any parts that are dirty need to be cleaned and any parts are damaged need to be replaced.
   (c) Auxiliary drains need to be opened on a regular basis to drain water that collects. In many cases, the water condenses out of the air that is placed in the system by the compressor or shop air system. Just how often you need to open the drains will depend on how much air is being placed in the system. When a standpipe system is new, the owner should frequently open these drains and pay attention to how much water comes out. A good frequency

can usually be determined within the first year of system operation. Prior to the onset of freezing weather, the drains should be opened and the water drained out. For systems that are holding air pressure, the drain needs to have two valves that need to be operated as an airlock. See the discussion on system design in Chap. 5 for more information on auxiliary drains in dry standpipe systems.

Internal Inspection and Flushing of Systems

Chapter 14 of NFPA 25 is dedicated to the concern of the interior conditions of the pipe. There are three items in the chapter that are important to standpipe systems: assessing the internal conditions of the pipe, recognizing signs of obstructions and figuring out what to do with them, and systems protecting freezers. Each of these problems will be discussed here.

Assessing Internal Conditions

On a regular basis, the internal condition of the pipe needs to be assessed. The default frequency is at least once every five years. NFPA 25 does allow the building owner to establish a different frequency using an approved risk analysis. The fact that the analysis has to be approved means that the local authority having jurisdiction needs to agree with the methodology and the conclusions used to justify some other frequency for evaluating the internal condition of the pipe.

While Section 14.2 of NFPA 25 spends a great deal of time talking about the frequency at which the assessment needs to be done, it does not actually include requirements for how to conduct the assessment. The intent is to look inside the system at a few strategic places by either taking the system apart or using some alternative method. Previous editions of NFPA 25 contained requirements for specific places to look inside of systems, but these locations have been eliminated from the 2014 edition. Instead, the standard contains some recommendations in the annex.

The recommendations start with an explanation of the purpose of the internal investigation. It is intended to provide reasonable assurance that destructive corrosion and obstruction issues are identified. More importantly, the internal assessment is not intended to find every speck of corrosion or every little obstruction that might be in a fire protection system. The annex contains some guidance for where to look in fire sprinklers systems, which certainly helps, but NFPA 25 applies to all water-based fire protection systems and it would be helpful to have examples that apply to standpipes. The best that the user can do is read through the examples and try and make parallels to where strategic places within standpipe systems might be. The following are some suggestions for where to conduct internal inspections in standpipe systems based on the annex material in NFPA 25:

- Open the system at a flushing connection at the end of the feed main (interconnection piping). Look inside for foreign material and the condition of the piping.
- For dry systems, open the system at the most remote hose connection on the most remote standpipe and look inside for foreign material and the condition of the piping.
- Any time the system is drained and opened for repair or normal maintenance, look inside at the location where the work is being performed for foreign material and the condition of the piping.

For building owners that are reluctant to take the pipe apart, NFPA 25 provides a discussion of other options in the annex. The building owner can consider using video inspection by inserting a small camera into the piping at a convenient point like a hose connection or a drain and running the camera as far into the piping as you can. Another option is ultrasonic technology that can help determine the thickness of the pipe wall and the presence of sediment or pipe scale build-up from the outside of the pipe. This has the advantage of not needing to drain the pipe down or take the system out of service.

A third option that is discussed by NFPA 25 is to take samples of the water and have it evaluated for the presence of microbes that are responsible for Microbiological Influenced Corrosion (MIC). By itself, this may be a questionable assessment technique. While this may help to pick up MIC in a system, it ignores other forms of corrosion and debris build-up. In conjunction with other methods of internal assessment, this can certainly be helpful.

Obstruction Investigations

The obstruction investigation is described in Section 14.3 of NFPA 25. This is very different from the internal assessment. While the internal assessment needs to be done on a regular basis, the obstruction investigation only needs to be performed when there is some evidence that there is a problem inside the piping system. NFPA 25 contains fifteen different specific conditions that, if they occur, warrant further investigation. These conditions are:

1. Defective (or missing) suction screen for pumps taking water from open sources like ponds or reservoirs.
2. Noticing obstructive material discharging from the system during any flow test.
3. Finding foreign material in pumps or dry-pipe valves.
4. Finding foreign material coming out of drains or inspector's test connections (also called alarm test connections and trip test connections).
5. Hearing unknown material banging around in the pipe during filling or flow testing.
6. Finding plugged sprinklers on a system.

7. Finding foreign material in pipe while performing system maintenance or repair.
8. You have reason to believe that public mains were not flushed correctly after they have been installed or repaired after a break.
9. There is a record of broken mains in the vicinity.
10. Dry-pipe valve false trip frequently.
11. A system is returned to service after being out of service for more than one year.
12. There is reason to believe that sodium silicate or similar chemicals were placed into the system.
13. The fire department pumped water from an open source (like a pond or reservoir) into the fire department connection.
14. The system experiences pinhole leaks.
15. In dry systems, there is a 50 % or more increase in the time that it takes water to get from the dry-pipe valve to the most remote portion of the system. Note that this is compared to the original system acceptance test, not the immediately previous test.

If any of the fifteen conditions occur, an internal obstruction investigation needs to be performed. This could be conducted by opening up the system or using an alternative method like video surveillance or some other nondestructive testing. NFPA 25 contains a list of four places where the system needs to be inspected. If anything is found in these places, then additional inspections should be conducted and/or back flushing of the pipe should be performed. Back flushing of the pipe involves putting water into the smaller pipes at the end of the system and flowing it in the direction of the bigger pipes so that obstructions are pushed towards larger openings rather than smaller ones. The four places that need to be inspected for an obstruction investigation are:

- At the main system valve.
- At the riser.
- In the cross main (or pick one of them if there are more than one).
- In one of the branch lines (pick one)

Note that these four locations are descriptive of sprinkler systems, not necessarily standpipe systems. The best way to translate these four locations to standpipe systems is probably to do the internal obstruction investigation at the main control valve, within one of the vertical standpipes, in the feed main (interconnection piping) that connects the standpipes, and in one of the lateral runs of pipes (branch line or horizontal standpipe) if there are any.

Assuming that some problem is observed and that some investigation and flushing of the system is conducted, the building owner and/or maintenance contractor needs to then pay attention to the system during the regular inspection and testing immediately after that. If the problem goes away after the obstruction investigation, then there are no additional requirements. But if the problems persist, NFPA 25 requires continued follow-up obstruction investigations and internal inspections every five years.

Ice Obstructions

There are very few standpipe systems that protect freezers, but there are some. Section 14.4 of NFPA 25 outlines some specific concerns for making sure that ice does not obstruct system piping, starting with an annual inspection at the location where the piping penetrates the freezer wall. There will always be some water suspended in the air that gets put into an automatic or semiautomatic standpipe system. The ability for air to hold the water goes down as the air temperature goes down, so it is very typical for water to come out of suspension right at the point where the pipe enters the freezer and turn to ice inside the pipe. For this reason, NFPA 25 requires this internal inspection at this location.

There is no specific information in NFPA 14 as to how to design a standpipe system for a freezer. In this situation, it is good to use the information in NFPA 13, which has a special section for sprinkler system design in freezers. There are really four important concepts in NFPA 13 that need to be brought over to standpipe systems:

1. The pipe that penetrates the freezer wall should be installed with easily removable pieces on both sides of the freezer wall. This will make it easy to perform the annual inspection discussed above.
2. The air lines from the compressor to the standpipe system should penetrate the freezer wall before they connect to the standpipe piping. This way, the water drops out of the air and blocks the air lines rather than the standpipe piping.
3. Two air lines get installed, but only one is in service at any given time. There is no way to stop the build-up of ice, so let the ice build-up in one line and when it becomes blocked, close that line and open the second until you can clean out the first.
4. Use two pressure gauges at different locations to indicate when an ice blockage has occurred in the air lines. If the gauges read the same pressure, the lines are open. If one gauge reads more than the other, a blockage has occurred and the air pressure cannot equalize, which is when the line needs to be closed and the second line put in service.

Impairment Program and Procedures

An impairment is defined in Section 3.3.20 of NFPA 25 as, "A condition where a fire protection system or unit or portion thereof is out of order, and the condition can result in the fire protection system or unit not functioning in a fire event." There are two possible types of impairments: emergency impairments and preplanned impairments. An emergency impairment is one that occurs without any planning or notice, like a leak in a pipe that causes the pipe to be shut down. The preplanned impairment is one that is known in advance, such as when a system is going to be shut down for an internal inspection. Regardless of whether the impairment is an emergency or it is preplanned, the procedures that need to be followed are the same and are laid out in Chap. 15 of NFPA 25.

Long before an impairment has occurred, the building owner needs to be prepared for an impairment to occur. Whether you like it or not, you have to realize that at some point in time, the fire protection system will become impaired, so you need to have a plan for how to deal with the situation when it occurs. The first thing that a building owner needs to do is appoint an Impairment Coordinator to plan for, and handle impairment, when they occur.

The job of an Impairment Coordinator starts with understanding the way that the building is being used and the risks created by this use regarding the potential for fire to start and the damage that a fire could do. The job of the Impairment Coordinator will be to minimize these risks when an impairment occurs and minimize the time of the impairment so that the system is returned to service as soon as possible. The role of the Impairment Coordinator is to do the following when an impairment occurs:

1. Figure out the extent of the impairment and the expected duration of the impairment.
2. The portion of the building affected by the impairment has been inspected and increased risks have been identified.
3. Recommendations are made to the building owner to decrease the risks associated with the potential for a fire or bad damage from a fire.
4. In the situation where a system is impaired for more than 10 h in a 24 h period, one of the following needs to be done:
    (a) The portion of the building affected by the impairment is evacuated.
    (b) A fire watch is established with trained people continuously circulating around the building looking for fire. Extra fire extinguishers are kept handy and the people performing the fire watch keep in communication with each other while having the ability to call the fire department quickly if a fire is spotted.
    (c) A temporary water supply is established so that the system can be used, even in its impaired state.
5. Notify all of the following individuals or entities that the impairment has occurred or is going to occur:
    (a) The fire department.
    (b) The insurance carrier.
    (c) The alarm monitoring company.
    (d) The property owner or the owner's specifically designated representative.
    (e) Other authorities having jurisdiction. This might include building officials, fire officials, or water utility representatives.
    (f) Supervisors of employees in the areas affected by the impairment.
6. Impairment Tags are placed on the control valve that is closed due to the impairment, the fire department connection that feeds the portion of the system affected,

and any other piece of equipment where people might need to know about the impairment.

7. Make sure all of the tools and materials are assembled to get the system back in service. Once the system has been returned to service, the Impairment Coordinator needs to perform the following steps:

   (a) Make sure that the inspections and tests get performed to make sure that the system is working properly after the impairment has been repaired.
   (b) Notify everyone that the system has been restored to service. This needs to be everyone that was notified when the system was taken out of service.
   (c) Remove all the Impairment Tags. With the control valve being closed during an impairment, a main drain test will need to be performed after the valve is opened. Many building owners like to record the information from the main drain test on the Impairment Tag and then file this with the system records in case the question is ever asked about whether the test was run.

A good impairment plan will make sure that preplanned impairments occur during times when the risk of fire is lowest in the building. Even in buildings with relatively low fire risk like office buildings, if preplanned impairments can be taken care of during the hours of 6:00 p.m. to 6:00 a.m., fewer people will be in the building and will be affected by the impairment. While scheduling work during these hours will increase the cost of the work, it may ultimately save the building owner money in not having to hire a fire watch or set up a temporary water supply.

Records of Inspection, Testing and Maintenance

There is a significant amount of inspection, testing and maintenance that needs to be performed in standpipe systems as has been outlined in this chapter of this text. Much of this work is required to be performed as a matter of law in most portions of the United States, so a building owner is going to need to provide documentation that shows that they complied with the law. The best way to do this is to maintain excellent records of the inspection, testing and maintenance work that is performed on the buildings that you own.

There are a variety of different forms that are available to keep track of the inspection, testing and maintenance performed on fire protection systems. The American Fire Sprinkler Association (AFSA) has developed some forms for their members to use. For more information on downloading these forms, go to their website at www.firesprinkler.org. The National Fire Sprinkler Association (NFSA) has a set of forms that they sell to members or nonmembers. For more information, go to their website at https://nfsa.site-ym.com/store/ListProducts.aspx?catid=129284. There are also a number of different jurisdictions, including the State of California, who have developed their own forms that they want contractors to use, so make sure that you understand the local requirements before finalizing the form that you are going to use for a specific job.

Fire protection contractors should get used to using two forms when performing inspection, testing and maintenance work. One form should be to report what they encountered when they were conducting the work that they were hired to do (as a part of the contract to perform the inspection, testing and maintenance). This report will be referred to here as the "NFPA 25 Report Form". The other form should be used to report observations that the contractor noticed while in the building that might affect the fire protection in the building, but are not within the nature of the reason that the firm was hired. This form will be called the "Observation Form" in this text.

The NFPA 25 Report Form needs to contain all of the information regarding the inspections, tests and maintenance that the contractor was hired to perform. This report needs to include any violations of NFPA 25 that were observed during the work. For example, if an employee of the contracted firm finds a broken control valve during an annual inspection of a standpipe system, then this needs to be reported on the form, because it is a problem that can be linked directly to a section of NFPA 25.

The Observation Form needs to be used to record important information that is outside the scope of NFPA 25. For example, if a contractor notices an exit door that is locked and chained shut, it should be reported on the Observation Form. This is outside the scope of NFPA 25, but it is a fire protection concern, so it should be brought to the attention of the building owner. It is also to bring to the attention of the building owner the fact that the information on the Observation Form is not a complete assessment of the fire protection that is outside the scope of NFPA 25. The information on the Observation Form is just an indication that there is something wrong that needs to be addressed. For example, if a contractor found a locked and chained exit door during a standpipe system inspection, they can report it on the Observation Form, but they need to make sure that the owner does not assume that they have conducted a complete analysis for the egress systems in the building by reporting this one concern. This is usually done through a legal disclaimer at the top of the Observation Form that warns the building owner that the information on the form is just something that the contractor noticed in the building and the form does not constitute a complete design review of the whole fire protection system for the building.

When a contractor sees a problem in a fire protection system, it isn't always clear at first which form needs to be used to report the problem. The issue comes down to the scope of NFPA 25, which is the inspection, testing and maintenance of the fire protection system that got installed, not a review of the design of the fire protection system. This is a very important and subtle distinction of NFPA 25. The goal of NFPA 25 is to make sure that the equipment that got installed continues to work properly. The point of NFPA 25 is NOT to make sure that the correct equipment got installed. The assumption that NFPA 25 makes is that someone agreed to allow the fire protection system to be installed in the first place. Once this system got installed, it needs to be maintained properly, but the critique of the design of the system is outside the scope of NFPA 25.

When asking whether a situation needs to be reported on the NFPA 25 form or the Observation Form, a person should ask the question, "Can you find a section in NFPA 25 that addresses the specific subject?" If the answer is "yes", then the situation needs to be reported on the NFPA Report Form. If the answer is "no", then you need to report the situation on the Observation Form.

NFPA 25 requires that the initial acceptance records be kept for the life of the system (Section 4.3.4). The records of inspection, testing and maintenance need to be kept for a period of one year past when the next inspection, test or maintenance of that type is required. So, a record of a quarterly inspection needs to be kept for 15 months (one year past the next quarterly inspection) whereas the record of a full flow trip test of an automatic dry standpipe system would need to be kept for four years (one year past the three year interval between tests). Holding on to the inspection, testing and maintenance reports this way insures that there will always be at least one previous result to compare the data to when an inspection or test is performed.

While the emphasis on test reports was on paper copies in the early days of NFPA 25, we are now seeing a shift to electronic storage for the information on inspection, testing and maintenance of systems. There are a number of advantages to electronic storage. It takes up less space, it is easier to back up, and it is easier to search for the information you need if the information has been cataloged correctly. NFPA 25 makes no mention of a preference, but leaves the decision up to the building owner regarding how they want to hold on to their information.

One popular way of keeping track of inspection, testing and maintenance information is to assign a bar code to every piece of equipment in a fire protection system. The information regarding this piece of equipment (make, model, serial number, etc.) can be entered into a computer database once and the bar code can be uniquely assigned to this piece of equipment. During any inspection, testing or maintenance, the person doing the work can scan the bar code and enter the data, which then allows a building owner to keep track of every individual piece of equipment in the system and what has happened to that equipment. See Fig. 14.10, which shows a Class III standpipe system with a bar code at the hose connection.

For dry-pipe and preaction valves, NFPA 25 specifies some additional records that need to be maintained. For preaction valves, Section 13.4.3.2.12 requires that records indicating the date that the preaction valve was last tripped, the tripping time, and the individual and organization conducting the test have to be maintained by the building owner at a location or in a manner readily available for review by the authority having jurisdiction. For dry-pipe systems, the rules get a little more stringent. Not only do records need to be kept by the owner, but Section 13.4.4.2.5 requires a tag or card be attached to the valve with the information as to when the valve was last tripped and who performed the trip test (individual and organization). Interestingly, this tag is not required to have any information on the performance of the valve. Separately, the owner needs to maintain records of initial air and water pressure, tripping air pressure, tripping time, water delivery time, and dry pipe valve operating conditions so that comparisons can be made the next time these tests are performed. This means that the records for the full flow trip test, which is performed every three years, need to be maintained for at least three years so that they can be referred to when the next test is done.

Fig. 14.10 Bar code used to keep track of inspection, testing and maintenance of a hose connection on a Class III standpipe system

Test Yourself

14.1 A control valve that is found in the open position is closed to see if the tamper switch notices the closing of the valve and sends a signal to the alarm panel. Which of the following correctly identify this procedure?

(a) This is an inspection of the tamper switch.
(b) This is a test of the tamper switch.
(c) This is preventative maintenance of the tamper switch.
(d) This is maintenance in the form of a repair of the tamper switch.

14.2 How often does the piping in a standpipe system need to be inspected?

(a) Monthly
(b) Quarterly
(c) Annually
(d) Once every three years

14.3 A standpipe system with no backflow preventer between the hose connections and the water supply has an annual main drain test conducted. Table 14.3 shows the results of this year's test as compared to last year's test. Does the system pass the main drain test?

14.4 An automatic dry standpipe system is protecting a high-rise parking garage. The system consists of three 100 ft tall vertical standpipe risers that are 4 in.

Table 14.3 Summary of main drain tests in question 14.3

Date	Static pressure (psi)	Residual pressure (psi)
23 August 2015	84	62
5 September 2016	82	48

Table 14.4 Summary of full flow trip tests in question 14.4

Date	Air pressure before test (psi)	Water pressure before test (psi)	Trip time for dry-pipe valve (s)	Trip pressure for dry-pipe valve (psi)	Water delivery time to remote hose connection (min)
19 May 2013	50	160	22	27	3.67
5 September 2016	52	170	24	28	3.5

Table 14.5 Summary of full flow trip tests in question 14.5

Date	Air pressure before test (psi)	Water pressure before test (psi)	Trip time for dry-pipe valve (s)	Trip pressure for dry-pipe valve (psi)	Water delivery time to remote hose connection (min)
5 June 2013	55	165	38	33	2.87
15 March 2016	65	155	72	31	4.25

Schedule 10 pipe and 250 ft of 6 in. Schedule 10 feed main interconnecting the standpipes. On May 19, 2013, a full flow trip test was conducted and the results are shown in Table 14.4. On June 27, 2016, a full flow trip test was conducted and the results are shown in Table 14.4. Does the system pass the full flow trip test?

14.5 An automatic dry standpipe system is protecting a high-rise parking garage. The system consists of five 100 ft tall vertical standpipe risers that are 4 in. Schedule 10 pipe and 500 ft of 6 in. Schedule 10 feed main interconnecting the standpipes. On June 5, 2013, a full flow trip test was conducted and the results are shown in Table 14.5. On March 15, 2016, a full flow trip test was conducted and the results are shown in Table 14.5. Does the system pass the full flow trip test?

Chapter 15
Solutions to "Test Yourself" Exercises

Following are the answers to the "Test Yourself" portions of each of the chapters. In many cases, these represent one possible answer. There are, in these cases, several other possible solutions. It is unrealistic to provide all of the possible answers, but in the case where there is more than one solution, sufficient information will be provided so that the reader should be able to determine whether their alternate solution is correct.

Note that there are no "Test Yourself" questions in Chaps. 1–3, so this portion of the text will begin with the answers to the Test Yourself questions in Chap. 4.

Answers to Test Yourself Questions for Chap. 4

4.1. A hose station includes hose connected to the outlet that can be used to immediately get water and fight a fire. A hose connection is just an outlet to which the fire department can connect their hose when they arrive at the scene of a fire.

4.2. 2.5 in.

4.3. 1.5 in.

4.4. Generally 2.5 in. with a 1.5 in. reducer, which may or not be connected to hose depending on whether the building is sprinklered or not.

4.5. The five places that Class I outlets are required are:

   (a) Exit stairwells
   (b) At horizontal exits
   (c) At exit passageways
   (d) At exterior entrances to shopping malls
   (e) On the roof of buildings where the exit stairwells do not provide access to the roof (only a single outlet required here)

K.E. Isman, *Standpipe Systems for Fire Protection*,
DOI 10.1007/978-3-319-47750-3_15

4.6. The allowable distances from any point in the building to a Class I standpipe system outlet are:

(a) Unsprinklered building (not parking garage) with manual wet standpipe system, 150 ft maximum travel distance.
(b) Sprinklered building (not parking garage) with automatic wet standpipe system, 200 ft maximum travel distance.
(c) Sprinklered parking garage with automatic dry standpipe system, maximum allowable travel distance is 200 ft. The restriction to 130 ft does not kick in unless the standpipe system is a manual dry one.
(d) Unsprinklered parking garage and manual dry standpipe system, maximum allowable travel distance is 130 ft. Note that this would have been the same answer even if the parking garage was sprinklered.

4.7. The answer is shown in Fig. 4.24a. Note that the figure is not drawn to scale and that the stairwell is 50 ft deep, which is not realistic, but has been drawn this way for simplicity and to aid the user in seeing some of the details.

At least four standpipes will be required for the building with one in each stairwell (positioned so that the outlets will be at the intermediate landings) and one on each side of the horizontal exits. The hose connections are shown in Fig. 4.24a as the black circles with the blue filling. After these four locations are chosen for the outlets, the travel distance needs to be checked from remote points in the building to make sure that each is less than or equal to 150 ft (since the building is unsprinklered and not a parking garage).

The green line in Fig. 4.24a (labeled as Path 1) shows one of the travel distances from a remote point in a room. The total travel distance along this path is 224.9 ft, which is calculated as follows:

1. 56 ft across the room (the room is 60 ft long, the remote point is one foot from the wall, the measurement continues to the point opposite from the middle of

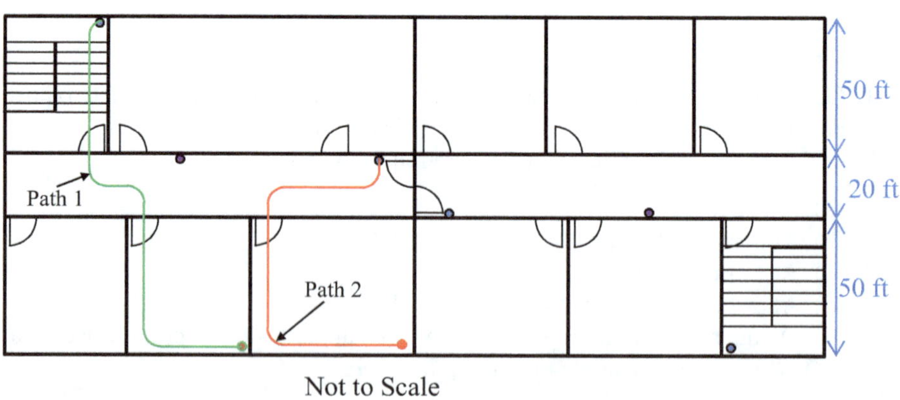

Fig. 4.24a Solution to question 4.7

the door, which is 2 ft from the wall, the arc starts 1 ft inside of this point, making the total length of this section 60 – 1 – 2 – 1 = 56 ft).

2. 1.6 ft of an arc turning the corner of the room (one-quarter of the circumference or a circle with a 1 ft radius)
3. 48 ft across the room to the door (the room is 50 ft wide, the path started 1 ft from the wall, the arc replaces another 1 ft, making the total length of this section 50 – 1 – 1 = 48 ft).
4. 9 ft to the center of the corridor (the corridor is 20 ft wide, but the arc starts 1 ft short of the center).
5. 1.6 ft of an arc in the center of the corridor.
6. 47 ft down the corridor. This is hard to see on the figure since it is not drawn to scale. The stairwell is 15 ft wide and the centerline of the door is 2 ft from the wall, so the centerline of the stairwell door is 13 ft from the end of the corridor. The doorway to the room in question is 62 ft from the end of the corridor (60 ft width of the room next door plus 2 ft to the centerline of the door in the room). Taking into account the arcs at each end of this measurement, the distance down the corridor 62 – 13 – 1 – 1 = 47.
7. 1.6 ft arc to bring the path in line with the center of the door to the stairwell.
8. 9 ft to the corridor wall.
9. 21.35 ft across the main floor landing. While this is unrealistic as an example, it is half the distance available to the two landings taking out the treads of the stairs. The treads on the stairs are 11 in. wide, making the total distance of 8 steps, 7.3 ft. The total width of the stairwell is 50 ft, so the landings are a total of 42.7 ft wide (50 – 7.3 = 42.7). Assuming that the landings are the same size, each would be 21.35 ft wide (42.7/2 = 21.35). Not all landings are equal, but the total distance is the only really important measurement.
10. 8.8 ft down the stairs, measured parallel to the slope of the stairs with a 7 in. riser and an 11 in. tread. The hypotenuse of that right triangle is 1.1 ft per step or 8.8 ft for eight steps.
11. 19.35 ft across the intermediate landing. This is the same distance as the main floor landing, but the hose connection is 1 ft from the wall and the arc starts 1 ft before that.
12. 1.6 ft arc to the hose connection.

Adding up all of these distances, 56 + 1.6 + 48 + 9 + 1.6 + 47 + 1.6 + 9 + 21.35 + 8.8 + 19.35 + 1.6 = 224.9 ft. Since this exceeds the limit of 150 ft, an extra standpipe needs to be added in a location accessible to the firefighters. A location in the corridor has been added as shown by the black circle with the purple filling. Since the design of the building is fairly symmetrical on both sides of the horizontal exit, a standpipe needs to be added there as well.

4.8. Assuming that 1.5 in. hose will be used, every point in the building must be within 130 ft of a hose connection. The solution shown in Fig. 4.25a is one possible answer with four Class II hose stations in the corridor, two on each side of the horizontal exit. There are certainly other possible solutions.

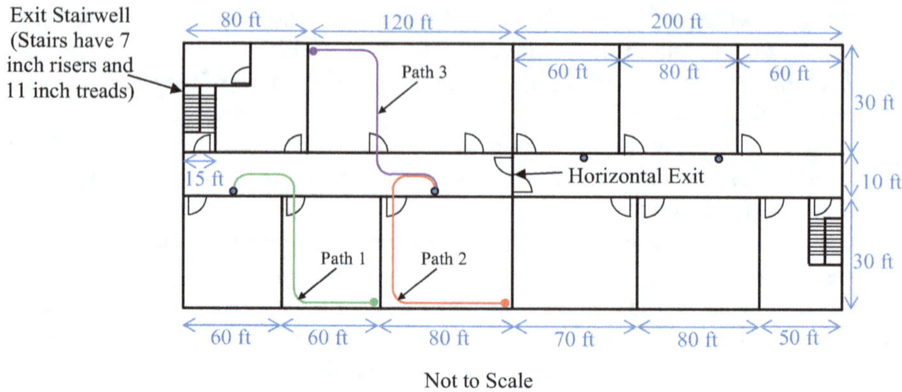

Fig. 4.25a Solution to question 4.8

There are three paths shown in Fig. 4.25a. In order to prove that the solution is correct, pathways for each of the rooms on the other side of the horizontal exit should be explored to make sure that each of them is less than or equal to 130 ft. In order to completely prove the solution to an authority having jurisdiction, more information should be provided on the plan like the exact location of the doors in the corridor.

Path 1 (the green path) is 127.8 ft with the hose station approximately halfway down the corridor of the 60 ft wide room. Tracing the distance from the most remote point, the lengths are (in feet) 57 + 1.6 + 28 + 4 + 1.6 + 30 + 1.6 + 4 = 127.8. This is within the maximum of 130 ft, so it is acceptable.

Path 2 (the red path) is 128.8 ft with the hose station only 11 ft to the right of the center of the door to the room. Tracing the distance from the remote point, the lengths are (in feet) 77 + 1.6 + 28 + 4 + 1.6 + 11 + 1.6 + 4 = 128.8. This is within the maximum of 130 ft, so it is acceptable.

Path 3 (the purple path) is a bit more difficult to pin down to an exact distance unless you know exactly where the door to the room is. Assuming that the center of the door is 42 ft to the right of the wall to the room, the distance to the hose station would be (in feet) 40 + 1.6 + 28 + 4 + 1.6 + 40 + 1.6 + 4 = 120.8 ft, which provides a bit of flexibility if the owner decides to break up this large room with partitions. The owner would need to be advised to have an evaluation down by an engineer if they wanted to renovate the space.

Answers to Test Yourself Questions for Chap. 5

5.1. The following three would be acceptable: ductile iron, steel and copper.

5.2. The correct answer is "a". Pipe using threaded fittings needs to have a minimum thickness of Schedule 40 in order to be used in a standpipe system unless the threading method and joining technique is listed for standpipe service.

5.3. The correct answer is "d". Looking back at Table 5.2 and reading the cast iron fitting rows, the standard weight fittings are limited to 175 psi. There is no such thing as a heavy weight fitting, so you have to go up to the extra heavy pattern fittings, which can handle a pressure up to 300 psi.

5.4. The correct answer is "b". Looking back at Table 5.2 and reading the malleable iron fitting rows, the standard weight fittings are allowed to go up to 300 psi, so the maximum pressure of 200 psi will be fine to use.

5.5. The correct answer is "b". Answer "a" would have been okay if the trapped section of pipe was less than 5 gallons, but in this case, the volume of the pipe is over 14 gallons (see Table 5.5 for pipe volume), so this option is not allowed. Answer "c" is for dry systems holding air pressure, so it is not applicable to a manual wet system. A full auxiliary drain needs to be installed complete with a place to put the water.

5.6. The volume of the system is 622 gallons. The answer was arrived at like this (see Table 5.5 for the volume of pipe of different sizes):

- 4 standpipes each 45 ft long is 180 ft of pipe
- 180 ft of 4 in. Schedule 40 is $180 \times 0.661 = 119$ gallons
- 250 ft of 4 in. interconnection piping has a volume of $250 \times 0.661 = 165$ gallons
- 225 ft of 6 in. interconnection piping has a volume of $225 \times 1.501 = 338$ gallons
- Total volume on the system side of the dry-pipe valve $= 119 + 165 + 338 = 622$ gallons

Note that the volume of pipe in the FDC connection or in the supply pipe between the water supply and the dry-pipe valve is not counted in this calculation because the effort here is to calculate the volume of the system on the standpipe side of the dry-pipe valve. This is the portion of the system that will need to be evacuated of air in order for the water to get to the open hose connection.

Given that the system volume is less than 750 gallons, the water delivery time has no specific requirement. The reality is that the water will get to the most remote hose connection (Node A) fairly quickly, so there is no reason to have a specific water delivery requirement.

Answers to Test Yourself Questions in Chap. 6

6.1. The answer is 153.3 psi. The elevation difference of 100 ft needs to be multiplied by the constant 0.433 to get an elevation loss of 43.3 psi ($100 \times 0.433 = 43.3$). This needs to be added to the original pressure demand of 110 psi at the top of the riser to obtain the final answer of 153.3 psi ($110 + 43 = 153.3$).

6.2. The answer is 0.2 psi. Using the Hazen-Williams formula with the flow of 250 gpm, the C-factor of 120 (wet system) and the internal diameter of 4.026, the friction loss per foot should come out to be 0.02 psi per foot. The pipe is 10 ft long, so the final friction loss is 0.2 psi ($0.02 \times 10 = 0.2$).

6.3. The answer is 6.5 psi. Using the Hazen-Williams formula with the flow of 500 gpm, the C-factor of 120 (wet system) and the internal diameter of 4.026, the friction loss per foot should come out to be 0.072 psi per foot. This needs to be multiplied by the 90 ft distance of the pipe yielding the answer 6.5 psi ($90 \times 0.072 = 6.5$).

6.4. The answer is 160 psi. You have to add up the pressure demands from elevation and friction loss. You start with a demand for 110 psi at the top of the riser, then you have to add 43.3 psi for elevation loss, 0.2 for friction loss in the top 10 ft of the riser, and 6.5 psi for friction loss in the bottom 90 ft of the riser. This all adds up to 160 psi ($110 + 43.3 + 0.2 + 6.5 = 160$).

6.5. The answer is 37.8 ft. Table 8.3.1.3 in NFPA 14 provides an equivalent length of 30 ft for Schedule 40 pipe. The value needs to be adjusted because the Schedule 10 pipe in our system is not the same as the Schedule 40 pipe in the table. The adjustment can be calculated by knowing the internal diameter of 6 in. Schedule 10 and Schedule 40 pipe as follows: $(6.357/6.065)^{4.87} = 1.26$. Now that we know the adjustment, we apply it to the base number from the table to obtain the answer ($30 \times 1.26 = 37.8$).

6.6. The answer is 26.9 ft. For a dry system, the value in Table 8.3.1.3 needs to be adjusted by an additional factor because the C-factor for the pipe is 100 rather than 120. The adjustment for C-factor pipe of 100 is 0.713, so the final answer is $30 \times 1.26 \times 0.713 = 26.9$ ft.

6.7. The answer is 24.7 psi. The Hazen-Williams friction loss per foot is 0.039 psi per foot for the 1000 gpm flowing through 6.357 in. pipe with a C-factor of 100. The total length of the pipe and fittings is 634.5 (500 ft of pipe + 5(26.9) = 634.5 ft). The total friction loss is 24.7 psi ($0.039 \times 634.5 = 24.7$).

6.8. The answers are as follows:

(a) The flow and pressure demand is 1250 gpm at 203.9 psi, which was determined as shown in Figs. 6.15a and b. The most remote riser needs to flow 500 gpm (250 gpm from the top two outlets) and the second most remote riser needs to flow 500 gpm as well because the building is more than 80,000 ft^2 per floor. The third riser only needs to flow 250 gpm from the top outlet. Once this flow of 1250 has been met, there is no additional flow required from the other standpipes because the standpipe system has reached its maximum flow requirement. For each of the fittings where the water flow changes direction, the equivalent length of the fittings was adjusted due to the fact that dry-pipe systems have a C-factor of 100 or 140. The fittings in the Schedule 30 and lined ductile iron pipe also had adjustments applied due to the different inside pipe diameter from Schedule 40 (the adjustment calculation is shown in the "Notes" column in each case).

(b) The fire department connection will need one 2.5 in. inlet for each 250 gpm of system demand. This means that the fire department connection will need five 2.5 in. inlets.

(c) The answer is "no". The city water main can only provide a pressure of about 87 psi at 1250 gpm. Since the system has a demand of 203.9 psi, this is insufficient.

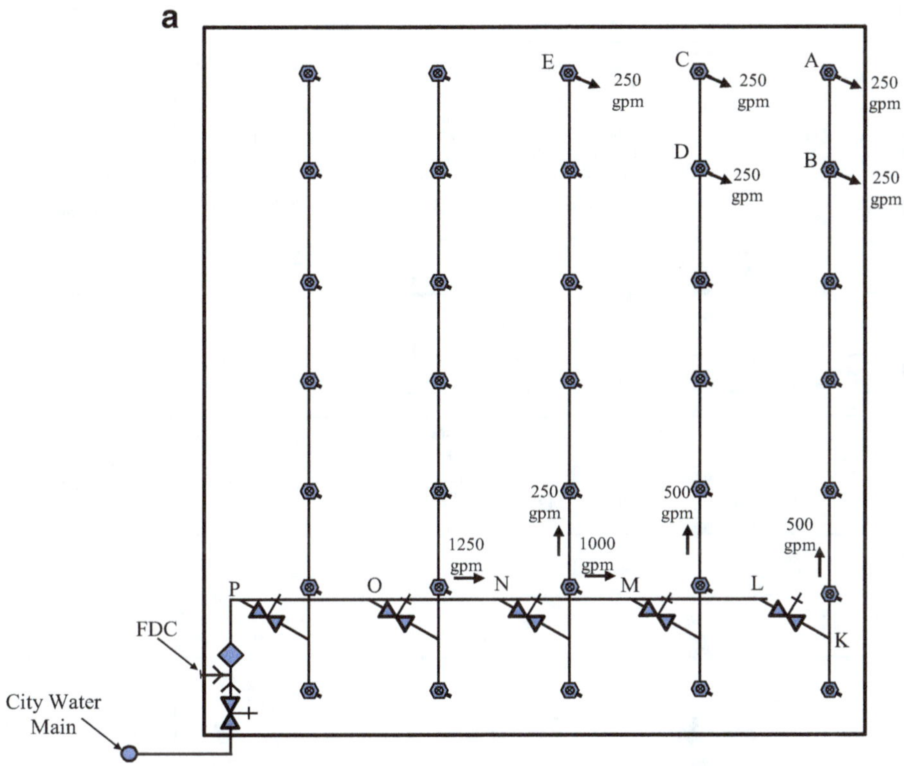

Fig. 6.15a Solution for question 6.8

b

Node 1 / Node 2	Elev 1 (ft) / Elev 2 (ft)	K-Factor	Flow added – this step (q) / Total flow (Q)	Nominal ID / Actual ID	Fittings–quantity and length	L ft / F ft / T ft	C Factor / P_f per foot (psi)	total (P_t) / elev (P_e) / frict (P_f)	Notes
AOUT	78		250	2½	Right angle valve (22.1)	1	100	100	Equivalent Length of Fittings 31 x 0.713 = 22.1 12 x 0.713 = 8.5
A	78		250	2.469	Tee (8.5)	30.6	0.302	0	
						31.6		9.5	
A	78		0	4	None	10	100	109.5	P_e = 0.433 x 10 P_e = 4.3
B	68		250	4.026		0	0.028	4.3	
						10		0.3	
B	68		250	4	2T and OS&Y (42 x 0.713 = 29.9)	260	100	114.1	P_e = 0.433 x 45 P_e = 19.5
M	23		500	4.026		29.9	0.101	19.5	
						289.9		29.3	
M	23		500	6		200	100	162.9	
N	23		1000	6.065		-	0.049	0	
						200		9.8	
N	23		250	6		200	100	172.7	
O	23		1250	6.065		-	0.074	0	
						200		14.8	
O	23			8	EL (18 x 0.713 x 1.06 = 13.6)	210	100	187.5	$(8.071/7.981)^{4.87} = 1.06$
DPV	13		1250	8.071		13.6	0.019	4.3	
						223.6		4.2	
DPV	13			8	DPV, GV, CV (64 x 1.06 = 67.8)	5	120	196.0	$(8.071/7.981)^{4.87} = 1.06$
FF	8		1250	8.071		67.8	0.013	2.2	
						72.8		0.9	
FF	8			8	EL and Tee (53 x 1.33 x 1.27 =	75	140	199.1	$(8.39/7.981)^{4.87} = 1.27$
CWM	0		1250	8.39		89.5	0.008	3.5	
						164.5		1.3	
CWM	0		1250					203.9	

Fig. 6.15b Hydraulic calculations for question 6.8

(d) The pump will need to provide a net pressure of 116.9 psi (203.9 – 87 = 116.9) at a flow of 1250 gpm. So, even though this is a very strong water supply, a significant fire pump is still required.

Note that the water pressure demand of 203.9 psi will mean that there is some need for pressure control in this system. That subject is covered in more detail in Chap. 8 of this text.

Answers to Test Yourself Questions in Chap. 7

7.1. Main Floor Landing—Advantageous

(a) The sprinkler system is easier to connect to the standpipe.
(b) The fire department uses less hose as compared to the connection at the intermediate landing.

7.2. Main Floor Landing—Disadvantage

(a) Harder for fire department to turn corner to get right out the door.
(b) Might be people who can't use stair waiting to be evacuated if stairwell is Area of Refuge.

7.3. Intermediate Floor Landing—Advantage

(a) Easier for fire department to unroll hose and advance it towards the door.
(b) Stays out of the way of people waiting to be rescued at main floor landing if that is an Area of Refuge.

7.4. Intermediate Floor Landing—Disadvantage

(a) Difficult to make connection for sprinkler system.
(b) More hose for firefighters to bring as compared to connecting at the main floor level.
(c) Less room for egress (since the landing is usually smaller than the main floor landing) so it is harder to keep the hose connection out of the clear width needed for egress.

7.5. Sprinklers supplied from two standpipes—Advantages

(a) Hydraulically, each sprinkler pipe carries less water, so the friction loss is decreased, which keeps the pressure demand down and allows smaller pipe sizes, all of which decreases cost.
(b) Reliability because the system will still get water, even if one of the standpipes is shut down for repair or maintenance.

7.6. The disadvantage to the sprinkler system being supplied from two standpipes is that the system is more complex, needing two valves to be closed in order to drain the system down and open it up. The person doing the work on the system needs to know that two valves need to be closed and they need to know where the second valve will be.

7.7. NFPA 14 requires a control valve and a check valve at each connection. In order to comply with NFPA 13 (and NFPA 72 regarding floor annunciation of a fire alarm) the waterflow switch will also need to be installed and then the test and drain will be needed in order to test the waterflow switch and drain the system from the control valve to the sprinklers. Technically, NFPA 14 only requires the check valve and the control valve. The other parts become necessary due to the check valve and the control valve to make the whole system work.

7.8. The total flow demand is 1000 gpm. The sprinkler demand of 200 gpm can be ignored because the sprinkler system meets NFPA 13.

7.9. The total flow demand will be 750 gpm. In this unusual situation, the standpipe demand is 750 gpm (500 from the first standpipe and 250 from the second). Again, the sprinkler demand can be ignored by the rules of NFPA 14, so the total demand is 750 gpm.

7.10. The total flow demand will be 1250 gpm. The standpipe system demand will be 1250 gpm (500 from the first standpipe and 250 from each of the other three standpipes). The sprinkler demand is not required to be added since the sprinkler system meets NFPA 13R.

7.11. The total flow demand will be 1250 gpm. The standpipe demand maxes out at 1250 gpm and the sprinkler demand is not required to be added.

7.12. The total flow demand will be 1430 gpm. The standpipe demand will be 1250 gpm (500 gpm for the most remote standpipe and 750 for the other three). The partial sprinkler system demand needs to be added, so $1250+180=1430$. Note that the sprinkler demand maxes out at 150 gpm for light hazard occupancies, but since this is an ordinary hazard occupancy, the maximum of 500 gpm has not yet been reached, so the sprinkler demand of 180 gpm needs to be added in its entirety.

Answers to Test Yourself Questions in Chap. 8

8.1. The correct answer is "c". NFPA 14 requires at least 100 psi at the Class I outlet.

8.2. The correct answer is "b". NFPA 14 requires at least 65 psi at the Class II outlet. Note that this is where the hose would be connected, making it a hose station, but the 65 psi is measured at the outlet of the standpipe, not the outlet of the nozzle on the hose.

8.3. The correct answer is "e". NFPA 14 allows the static pressure to go up to 175 psi in Class I systems, but no higher at the outlet to which the firefighter will connect a hose.

8.4. The correct answer is "c". NFPA 14 allows the residual pressure to go up to 100 psi in Class II systems, but no higher at the outlet to which the hose is connected. The static pressure is not explicitly controlled, but with the flow only at 100 gpm, there should not be too much difference between the static and residual pressures.

8.5. The correct answer is "d". NFPA 14 allows the pressure anywhere within the pipe that there is a hose connection to be a maximum of 350 psi. This is allowed all the way to the hose connection where it would need to be reduced to 175 psi before it gets to the outlet as explained in question 8.3.

8.6. The correct answer is "e". NFPA 14 allow the pressure anywhere within the pipe to be any number (as long as the pipe and fittings are rated for the pressure) at any point in the system that does not have hose connections attached. This would be in express risers or water supply pipes.

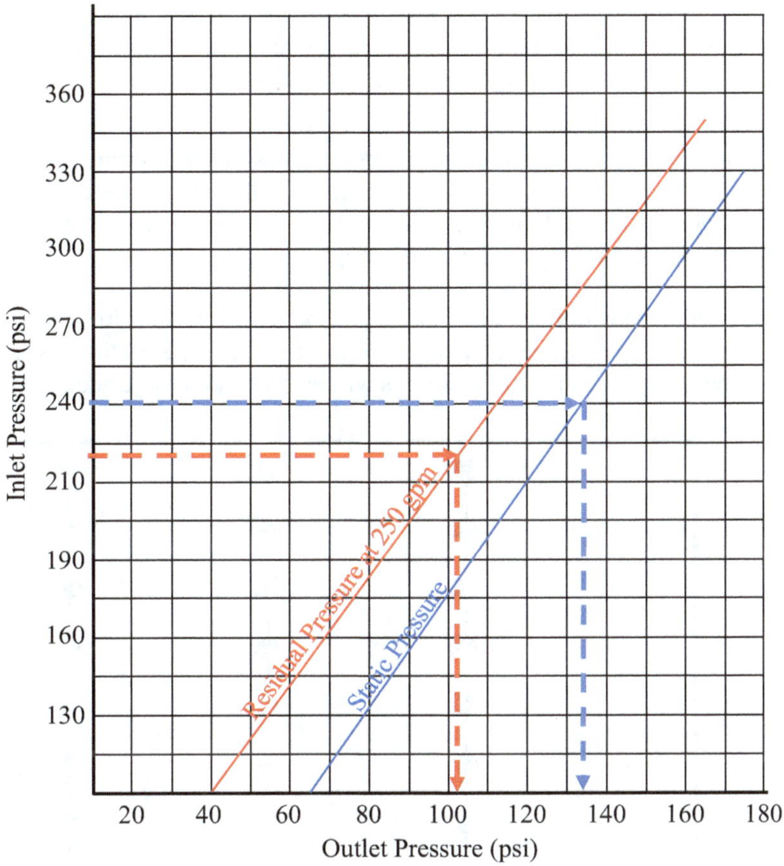

Fig. 8.24a Determination of outlet pressure at specific conditions in questions 8.9 and 8.10 when using model MD1 direct acting pressure reducing valve

8.7. A pressure reducing valve is a device that decreases both the static and residual pressure on the downstream side. A pressure restricting device only decreases the residual pressure and will have no effect on the static pressure.

8.8. A pilot operated pressure reducing valve senses the pressure on the upstream side and adjusts so that the discharge pressure (or outlet pressure) is constant. The direct acting valve is set so that there is always a pressure drop across the device and the outlet pressure will depend on the inlet pressure. As the inlet pressure goes up and down, the outlet pressure goes up and down.

8.9. The static pressure on the outlet side of the pressure reducing valve will be about 134 psi (depending on how precisely you try to read the graph). Yes, this value will be acceptable to NFPA 14 because it is between 100 and 175 psi. See the blue dashed line on Fig. 8.24a for how this value was obtained from the static pressure curve.

8.10. The residual pressure will be about 103 psi. Yes, this value will be acceptable to NFPA 14 because it is between 100 and 175 psi. See the red dashed line on Fig. 8.24a for how this value was obtained from the residual pressure curve.

8.11. There are multiple correct answers for this question. The information shown in Fig. 8.25a is one correct answer. The methodology as to how these answers were obtained is explained so that the reader can follow along and determine whether their particular solution is acceptable.

(a) While many options will probably work, 6 in. Schedule 40 steel has been selected here because the pipe size will minimize the friction loss and the equivalent lengths of the fittings will not need to be adjusted.

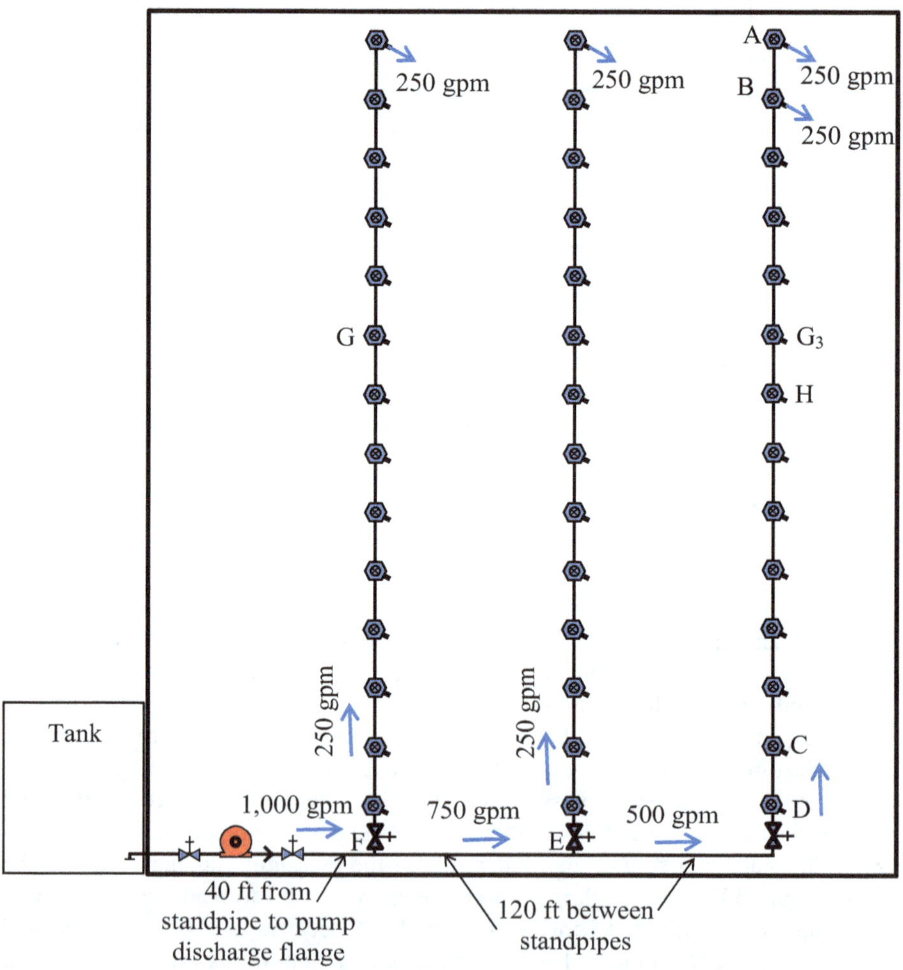

Fig. 8.25a Nodes and flows for Class I standpipe system in question 8.11

Node 1 / Node 2	Elev 1 (ft) / Elev 2 (ft)	K-Factor	Flow added – this step (*q*) / Total flow (*Q*)	Nominal ID / Actual ID	Fittings – quantity and length	L ft / F ft / T ft	C Factor / P_f per foot (psi)	total (P_t) / elev (P_e) / frict (P_f)	Notes
Outlet	135		250	2½	Right angle hose valve (31) + tee (12)	1	120	100	
A	135		250	2.469		43 44	0.215	0 9.5	
A	135		0	6		10	120	109.5	
B	125		250	6.065		0 10	0.003	4.3 0.0	
B	125		250	6	GV (3) T (30)	245	120	113.8	
E	0		500	6.065		33 278	0.010	54.1 2.8	
E	0		250	6		120	120	170.7	
F	0		750	6.065		0 120	0.021	0 2.5	
F	0		250	6	CV (32) GV (3)	40	120	173.2	
Dis. Flange	0		1,000	6.065		35 75	0.035	0 2.6	
Dis. Flange								175.8	
			1,000						

Fig. 8.27a Hydraulic calculations for Class I standpipe system in question 8.11

(b) The demand at the discharge flange of the pump will be 1000 gpm at 175.8 psi. See Fig. 8.27a to see how these values were determined. Note that NFPA 14 would also require a calculation to the FDC, which would be very similar in this case; however, the fire department will have to pump into the FDC as a pressure closer to the pump rating due to the pressure reducing valves that will probably need to be installed.

(c) The maximum pressure that will come from the pump is the elevation pressure created by the high water level in the tank ($24 \times 0.433 = 10.4$ psi) plus the churn pressure of the pump (200 psi), which equals 210.4 psi. The hose connections closest to the pump are 5 ft above the elevation of the pump, so they will experience a static pressure of 208.2 psi. Since this exceeds 175, they will need pressure reducing valves (PRV's). The ninth level of valves above the elevation of the pump (Node G in the figure) is 85 ft above the level of the pump. These valves will experience a static pressure of 173.6 psi ($210.4-(85 \times 0.433)=173.6$). This is just barely below the maximum allowed by NFPA 14, which means that every hose connection below this needs a PRV. The hose connection immediately below the level at Node G will operate at 177.9 psi (173.6 + 4.3 = 177.9). This will mean a total of 24 hose connections (8 on each standpipe) will need PRV's.

(d) For the hose connections that will see a static pressure of 177.9 psi, the Model A PRV shown in Fig. 8.28 (represented by the blue line) will work. With this inlet pressure, the outlet pressure will be about 141 psi, which is well within the limits of NFPA 14.

For the hose connections at the bottom of the system that will see a static pressure of 208.2 psi at the inlet of the PRV, the outlet static pressure will be about 165 psi using the Model A PRV, which is a little on the high side, but is still acceptable under the rules of NFPA 14.

Next, the effect of the PRV on the system during low pressure situations also needs to be calculated. The best way to do this is to work forward through the system from the fire pump. At a flow of 1000 gpm, the suction pressure from the fire pump is allowed to be as low as −3 psi, and the net pressure is 180 psi, so the worse-case discharge pressure will be 177 psi. The friction loss to Node F will be 2.6 psi (see Fig. 8.27a). The friction loss to Node E will be 2.5 psi (also see Fig. 8.27a). The friction loss to Node H (the highest level of hose connection with a PRV) when 250 gpm is flowing from Node H and Node G_3 will be 2.2 psi (0.01 × 218 = 2.2) while the elevation loss is 32.5 psi (75 × 0.433 = 32.5). This makes the total pressure loss from the pump to Node H 39.8 psi (2.6 + 2.5 + 2.2 + 32.5 = 39.8). The inlet residual pressure at the PRV at Node H will be 137.2 psi (177 − 39.8 = 137.2). At this inlet pressure, the Model A PRV will reduce the pressure to exactly 100 psi (see the blue line on Fig. 8.29). While this is the low limit of what is permitted by NFPA 14, it is within the permitted limits. Each of the PRV's below the one at Node H will have a higher inlet pressure, and therefore an acceptable outlet pressure.

Answers to Test Yourself Questions in Chap. 9

9.1. The answer is 157 psi (30 + 127 = 157).

9.2. The answer is 118 psi (162 − 44 = 118).

9.3. (a) The answer is 75 psi represented by the blue "X" on Fig. 9.12a.

(b) The answer is around 37 psi represented by the red "X" on Fig. 9.12a.

(c) The answer is around 34 psi represented by the green "X" on Fig. 9.12a. Note that you must evaluate the maximum flow at 150 % of rated flow, so for a pump rated at 750 gpm, the maximum flow is 1125 gpm (750 × 1.5 = 1125).

(d) The answer is around 22 psi represented by the purple "X" on Fig. 9.12a. Note that you must evaluate the maximum flow at 150 % of rated flow, so

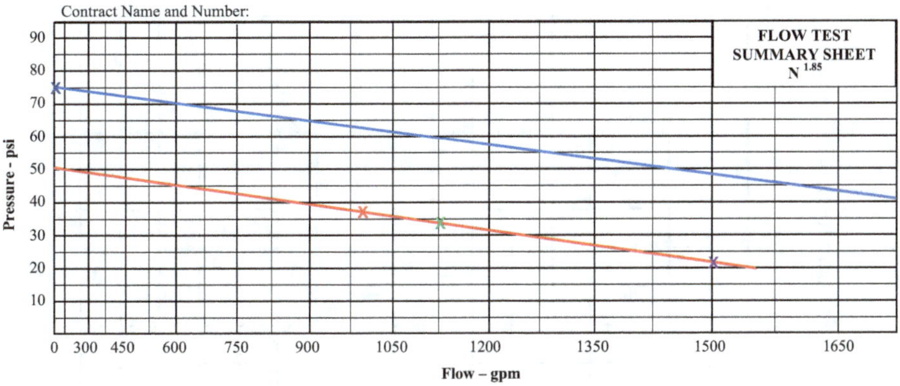

Fig. 9.12a Answers regarding the water supply in question 9.3

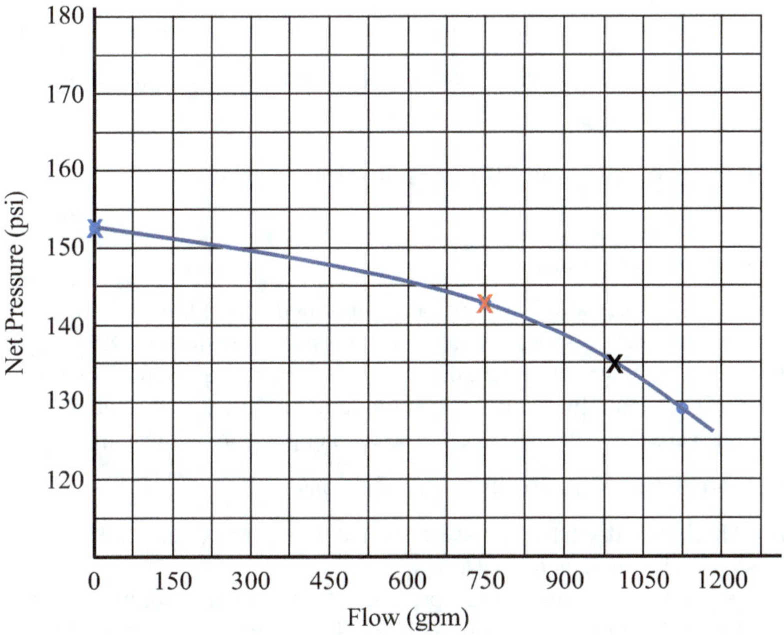

Fig. 9.13a Answer regarding the 750 gpm pump in question 9.4

for a pump rated at 1000 gpm, the maximum flow is 1500 gpm $(1000 \times 1.5 = 1500)$.

9.4. For the 750 gpm pump:
- The churn pressure is 152 psi as represented by the blue "X" in Fig. 9.13a.
- The rated net pressure is 143 psi as represented by the red "X" in Fig. 9.13a.
- The net pressure at system demand is 135 psi as represented by the black "X" in Fig. 9.13a.

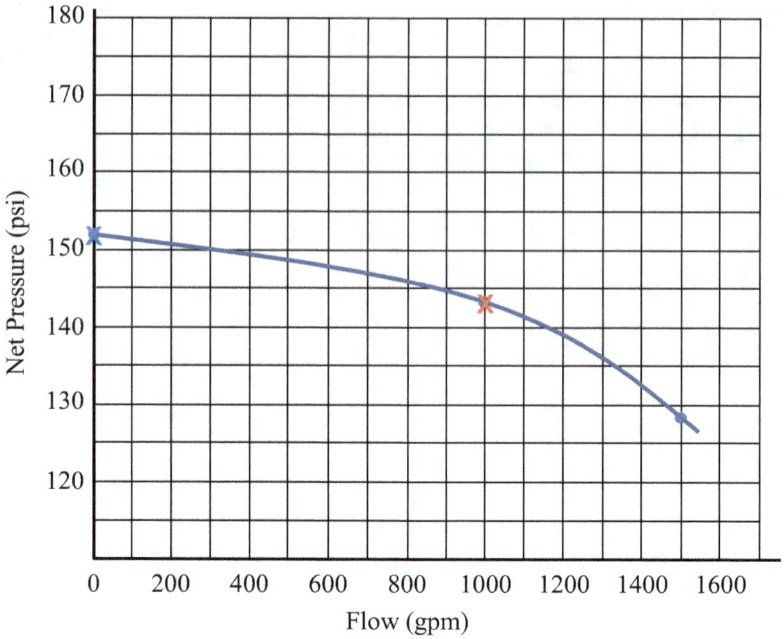

Fig. 9.14a Answers regarding the 1000 gpm pump in question 9.4

For the 1000 gpm pump:

- The churn pressure is 152 psi as represented by the blue "X" in Fig. 9.14a.
- The rated net pressure is 143 psi as represented by the red "X" in Fig. 9.14a.
- The net pressure at system demand is 143 psi as represented by the red "X" in Fig. 9.14a. Note that the system demand flow and the rated flow of the pump are the same in this case, so the net pressure is the same.

9.5. Here are the answers to each of the questions:

(a) Maximum discharge pressure for the 750 gpm pump is 231.3 psi, which was calculated as follows:

Add maximum static pressure from water supply (75 psi from Fig. 9.12a) to elevation pressure ($0.433 \times 10 = 4.3$, note that this is added because the pump is downhill from the location where the water main pressure was obtained) and pump churn pressure (152 from Fig. 9.13a).

75 psi + 4.3 + 152 = 231.3 psi

(b) The discharge pressure at the standpipe system demand for the 750 gpm pump will be 172.3 psi, which was calculated as follows:

Subtract the friction loss in the underground piping and the friction loss in the pump room piping from the residual pressure at the water supply. Then add the elevation pressure and the net pressure from the pump at the demand flow.

In order to calculate the friction loss, you will need to know the friction loss per ft (using the Hazen-Williams formula) and the equivalent length of the fittings. The equivalent length of the underground fittings was already calculated in Chap. 9 as 127.8 ft and the equivalent length of the pump room piping was already calculated in Chap. 9 as 23.3 ft.

The friction loss in the underground pipe will be 3.7 psi $(0.021 \times (50 + 127.8) = 3.7)$.

The friction loss in the pump room piping will be 0.3 psi $(0.009 \times (10 + 23.3) = 0.3)$.

Discharge Pressure $= 37 - 3.7 - 0.3 + 4.3 + 135 = 172.3$ psi.

(c) The low suction pressure for the 750 gpm pump will be 32.1 psi, which was calculated as follows:

Take the residual pressure from the water supply, subtract the friction loss in the underground pipe and the friction loss in the pump room, then add the elevation pressure.

Friction loss in the underground at maximum flow (1125 gpm) $= 0.027 \times 177.8 = 4.8$ psi.

Friction loss in the pump room pipe at maximum flow $= 0.011 \times 33.3 = 0.4$ psi

Suction pressure $= 33 - 4.8 - 0.4 + 4.3 = 32.1$ psi

(d) Maximum discharge pressure for the 1000 gpm pump is 231.3 psi, which was calculated as follows:

Add maximum static pressure from water supply (75 psi from Fig. 9.12a) to elevation pressure $(0.433 \times 10 = 4.3$, note that this is added because the pump is downhill from the location where the water main pressure was obtained) and pump churn pressure (152 from Fig. 9.14a).

75 psi $+ 4.3 + 152 = 231.3$ psi

(e) The discharge pressure at the standpipe system demand for the 1000 gpm pump will be 180.3 psi, which was calculated as follows:

Subtract the friction loss in the underground piping and the friction loss in the pump room piping from the residual pressure at the water supply. Then add the elevation pressure and the net pressure from the pump at the demand flow.

In order to calculate the friction loss, you will need to know the friction loss per ft (using the Hazen-Williams formula) and the equivalent length of the fittings. The equivalent length of the underground fittings was already calculated in Chap. 9 as 127.8 ft and the equivalent length of the pump room piping was already calculated in Chap. 9 as 23.3 ft.

The friction loss in the underground pipe will be 3.7 psi $(0.021 \times (50 + 127.8) = 3.7)$.

The friction loss in the pump room piping will be 0.3 psi $(0.009 \times (10 + 23.3) = 0.3)$.

Discharge Pressure $= 37 - 3.7 - 0.3 + 4.3 + 143 = 180.3$ psi.

(f) The low suction pressure for the 1000 gpm pump will be 17.5 psi, which was calculated as follows:

Take the residual pressure from the water supply, subtract the friction loss in the underground pipe and the friction loss in the pump room, then add the elevation pressure.

Friction loss in the underground at maximum flow (1500 gpm) = 0.046 × 177.8 = 8.2 psi.

Friction loss in the pump room pipe at maximum flow = 0.019 × 33.3 = 0.6 psi

Suction pressure = 22 − 8.2 − 0.6 + 4.3 = 17.5 psi

(g) The 1000 gpm pump is my choice.

The 1000 gpm pump is the better pump if no changes can be made to the standpipe system. As it stands right now, the standpipe system has a demand of 1000 gpm at 180 psi. The 750 gpm pump delivers 172.3 psi at 1000 gpm (answer b) while the 1000 gpm pump delivers 180.3 psi at 1000 gpm (answer e). The 750 gpm pump does not meet the system demand while the 1000 gpm pump does.

Both pumps produce a positive gauge pressure at the suction flange during maximum flow, so that criterion should not have much of an effect on this decision. Both pumps produce the same maximum pressure, so that criterion should not have much of an effect on this decision.

Some people prefer to install a smaller pump to make testing easier, as well as control cost. So it might be worth seeing if you could redesign the standpipe system to get the demand down (increase the pipe size in the standpipes or interconnecting pipe) and/or to reduce the friction loss in the underground. But if you had to pick between these two pumps with no changes to the system, the 1000 gpm pump is the only one that meets the demand.

The 1000 gpm pump discharge pressure only exceeds the demand pressure by 0.3 psi, which is not much of a margin. But remember that the calculation was already performed using the reasonable worse case water supply already taking daily and seasonal fluctuations into account, so no additional margin should be required.

(h) Most likely all of the hose connections will need pressure reducing valves. The maximum static pressure from either pump will be 231.3 psi (answer a and answer d). This exceeds the allowable outlet pressure at a hose connection, so at least some of the valves low down in this system will need pressure reducing valves.

If you subtract the maximum allowable pressure at the outlet of the hose connection from the maximum static pressure, then divide by the relationship between pressure and elevation, you get the height at which pressure reducing valves will no longer be needed. In this case, that formula looks like this:

(231.3 − 175)/0.433 = 130 ft

You have not been given information as to how tall this building is, but it is likely that the building does not exceed 130 ft given that the standpipe system demand is only 180 psi. Recognizing that you need 100 psi at the most remote hose connection and that some of the rest of the demand has to be from friction loss, it is unlikely that this system exceeds 80 or 90 ft above the level of the fire pump. So, it is likely that all of the hose connections will need a pressure reducing valve.

Answers to Test Yourself Questions in Chap. 10

10.1. The answer is “b”. All of the model codes have the same definition of a high rise building.
10.2. The answer is “e”, at 420 ft the code requires the additional connection.
10.3. The answer is “c”.
10.4. Pick two of the following: air traffic control towers, open parking garages, portions of buildings that are A-5 occupancies (outdoor activities), special industrial occupancies, and high hazard.
10.5. The correct answer is “d”. With all of the other options, if you take the largest tank out of service, you take more than 50 % of the duration demand away. With option d, if you take the largest tank out of service, you still have 17,000 gallons, which is more than 50 % of the duration demand.

Answers to Test Yourself Questions in Chap. 11

11.1. The answer is 2626.75 lb. The hanger needs to support 15 ft of pipe, which will weigh 475.35 ($31.69 \times 15 = 475.35$). The hanger needs to support 5 times the weight of the pipe plus 250 lbs, so it will need to hold 2626.75 ($475.35 \times 5 + 250 = 2626.75$),
11.2. The section modulus for the 4 in. Schedule 40 steel pipe in the standpipe system is 0.79. The smallest Schedule 40 pipe that you could use is 2.5 in., which has a section modulus of 1.06. Smaller pipes have a smaller section modulus than 0.79, and therefore could not be used for this situation.
11.3. Using the formula, $L = 4ab/(a+b) = 4(1)(7)/(1+7) = 48/8 = 3.5$ ft. For a 3.5 ft span, the section modulus is 0.35 to support 4 in. Schedule 40. This can be supported by a trapeze span that is 2 in. Schedule 40 pipe with a section modulus of 0.56.
11.4. The answers are as follows with Fig. 11.18a showing the parts that it can:

   (a) There are no flexible couplings required on the horizontal pipe. The lateral braces need to be installed at 25 ft spacing if they are to be connected to the major structural members, which they should be. If you were to go every other one, the braces would be 50 ft apart, which is too far. There will need to be two longitudinal braces since the distance between the standpipes is 125 ft and the maximum allowable spacing between longitudinal braces is 80 ft. The two braces can be installed as shown with 75 ft between the braces or in other locations.

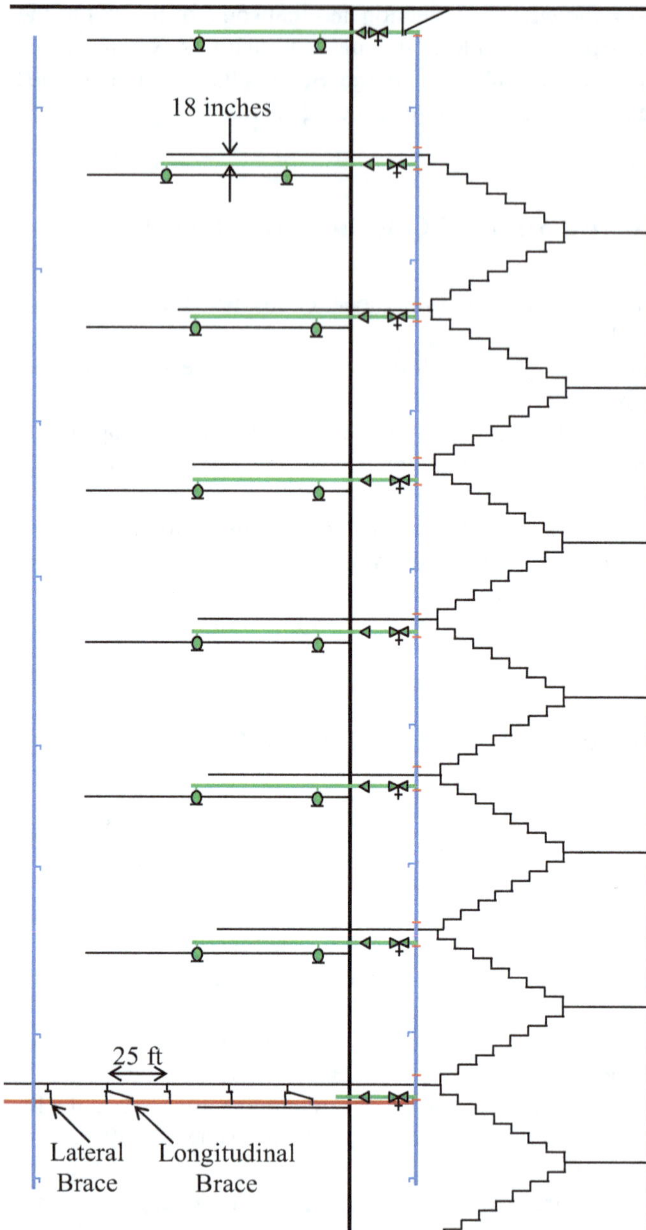

Fig. 11.18a Answers to question 11.4

With the S_S value of 0.5, the C_P will be 0.4. That makes the load that the braces need to handle $0.46W_P$ ($1.15\times0.4=0.46$). Six inch Schedule 10 pipe weighs 23.03 lb per ft (see Table A.9.3.5.9 in NFPA 13). The lateral brace will need to handle 265 lb ($0.46\times25\times23.03=265$). This is permitted by Table 9.3.5.5.2(a) since it will allow a load up to 3231 lb for braces spaced at 25 ft apart. The longitudinal brace will need to handle 795 lb ($0.46\times75\times23.03=795$).

(b) Flexible couplings are required at the top and bottom of the standpipe. The flexible couplings are represented by the bright red short lines perpendicular to the pipe. The one at the bottom is at the feed main piping, which counts as the bottom of the standpipe given where the support for the feed main is. Flexible couplings are also required within 12 in. above each floor in the building. Flexible couplings are required within 24 in. under each floor. In this case, they are placed 24 in. below the floor, which puts them below the sprinkler system connection, so no additional flexible couplings are required on the sprinkler piping.

There is only one brace required on the standpipe, at the top. The four-way brace is on the horizontal sprinkler piping within 24 in. of the centerline of the standpipe. This assumes that the clearance around the floor penetrations is the minimum necessary and not any more.

11.5. The following are the answers for this question. Remember that the requirements in NFPA 13 are all based on the brace angle being measured from the vertical, and in this example, it has been provided in relation to the horizontal, so you need to convert the angle by subtracting the angle from the horizontal from 90 degrees. The angle was provided as 34 degrees from the horizontal, so it is 56 degrees from the vertical ($90-34=56$).

(a) Will a sway brace fitting rated for 800 lb be sufficient?

A sway brace rated for 800 lb has been rated in the horizontal direction. When used at an angle of 56 degrees to the vertical, it needs to be divided by 1.414, so the brace can handle a load of 566 lb ($800/1.414=566$).

The load on the lateral brace is only 265 lb (see the answer for Question 11.4), so that will not be a problem. But the load for the longitudinal brace covering 75 ft of pipe is 795 lb, which exceeds the ability of the fitting. A different fitting would need to be found that can hold up to 795 lb, or the longitudinal braces would need to be spaced at every other structural member (50 ft apart) so that they had a total load of 530 lb, which could be held by the fitting.

(b) What is the smallest all thread rod that can be used as the brace member?

For both braces, 3/4 in. all thread rod is the smallest diameter that can be used. If you look at the $l/r=100$ table, none of the all thread rod can be used because none of the lengths allow 30 in. (2 ft – 6 in.). Using the $l/r=200$ table, the 3/4 in. in. all thread rod can be used for a length of

2 ft – 7 in. and can handle a load of 819 lb at an angle of 56 degrees to the vertical. This would work for any of the lateral and longitudinal brace options we have discussed here. Looking at the l/r = 300 table, none of the rods smaller than ¾ can handle the loads, so the ¾ in. rod is the smallest all thread rod that can be used.

(c) What is the smallest size unfinished bolt that could be used to fasten the brace to the steel structural member?

The fastener will go up into the steel structural member vertically, making the combination of the fastener direction and brace angle match situation B for the fastener tables. Since the fastener is going to be a bolt through the steel, Table 9.3.5.12.2(g) is the appropriate table to use from NFPA 13. This table allows for a ¼ in. bolt to handle a load of 500 lb, which will work for the lateral brace. A 3/8 in. bolt can handle a load of 1200 lb, which will handle any of the longitudinal braces that have been discussed here.

Answers to Test Yourself Questions in Chap. 12

12.1. Both consist primarily of horizontal piping, but a branch line serves only a single hose connection while a horizontal standpipe serves two or more hose connections.

12.2. The answer is "No." A separate control valve for a branch line only needs to be installed if the branch line exceeds 40 ft in length.

12.3. The answer is 20.9 psi. Using the Hazen-Williams formula, the friction loss per length of pipe with 250 gpm flowing through 2.5 in. Schedule 40 pipe in a wet pipe system is 0.215 psi per ft. The equivalent length of four elbows is 24 ft. The equivalent length of a tee is 12 ft. The equivalent length of the OS&Y gate valve is 1 ft. With 60 ft of pipe, the total length of pipe and fittings is 97 ft ($60+24+12+1=97$). The total friction loss is then 20.9 psi ($0.215\times97=20.9$).

12.4. The answer is 17.4 psi. Using the Hazen-Williams formula, the friction loss per length of pipe with 250 gpm flowing through the 2.5 in. Schedule 10 pipe in a wet pipe system is 0.157 psi per ft. While the length of pipe has not changed, the equivalent length of the fittings does because Table 8.3.1.3 in NFPA 14 was written assuming that the pipe being used was Schedule 40. An adjustment to the fitting length for Schedule 10 needs to be calculated using the formula in Section 8.3.2.5 as follows: $(2.635/2.469)^{4.87}=1.37$. The equivalent lengths of the fittings from Table 8.3.1.3 need to be revised by 1.37 as follows: $1.37(24+12+1)=50.7$ ft. The total length of pipe and fittings is 110.7 ft ($60+50.7=110.7$) making the total friction loss 17.4 psi ($110.7\times0.157=17.4$).

12.5. The answer is 1000 gpm at 144 psi, which any fire department should be able to provide. See Fig. 12.7a for how this answer was obtained.

Node 1 Node 2	Elev 1 (ft) Elev 2 (ft)	K-Factor	Flow added – this step (*q*) Total flow (*Q*)	Nominal ID Actual ID	Fittings–quantity and length	L ft F ft T ft	C Factor P_f per foot (psi)	total (P_t) elev (P_e) frict (P_f)	Notes
Outlet	23	-	250	2.5	Angle hose valve (31) 2T (24) 1.37 * 55 = 75.3	1	120	100	
A	24		250	2.635		75.3 76.3	0.157	-0.4 12.0	
A	24	-	-	4		125	120	111.6	
B	24		250	4.26		- 125	0.015	0 1.9	
B	24	-	250	4	EL(10) 10 * 1.32 = 13.2	100	120	113.5	$(4.26/4.026)^{4.87} = 1.32$
C	24		500	4.26		13.2 113.2	0.055	0 6.2	
C	24	-	250	6	3EL (42) GV (3) T (30) 75 * 1.26 = 94.5	470	120	119.7	$(6.357/6.065)^{4.87} = 1.26$ 20 * 0.433 = 8.7
D	4		750	6.357		94.5 564.5	0.016	8.7 9.0	
D	4	-	250	6	3T(90) CV(32) 122 * 1.26 = 153.7	20	120	137.4	4 * 0.433 = 1.7
FDC	0		1000	6.357		153.7 173.7	0.028	1.7 4.9	
FDC								144	

Fig. 12.7a Hydraulic calculation for horizontal standpipe in question 12.5

Answers to Test Yourself Questions in Chap. 13

13.1. The system needs to be tested at 200 psi at the lowest point in the system. You only add 50 psi to the system static pressure if the static pressure is over 150 psi.

13.2. The pressure that will be created by the elevated tank is 65 psi ($150 \times 0.433 = 65$). The leakage (or weepage) would be ($700 \times 8 \times (65)^{0.5}/148{,}000 = 0.305$ gallons per hour. Since the test will be run for two hours, the total amount of leakage (weepage) will be 0.61 gallons ($0.305 \times 2 = 0.61$).

13.3. The maximum discharge pressure from the pump will be 190 psi ($50 + 140 = 190$). This means that the test will need to be run at 50 psi over this value, which will be 240 psi ($190 + 50 = 240$). The elevation loss 90 ft above the pump will be 39 psi (90 × 0.433 = 39). So, the pressure at the top of the standpipe will be 201 psi (240 − 39 = 201).

13.4. The flow will need to be 1000 gpm with 250 gpm from the most remote hose connection, 250 from the second most remote hose connection, then 250 gpm from a convenient place on the second most remote standpipe and finally 250 gpm from a convenient place on the third most remote standpipe. Even though there are many other standpipes in this building, the test only needs to be conducted at 1000 gpm because the building is fully sprinklered and the maximum flow demand for the standpipe system is 1000 gpm.

13.5. The results are not acceptable. With 250 gpm flowing and in inlet pressure of 230 psi, the chart shows that the pressure should be reduced to 100 psi. The fact that the valve is reducing the pressure only to 148 psi means that it is either not set correctly or the wrong valve was installed. It is possible that a Model B valve was installed because this is how the pressure would be reduced with an inlet pressure of 230 psi to that valve. More information is needed to fix this situation, but it is not acceptable.

Answers to Test Yourself Questions in Chap. 14

14.1. The correct answer is "b". This is a test. There is physical hands-on work being performed and the condition of a piece of equipment is changed. The goal is to see whether the device being tested recognizes the change in position. This is clearly a test.

14.2. The correct answer is "c". See part 20 of the list in Chap. 14 of inspection items from NFPA 25. Also see Table 6.2.1 of NFPA 25.

14.3. The answer is "no". There is something wrong with the system that needs investigating. Although the static pressures are very similar, the residual pressures are very different. During the 2015 test, the difference between the static and residual pressures was 22 psi (84 − 62 = 22), but during the 2016 test, the difference between static and residual pressures was 34 psi (82 − 48 = 34). The additional 12 psi loss needs to be explained. It is possible that a valve is partially closed, an obstruction is in the system, or extra friction loss has been created with new piping or equipment between the main drain and the water supply. Investigation is necessary. It could even be that a backflow device was added since there was not one in the past. It is common for such a device to be added without consulting a fire protection professional to evaluate the effect on the standpipe system. If this is the case, the hydraulics of the standpipe system need to be investigated to make sure that the standpipe system still works properly.

14.4. Before analyzing any of the information from the test, you will need to determine the system volume. With no information from the records, you will need to figure it out from the amount of pipe. The system volume turns out to be 634 gallons, calculated as follows:

- 3 (standpipes) × 100 ft × 0.74 gal/ft = 222 gallons
- 1 (feed main) × 250 ft × 1.649 gal/ft = 412 gallons
- Total system volume = 222 + 412 = 634 gallons

Since the system volume does not exceed 750 gallons, there is no specific water delivery time requirement from NFPA 14 or NFPA 25. The two tests that we have data for are extremely comparable. There is no cause for concern. The system should be considered as passing the test.

14.5. Before analyzing any of the information from the test, you will need to determine the system volume. With no information from the records, you will need to figure it out from the amount of pipe. The system volume turns out to be 1194 gallons, calculated as follows:

- 5 (standpipes) × 100 ft × 0.74 gal/ft = 370 gallons
- 1 (feed main) × 500 ft × 1.649 gal/ft = 824 gallons
- Total system volume = 370 + 824 = 1194 gallons

Since the system volume does exceed 750 gallons, the water delivery time needs to be within 3 min, which it was for the test in 2013. For the test in 2016, there is clearly a problem. The water delivery time to the most remote hose connection was over 4 min, which violates the 3 min delivery time of NFPA 14. Further investigation is needed to determine what the problem might be.

One of the suspected problems is that the air pressure has been set much higher. A 10 psi increase over the last time the test was done is suspicious. Depending on the differential of the dry-pipe valve, this may or may not be a good idea. With the higher starting air pressure and the lower starting water pressure, it took the valve longer to trip, and then the lower water pressure took longer to push the water out of the remote hose connection.

One way to determine if this was the issue would be to reset the system with an air pressure of 55 psi and repeat the test. If the water delivery is under 3 min and closer to the previous test, this may have been the only issue. However, the reason that the air pressure was increased would need to be examined. If the air pressure was increased in order to solve multiple false trip problems, then the air pressure/water pressure situation needs to be re-examined or an obstruction investigation needs to be performed. If the air pressure was increased because someone was misinformed about the correct air pressure, then some education regarding correct air pressure settings is in order. Typically, the air pressure is set at 20 psi in excess of the maximum water pressure divided by the valve differential. So, if a valve had a 5:1 differential and the maximum expected water pressure was 175 psi, the air pressure should be set at 55 psi (175/5 + 20 = 55).

www.ingramcontent.com/pod-product-compliance
Lightning Source LLC
Chambersburg PA
CBHW070417010726
47387CB00003BA/210
9783319477497